Interactive Student Edition

# Reveal
# GEOMETRY™
Volume 2

Mc
Graw
Hill
Education

www.my.mheducation.com

Copyright © 2020 McGraw-Hill Education

All rights reserved. No part of this publication may be
reproduced or distributed in any form or by any means,
or stored in a database or retrieval system, without
the prior written consent of McGraw-Hill Education,
including, but not limited to, network storage or
transmission, or broadcast for distance learning.

Cover: (t to b, l to r) Kenny McCartney/Moment Open/Getty Images; YinYang/E+/Getty Images;
nycshooter/Vetta/Getty Images; michaelgzc/E+/Getty Images

Send all inquiries to:
McGraw-Hill Education
8787 Orion Place
Columbus, OH 43240

ISBN: 978-0-07-662601-4 (*Interactive Student Edition*, Volume 1)
MHID: 0-07-662601-6 (*Interactive Student Edition*, Volume 1)
ISBN: 978-0-07-899749-5 (*Interactive Student Edition*, Volume 2)
MHID: 0-07-899749-6 (*Interactive Student Edition*, Volume 2)

Printed in the United States of America.

1 2 3 4 5 6 7 8 9 10 QVS 27 26 25 24 23 22 21 20 19 18

Common Core State Standards © Copyright 2010. National
Governors Association Center for Best Practices and Council of
Chief State School Officers. All rights reserved.

# Contents in Brief

Copyright © McGraw-Hill Education

# Reveal AGA™ Makes Math Meaningful...

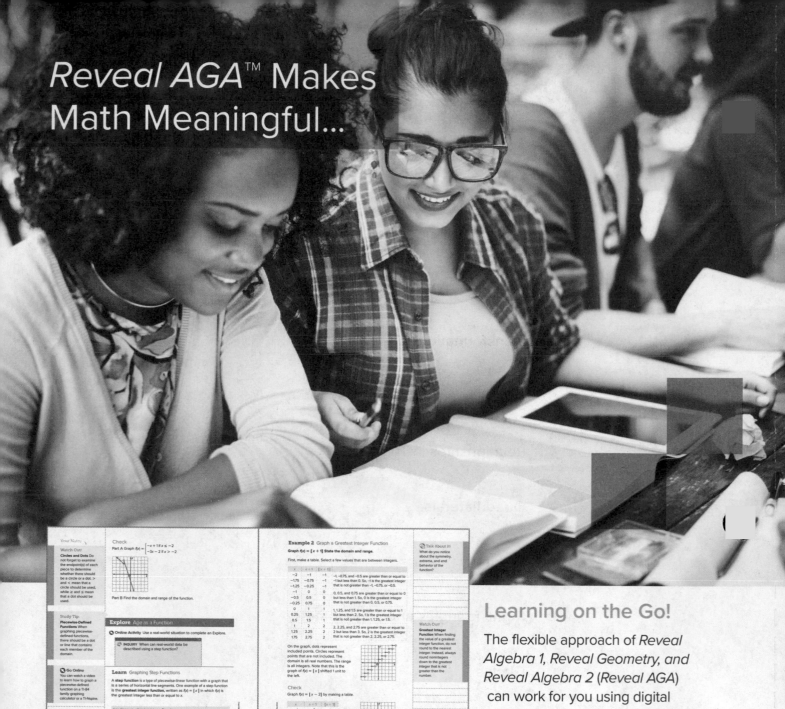

**Interactive Student Edition**

## Learning on the Go!

The flexible approach of *Reveal Algebra 1, Reveal Geometry, and Reveal Algebra 2 (Reveal AGA)* can work for you using digital only or digital and your *Interactive Student Edition* together.

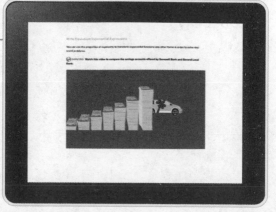

**Student Digital Center**

Copyright © McGraw-Hill Education. Rawpixel/Shutterstock

# ...to Reveal YOUR Full Potential!

## *Reveal AGA*™ Brings Math to Life in Every Lesson

*Reveal AGA* is a blended print and digital program that supports access on the go. You'll find the *Interactive Student Edition* mirrors the Student Digital Center, so you can record your digital observations in class and reference your notes later, or access just the digital center, or a combination of both! The Student Digital Center provides access to the interactive lessons, interactive content, animations, videos and technology-enhanced practice questions.

Write down your username and password here

Username: _____

Password: _____

## Go Online!
## my.mheducation.com

**WebSketchpad® Powered by The Geometer's Sketchpad®**- Dynamic, exploratory, visual activities embedded at point of use within the lesson.

**Animations and Videos** – Learn by seeing mathematics in action.

**Interactive Tools** – Get involved in the content by dragging and dropping, selecting, highlighting, and completing tables.

**Personal Tutors** – See and hear a teacher explain how to solve problems.

**eTools** – Math tools are available to help you solve problems and develop concepts.

Copyright © McGraw-Hill Education. dolgachov/123RF

## Module 1
# Tools of Geometry

Copyright © McGraw-Hill Education

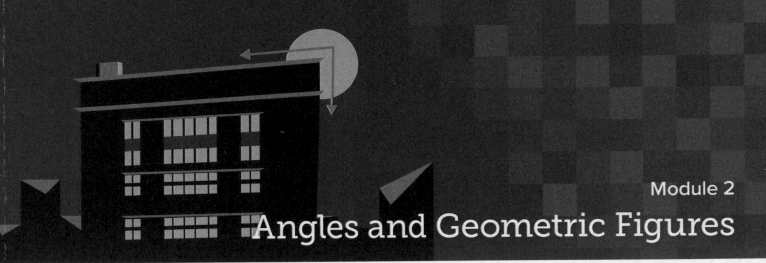

Module 2

# Angles and Geometric Figures

Copyright © McGraw-Hill Education

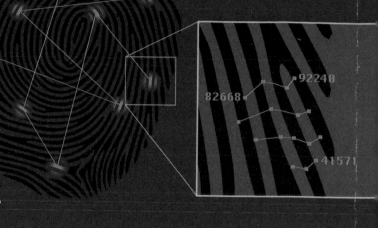

# Module 3
# Logical Arguments and Line Relationships

Copyright © McGraw-Hill Education

Module 4

# Transformations and Symmetry

Copyright © McGraw-Hill Education

# TABLE OF CONTENTS

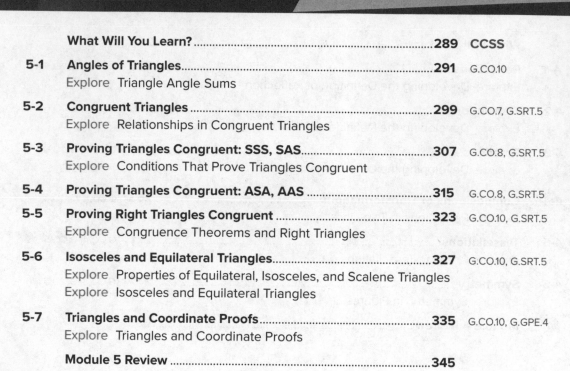

## Module 5
# Triangles and Congruence

Copyright © McGraw-Hill Education

Copyright © McGraw-Hill Education

Module 7
# Quadrilaterals

Copyright © McGraw-Hill Education

Module 8

# Similarity

Copyright © McGraw-Hill Education

Module 9

# Right Triangles and Trigonometry

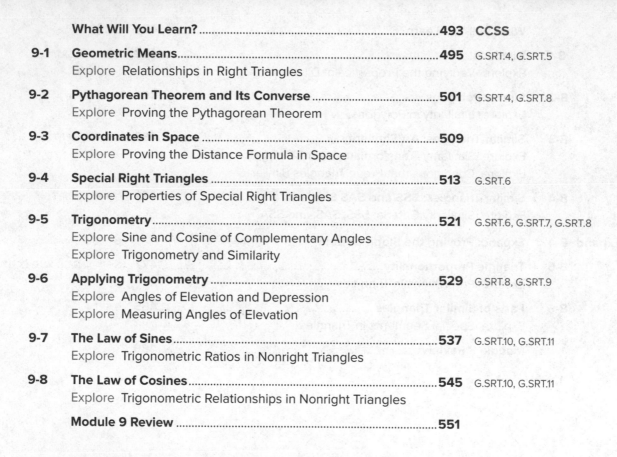

Copyright © McGraw-Hill Education

Module 10
# Circles

Copyright © McGraw-Hill Education

Module 11

# Measurement

Copyright © McGraw-Hill Education

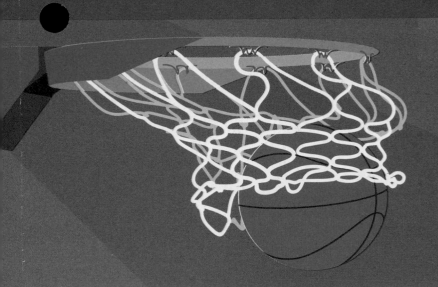

Module 12
# Probability

Copyright © McGraw-Hill Education

## Geometry

### Congruence G-CO

**G.CO.1** Experiment with transformations in the plane. Know precise definitions of angle, circle, perpendicular line, parallel line, and line segment, based on the undefined notions of point, line, distance along a line, and distance around a circular arc.

**G.CO.2** Represent transformations in the plane using, e.g., transparencies and geometry software; describe transformations as functions that take points in the plane as inputs and give other points as outputs. Compare transformations that preserve distance and angle to those that do not (e.g., translation versus horizontal stretch).

**G.CO.3** Given a rectangle, parallelogram, trapezoid, or regular polygon, describe the rotations and reflections that carry it onto itself.

**G.CO.4** Develop definitions of rotations, reflections, and translations in terms of angles, circles, perpendicular lines, parallel lines, and line segments.

**G.CO.5** Given a geometric figure and a rotation, reflection, or translation, draw the transformed figure using, e.g., graph paper, tracing paper, or geometry software. Specify a sequence of transformations that will carry a given figure onto another.

**G.CO.6** Understand congruence in terms of rigid motions. Use geometric descriptions of rigid motions to transform figures and to predict the effect of a given rigid motion on a given figure; given two figures, use the definition of congruence in terms of rigid motions to decide if they are congruent.

**G.CO.7** Use the definition of congruence in terms of rigid motions to show that two triangles are congruent if and only if corresponding pairs of sides and corresponding pairs of angles are congruent.

**G.CO.8** Explain how the criteria for triangle congruence (ASA, SAS, and SSS) follow from the definition of congruence in terms of rigid motions.

**G.CO.9** Prove geometric theorems.
Prove theorems about lines and angles.

**G.CO.10** Prove theorems about triangles.

**G.CO.11** Prove theorems about parallelograms.

**G.CO.12** Make geometric constructions.
Make formal geometric constructions with a variety of tools and methods (compass and straightedge, string, reflective devices, paper folding, dynamic geometric software, etc.).

**G.CO.13** Construct an equilateral triangle, a square, and a regular hexagon inscribed in a circle.

### Similarity, Right Triangles, and Trigonometry G-SRT

**G.SRT.1** Understand similarity in terms of similarity transformations.
Verify experimentally the properties of dilations given by a center and a scale factor:
a.   A dilation takes a line not passing through the center of the dilation to a parallel line, and leaves a line passing through the center unchanged.

b.   The dilation of a line segment is longer or shorter in the ratio given by the scale factor.

**G.SRT.2** Given two figures, use the definition of similarity in terms of similarity transformations to decide if they are similar; explain using similarity transformations the meaning of similarity for triangles as the equality of all corresponding pairs of angles and the proportionality of all corresponding pairs of sides.

**G.SRT.3** Use the properties of similarity transformations to establish the AA criterion for two triangles to be similar.

**G.SRT.4** Prove theorems involving similarity.
Prove theorems about triangles.

**G.SRT.5** Use congruence and similarity criteria for triangles to solve problems and to prove relationships in geometric figures.

**G.SRT.6** Define trigonometric ratios and solve problems involving right triangles.
Understand that by similarity, side ratios in right triangles are properties of the angles in the triangle, leading to definitions of trigonometric ratios for acute angles.

**G.SRT.7** Explain and use the relationship between the sine and cosine of complementary angles.

**G.SRT.8** Use trigonometric ratios and the Pythagorean Theorem to solve right triangles in applied problems. ★

**G.SRT.9** Apply trigonometry to general triangles.
(+) Derive the formula $A = \frac{1}{2} ab \sin (C)$ for the area of a triangle by drawing an auxiliary line from a vertex perpendicular to the opposite side.

**G.SRT.10** (+) Prove the Laws of Sines and Cosines and use them to solve problems.

**G.SRT.11** (+) Understand and apply the Law of Sines and the Law of Cosines to find unknown measurements in right and non-right triangles (e.g., surveying problems, resultant forces).

## Circles G-C

**G.C.1** Understand and apply theorems about circles. Prove that all circles are similar.

**G.C.2** Identify and describe relationships among inscribed angles, radii, and chords.

**G.C.3** Construct the inscribed and circumscribed circles of a triangle, and prove properties of angles for a quadrilateral inscribed in a circle.

**G.C.4** (+) Construct a tangent line from a point outside a given circle to the circle.

**G.C.5** Find arc lengths and areas of sectors of circles. Derive using similarity the fact that the length of the arc intercepted by an angle is proportional to the radius, and define the radian measure of the angle as the constant of proportionality; derive the formula for the area of a sector.

## Expressing Geometric Properties with Equations G-GPE

**G.GPE.1** Translate between the geometric description and the equation for a conic section. Derive the equation of a circle of given center and radius using the Pythagorean Theorem; complete the square to find the center and radius of a circle given by an equation.

**G.GPE.2** Derive the equation of a parabola given a focus and directrix.

**G.GPE.4** Use coordinates to prove simple geometric theorems algebraically.

**G.GPE.5** Prove the slope criteria for parallel and perpendicular lines and use them to solve geometric problems (e.g., find the equation of a line parallel or perpendicular to a given line that passes through a given point).

**G.GPE.6** Find the point on a directed line segment between two given points that partitions the segment in a given ratio.

**G.GPE.7** Use coordinates to compute perimeters of polygons and areas of triangles and rectangles, e.g., using the distance formula. ★

## Geometric Measurement and Dimension G-GMD

**G.GMD.1** Explain volume formulas and use them to solve problems. Give an informal argument for the formulas for the circumference of a circle, area of a circle, volume of a cylinder, pyramid, and cone.

**G.GMD.3** Use volume formulas for cylinders, pyramids, cones, and spheres to solve problems. ★

**G.GMD.4** Visualize relationships between two-dimensional and three-dimensional objects. Identify the shapes of two-dimensional cross-sections of three-dimensional objects, and identify three-dimensional objects generated by rotations of two-dimensional objects.

## Modeling with Geometry G-MG

**G.MG.1** Apply geometric concepts in modeling situations. Use geometric shapes, their measures, and their properties to describe objects (e.g., modeling a tree trunk or a human torso as a cylinder). ★

**G.MG.2** Apply concepts of density based on area and volume in modeling situations (e.g., persons per square mile, BTUs per cubic foot). ★

**G.MG.3** Apply geometric methods to solve problems (e.g., designing an object or structure to satisfy physical constraints or minimize cost; working with typographic grid systems based on ratios). ★

## Statistics and Probability

### Conditional Probability and the Rules of Probability S-CP

**S.CP.1** Understand independence and conditional probability and use them to interpret data. Describe events as subsets of a sample space (the set of outcomes) using characteristics (or categories) of the outcomes, or as unions, intersections, or complements of other events ("or," "and," "not").

**S.CP.2** Understand that two events $A$ and $B$ are independent if the probability of $A$ and $B$ occurring together is the product of their probabilities, and use this characterization to determine if they are independent.

**S.CP.3** Understand the conditional probability of $A$ given $B$ as $\frac{P(A \text{ and } B)}{P(B)}$, and interpret independence of $A$ and $B$ as saying that the conditional probability of $A$ given $B$ is the same as the probability of $A$, and the conditional probability of $B$ given $A$ is the same as the probability of $B$.

★ Mathematical Modeling Standards

**S.CP.4** Construct and interpret two-way frequency tables of data when two categories are associated with each object being classified. Use the two-way table as a sample space to decide if events are independent and to approximate conditional probabilities.

**S.CP.5** Recognize and explain the concepts of conditional probability and independence in everyday language and everyday situations.

**S.CP.6** Use the rules of probability to compute probabilities of compound events in a uniform probability model.
Find the conditional probability of $A$ given $B$ as the fraction of $B$'s outcomes that also belong to $A$, and interpret the answer in terms of the model.

**S.CP.7** Apply the Addition Rule, $P(A \text{ or } B) = P(A) + P(B) - P(A \text{ and } B)$, and interpret the answer in terms of the model.

**S.CP.8** (+) Apply the general Multiplication Rule in a uniform probability model, $P(A \text{ and } B) = P(A)P(B|A) = P(B)P(A|B)$, and interpret the answer in terms of the model.

**S.CP.9** (+) Use permutations and combinations to compute probabilities of compound events and solve problems.

**Using Probability to Make Decisions S-MD**

**S.MD.6** Use probability to evaluate outcomes of decisions. (+) Use probabilities to make fair decisions (e.g., drawing by lots, using a random number generator).

**S.MD.7** (+) Analyze decisions and strategies using probability concepts (e.g., product testing, medical testing, pulling a hockey goalie at the end of a game).

# Quadrilaterals

## ℮ Essential Question

What are the different types of quadrilaterals, and how can their characteristics be used to model real-world situations?

G.CO.11, G.GPE.4, G.MG.1
**Mathematical Practices:** MP1, MP2, MP3, MP4, MP5, MP6, MP7, MP8

## What will you learn?

Place a checkmark (✓) in each row that corresponds with how much you already know about each topic **before** starting this module.

KEY

👎 — I don't know.     ✋ — I've heard of it.     👍 — I know it!

| | Before | | | After | | |
|---|---|---|---|---|---|---|
| | 👎 | ✋ | 👍 | 👎 | ✋ | 👍 |
| solve problems involving the interior angles of polygons | | | | | | |
| solve problems involving the exterior angles of polygons | | | | | | |
| solve problems using the properties of parallelograms | | | | | | |
| solve problems involving the diagonals of parallelograms | | | | | | |
| solve problems using the properties of rectangles | | | | | | |
| solve problems using the properties of rhombi | | | | | | |
| solve problems using the properties of squares | | | | | | |
| solve problems using the properties of trapezoids | | | | | | |
| solve problems using the properties of kites | | | | | | |

📖 **Foldables** Make this Foldable to help you organize your notes about quadrilaterals. Begin with one sheet of notebook paper.

1. **Fold** widthwise.

2. **Fold** along the width of the paper twice and unfold the paper.

3. **Cut** along the fold marks on the left side of the paper to the center.

4. **Label** as shown.

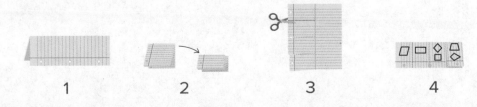

1          2          3          4

Copyright © McGraw-Hill Education

# What Vocabulary Will You Learn?

Check the box next to each vocabulary term that you may already know.

- ☐ base angle of a trapezoid
- ☐ bases of a trapezoid
- ☐ diagonal
- ☐ isosceles trapezoid
- ☐ kite
- ☐ legs of a trapezoid
- ☐ midsegment
- ☐ parallelogram
- ☐ rectangle
- ☐ rhombus
- ☐ square
- ☐ trapezoid

## Are you ready?

Complete the Quick Review to see if you are ready to start this module.
Then complete the Quick Check.

### Quick Review

**Example 1**

**Find the measure of each numbered angle.**

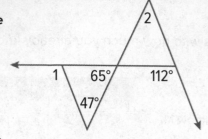

**a. m∠1**

| | |
|---|---|
| m∠1 = 65 + 47 | Exterior Angle Theorem |
| m∠1 = 112° | Add. |

**b. m∠2**

| | |
|---|---|
| 180 = m∠2 + 68 + 65 | Triangle Angle Sum Th. |
| 180 = m∠2 + 133 | Simplify. |
| m∠2 = 47° | Subtract. |

**Example 2**

**Find the lengths of the sides of isosceles △XYZ.**

| | |
|---|---|
| XY = YZ | Given |
| 2x + 3 = 4x − 1 | Substitution |
| −2x = −4x | Subtract |
| x = 2 | Simplify. |
| XY = 2x + 3 | Given |
| = 2(2) + 3, or 7 | Substitute x = 2. |
| YZ = 4x − 1 | Given |
| = 4(2) − 1, or 7 | Substitute x = 2. |
| XZ = 8x − 4 | Given |
| = 8(2) − 4, or 12 | Substitute x = 2. |

### Quick Check

**Find the value of x to the nearest tenth.**

**1.**     **2.**

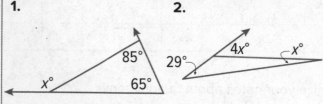

**Find the value of x to the nearest tenth.**

**3.**     **4.**

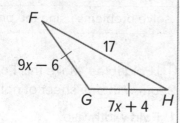

### How did you do?

Which exercises did you answer correctly in the Quick Check? Shade those Exercise numbers below.

① ② ③ ④

Copyright © McGraw-Hill Education

# Angles of Polygons

## Explore Angles of Polygons

 **Online Activity** Use dynamic geometry software to complete the Explore.

> ⊘ ×
> ② **INQUIRY** How can you find the sum of the interior angle measures of a polygon?

## Learn Interior Angles of Polygons

**Theorem 7.1: Polygon Interior Angles Sum Theorem**

The sum of the interior angle measures of an *n*-sided convex polygon is $(n - 2) \cdot 180°$.

A **diagonal** of a polygon is a segment that connects any two nonconsecutive vertices.

## Example 1 Find the Interior Angles Sum of a Polygon

**Find the measure of each interior angle of pentagon *HJKLM*.**

**Step 1 Find the sum.**
A pentagon has 5 sides. Use the Polygon Interior Angles Sum Theorem to find the sum of its interior angle measures.

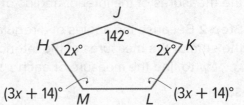

Sum of the interior angle measures $= (n - 2) \cdot 180°$    Polygon Interior Angles Th.

$= (\underline{\phantom{0}} - \underline{\phantom{0}}) \cdot 180°$    Substitute.

$= \underline{\phantom{0000}}$    Solve.

**Step 2 Find the value of *x*.**
Use the sum of the interior angle measures to determine the value of *x*.

$2x° + \underline{\phantom{00}}° + (3x + 14)° + (\underline{\phantom{0000}})° + \underline{\phantom{00}}° = 540°$   Write an equation.

$x = \underline{\phantom{000}}$   Solve.

**Step 3 Find the measure of each angle.**
Use the value of *x* to find the measure of each angle.

$m\angle J = \underline{\phantom{00}}°$   $m\angle K = 2(\underline{\phantom{0}})°$ or $\underline{\phantom{00}}°$   $m\angle L = (\underline{\phantom{0}}(37) + 14)°$ or $\underline{\phantom{00}}°$

$m\angle M = (3(37) + \underline{\phantom{0}})°$ or $125°$        $m\angle H = \underline{\phantom{0}}x° = 2(\underline{\phantom{0}})°$ or $\underline{\phantom{00}}$

 **Go Online** You can complete an Extra Example online.

Copyright © McGraw-Hill Education

**Today's Standards**
G.MG.1

MP1, MP2, MP4, MP5, MP6, MP8

**Today's Vocabulary**
diagonal

## Check

What is $m\angle E$?

**A.** 108°

**B.** 120°

**C.** 122°

**D.** 126°

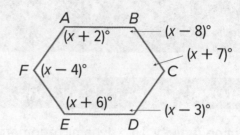

$A$   $B$
$(x + 2)°$   $(x - 8)°$
$F$   $(x - 4)°$   $(x + 7)°$
$C$
$(x + 6)°$   $(x - 3)°$
$E$   $D$

## 🌐 Example 2 Interior Angle Measures of a Regular Polygon

**FLOOR PLANS Penny is building a house using a floor plan that she designed. What is the measure of $\angle ABC$?**

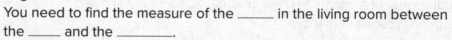

Deck | Living Room | Kitchen
Breezeway
Bedroom | Garage

**Understand**

You know the living room can be modeled by a _____.
The living room has ____ sides and ____ angles.
You need to find the measure of the _____ in the living room between the _____ and the _____.

**Plan**

**Step 1** Use the _____ Theorem to find the sum of the measures of the interior angles of the living room.

**Step 2** Because the angles of a regular polygon are _____, divide the sum of the measures of the interior angles by the total number of _____ to find the measure of each _____.

**Solve**

**Step 1** Find the sum of the interior angle measures.

$(n - 2) \cdot 180° = (\_\_\_ - 2) \cdot 180°$     $n = 6$

$= \_\_\_°$     Solve.

**Step 2** Find the measure of one interior angle.

$$\frac{\text{sum of interior angle measures}}{\text{number of congruent angles}} = \frac{\quad\quad}{6}$$     Substitute.

$= _____$     Solve.

The measure of $\angle ABC$ is 120°.

**Check**

How do you know your solution is reasonable?

Sample answer: _____

_____

🅑 **Go Online** You can complete an Extra Example online.

Copyright © McGraw-Hill Education

## Check

**PONDS** Miguel has commissioned a pentagonal koi pond to be built in his backyard. He wants the pond to have a deck of equal width around it. The lengths of the interior deck sides are the same length, and the lengths of the exterior sides are the same.

The measure of the angle of the pond formed by two sides of the deck is _____.

## Example 3 Identify the Polygon Given Interior Angle Measure

**The measure of an interior angle of a regular polygon is 144°. Find the number of sides in the polygon.**

Let $n$ = the number of sides in the polygon. Because all angles of a regular polygon are congruent, the sum of the interior angle measures is $144n°$. By the Polygon Interior Angles Sum Theorem, the sum of the interior angle measures can also be expressed as $(n - 2) \cdot 180°$.

$$\underline{\hspace{1cm}}n° = (n - 2) \cdot 180° \qquad \text{Write an equation.}$$

$$n = \underline{\hspace{1cm}} \qquad \text{Solve.}$$

The polygon has _____ sides, so it is a regular decagon.

## Check

The measure of an interior angle of a regular polygon is 150°.

The polygon has _____ sides.

---

## Learn Exterior Angles of Polygons

**Theorem 7.2 Polygon Exterior Angles Sum Theorem**

The sum of the exterior angle measures of a convex polygon, one angle at each vertex, is 360°.

## Example 4 Find Missing Values

**Find the value of x.**

Use the _____
Theorem to write an equation. Then solve for x.

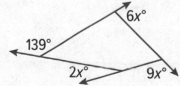

$$6x° + \underline{\hspace{0.7cm}}° + 2x° + \underline{\hspace{0.7cm}}° = \underline{\hspace{0.7cm}}° \qquad \text{Write an equation.}$$

$$x = \underline{\hspace{1cm}} \qquad \text{Solve.}$$

 **Go Online** You can complete an Extra Example online.

Copyright © McGraw-Hill Education

### Study Tip

**Naming Polygons**
Remember, a polygon with *n*-sides in an *n*-gon, but several polygons have special names.

| Number of Sides | Polygon |
|---|---|
| 3 | triangle |
| 4 | quadrilateral |
| 5 | pentagon |
| 6 | hexagon |
| 7 | heptagon |
| 8 | octagon |
| 9 | nonagon |
| 10 | decagon |
| 11 | hendecagon |
| 12 | dodecagon |
| $n$ | $n$-gon |

**Go Online**

An alternate solution method is available for this example.

## Check

What is the value of *x*?

A. 45

B. 52

C. 93

D. 97

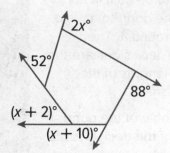

---

## Example 5 Find Exterior Angle Measures of a Polygon

**Find the measure of each exterior angle of a regular dodecagon.**

A regular dodecagon has 12 congruent sides and 12 congruent interior angles. The exterior angles are also congruent, because angles supplementary to congruent angles are congruent. Let *n* = the measure of each exterior angle and write and solve an equation.

$$\underline{\qquad}n = \underline{\qquad} \qquad \text{Polygon Exterior Angles Sum Theorem}$$

$$n = \underline{\qquad} \qquad \text{Solve.}$$

The measure of each exterior angle of a regular dodecagon is 30°.

## Check

The measure of each exterior angle of a regular octagon is _____.

---

## Pause and Reflect

Did you struggle with anything in this lesson? If so, how did you deal with it?

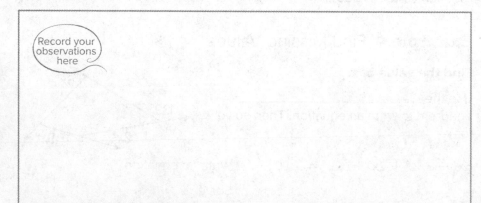

Record your observations here

🅡 **Go Online** You can complete an Extra Example online.

Copyright © McGraw-Hill Education

# Practice

🔴 **Go Online** You can complete your homework online.

**Example 1**

**Find the sum of the measures of the interior angles of each convex polygon.**

**1.** 16-gon

**2.** 30-gon

**Find the measure of each interior angle.**

**3.**

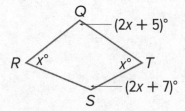

**4.**

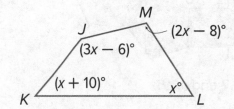

**5.**

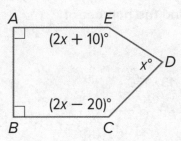

**6.**

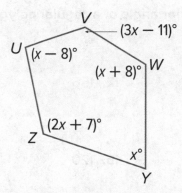

**Example 2**

**7.** ARCHITECTURE  In the Uffizi gallery in Florence, Italy, there is a room built by Buontalenti called the Tribune (*La Tribuna* in Italian). This room is shaped like a regular octagon. What angle do consecutive walls of the Tribune make with each other?

La Tribuna

**8.** THEATER  A theater floor plan is shown in the figure. The upper five sides are part of a regular dodecagon. Find $m\angle 1$.

Stage 1

Copyright © McGraw-Hill Education

**9.** An animal pen is in the shape of a regular heptagon. What is the measure of each interior angle of the animal pen? Round to the nearest tenth.

**10.** POLYGON PATH In Ms. Rickets' math class, students made a "polygon path" that consists of regular polygons of 3, 4, 5, and 6 sides joined together as shown.

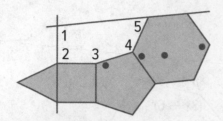

    **a.** Find $m\angle 2$ and $m\angle 5$.

    **b.** Find $m\angle 3$ and $m\angle 4$.

    **c.** What is $m\angle 1$?

Example 3

**The measure of an interior angle of a regular polygon is given. Find the number of sides in the polygon.**

**11.** 144

**12.** 156

**13.** 160

**14.** 108

**15.** 120

**16.** 150

Example 4

**Find the value of $x$ in each diagram.**

**17.**

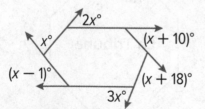

**18.**

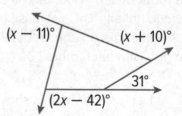

**19.**

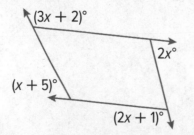

**20.**

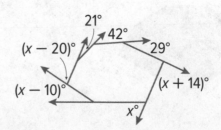

Copyright © McGraw-Hill Education

**Example 5**

**Find the measure of each exterior angle of each regular polygon.**

**21.** pentagon

**22.** 15-gon

**23.** hexagon

**24.** octagon

**25.** nonagon

**26.** 12-gon

**Mixed Exercises**

**Find the measures of an exterior angle and an interior angle given the number of sides of each regular polygon. Round to the nearest tenth, if necessary.**

**27.** 7

**28.** 13

**29.** 14

**For Exercises 30 and 31, find the value of $x$.**

**30.** A convex octagon has interior angles with measures $(x + 55)°$, $(3x + 20)°$, $4x°$, $(4x - 10)°$, $(6x - 55)°$, $(3x + 52)°$, $3x°$, and $(2x + 30)°$.

**31.** A convex hexagon has interior angles with measures $x°$, $(5x - 103)°$, $(2x + 60)°$, $(7x - 31)°$, $(6x - 6)°$, and $(9x - 100)°$.

**For Exercises 32 and 33, find the measure of each interior angle.**

**32.** decagon, in which the measures of the interior angles are $x + 5$, $x + 10$, $x + 20$, $x + 30$, $x + 35$, $x + 40$, $x + 60$, $x + 70$, $x + 80$, and $x + 90$

**33.** polygon $ABCDE$, in which the measures of the interior angles are $6x$, $4x + 13$, $x + 9$, $2x - 8$, $4x - 1$

**34.** Find the measure of each exterior angle of a regular 2x-gon.

**35.** Find the sum of the measures of the exterior angles of a convex 65-gon.

**36. ARGUMENTS** Write a paragraph proof to prove the Polygon Interior Angles Sum Theorem for octagons.

**37. ARGUMENTS** Use algebra to prove the Polygon Exterior Angles Sum Theorem.

**38. REASONING** The measure of each interior angle of a regular polygon is 24 more than 38 times the measure of each exterior angle. Find the number of sides of the polygon.

Copyright © McGraw-Hill Education

39. **ARCHEOLOGY** Archeologists unearthed parts of two adjacent walls of an ancient castle. Before it was unearthed, they knew from ancient texts that the castle was shaped like a regular polygon, but nobody knew how many sides it had. Some said 6, others 8, and some even said 100. From the information in the figure, how many sides did the castle really have?

24°

40. **MODELING** Jasmine is designing boxes she will use to ship her jewelry. She wants to shape the box like a regular polygon. In order for the boxes to pack tightly, she decides to use a regular polygon that has the property that the measure of its interior angles is half the measure of its exterior angles. What regular polygon should she use?

41. **CRYSTALLOGRAPHY** Crystals are classified according to seven crystal systems. The basis of the classification is the shapes of the faces of the crystal. Turquoise belongs to the triclinic system. Each of the six faces of turquoise is in the shape of a parallelogram. Find the sum of the measures of the interior angles of one such face.

42. **STRUCTURE** If three of the interior angles of a convex hexagon each measure 140°, a fourth angle measures 84, and the measure of the fifth angle is 3 times the measure of the sixth angle, find the measure of the sixth angle.

43. **FIND THE ERROR** Marcus says that the sum of the exterior angles of a decagon is greater than that of a heptagon because a decagon has more sides. Liang says that the sum of the exterior angles for both polygons is the same. Is either of them correct? Explain your reasoning.

44. **WRITE** Explain how triangles are related to the Interior Angles Sum Theorem.

45. **CREATE** Sketch a polygon and find the sum of its interior angles. How many sides does a polygon with twice this interior angles sum have? Justify your answer.

46. **PERSEVERE** Find the values of $a$, $b$, and $c$ if QRSTVX is a regular hexagon. Justify your answer.

47. **ANALYZE** If two sides of a regular hexagon are extended to meet at a point in the exterior of the polygon, will the triangle formed *always*, *sometimes*, or *never* be equilateral? Justify your answer.

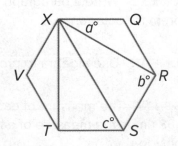

Copyright © McGraw-Hill Education

# Parallelograms

## Explore Properties of Parallelograms

🔗 **Online Activity** Use dynamic geometry software to complete the Explore.

⊚ **INQUIRY** What special properties do parallelograms have?

## Learn Properties of Parallelograms

A **parallelogram** is a quadrilateral with both pairs of opposite sides parallel. To name a parallelogram, use the symbol ▱. In ▱ABCD, $\overline{BC} \parallel \overline{AD}$ and $\overline{AB} \parallel \overline{DC}$ by definition.

**Theorem 7.3: Properties of Parallelograms**

If a quadrilateral is a parallelogram, then its opposite sides are congruent.

**Theorem 7.4: Properties of Parallelograms**

If a quadrilateral is a parallelogram, then its opposite angles are congruent.

**Theorem 7.5: Properties of Parallelograms**

If a quadrilateral is a parallelogram, then its consecutive angles are supplementary.

**Theorem 7.6: Properties of Parallelograms**

If a parallelogram has one right angle, then it has four right angles.

## Example 1 Use Properties of Parallelograms

**Find CD.**

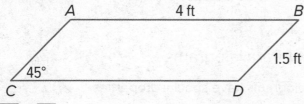

$\overline{CD} \cong \overline{AB}$       Opposite sides of a ▱ are ≅.

$CD =$ _____       Definition of congruent

$= $ _____ ft       Substitution

**Find m∠A.**

$m\angle A + m\angle$ _____ $= 180°$       Consecutive ∠s in a ▱ are supplementary.

$m\angle A + $ _____° $= 180°$       Substitution

$m\angle A = $ _____°       Solve.

Copyright © McGraw-Hill Education

---

**Today's Standards**
G.CO.11, G.GPE.4
MP2, MP3, MP4, MP6

**Today's Vocabulary**
parallelogram

**Watch Out!**

**Parallelograms**
Theorems 7.3 through 7.6 apply only if you already know that the figure is a parallelogram.

**Study Tip**

**Including a Figure**
Theorems are presented in general terms. In a proof, you must include a drawing so that you can refer to segments and angles specifically.

Copyright © McGraw-Hill Education

**💬 Talk About It!**

Thiago states that because all parallelograms are quadrilaterals, all quadrilaterals are parallelograms. Do you agree? Justify your answer.

## Check

Find each measure.

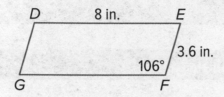

**a.** $m\angle D$ _____ °

**b.** $FG$ _____ in.

## Example 2 Proofs Using the Properties of Parallelograms

You can use the properties of parallelograms to write proofs.

**Write a two-column proof.**

**Given:** ▱HJKP and ▱PKLM

**Prove:** $\overline{HJ} \cong \overline{ML}$

| Statements | Reasons |
|---|---|
| 1. ▱HJKP and ▱PKLM | 1. |
| 2. $\overline{HJ} \cong \overline{PK}$, $\overline{PK} \cong \overline{ML}$ | 2. |
| 3. $\overline{HJ} \cong \overline{ML}$ | 3. |

## Check

Write the correct statements and reasons to complete the proof.

**Given:** ▱JKLM, $\overline{KN} \cong \overline{KL}$

**Prove:** $\angle J \cong \angle KNL$

| Statements | Reasons |
|---|---|
| 1. ▱JKLM, $\overline{KN} \cong \overline{KL}$ | 1. |
| 2. | 2. |
| 3. | 3. Isos. △Thm. |
| 4. $\angle J \cong \angle KNL$ | 4. |

## Learn Diagonals of Parallelograms

The diagonals of parallelograms have special properties.

**Theorem 7.7: Diagonals of Parallelograms**

If a quadrilateral is a parallelogram, then its diagonals bisect each other.

**Theorem 7.8: Diagonals of Parallelograms**

If a quadrilateral is a parallelogram, then each diagonal separates the parallelogram into two congruent triangles.

🡢 **Go Online** You can complete an Extra Example online.

## Example 3 Use Properties of Parallelograms and Algebra

**Find the values of *x* and *z* in □ABCD.**

$m\angle ADC = 4x°$ and $m\angle DAB = (2x - 6)°$

**Part A  Find the value of *x*.**

| | |
|---|---|
| $180° = m\angle ADC + m\angle DAB$ | Consec. ∠s in a □ are _____ |
| $180 = \underline{\quad} + (2x - 6)$ | Substitution |
| $x = \underline{\quad}$ | Solve. |

**Part B  Find the value of *z*.**

| | |
|---|---|
| $\overline{AE} \cong \overline{CE}$ | Diagonals of a □ bisect each other. |
| $AE = CE$ | Def. of congruent |
| $3z - 4 = \underline{\quad}$ | Substitute |
| $z = \underline{\quad}$ | Solve. |

## 🌐 Example 4 Parallelograms and Coordinate Geometry

**SCRAPBOOKING Tomas is making envelopes to sell with handmade cards at craft fairs. He uses a different style of paper to create the flap of the envelope, and he edges the envelopes with washi tape. The envelopes are parallelograms, and the edges of the flaps lie along the diagonals of the parallelograms. Find the area of the flap and the perimeter of the envelope.**

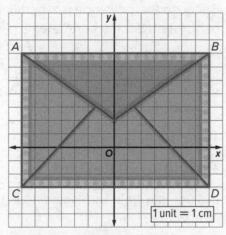

1 unit = 1 cm

**Part A  Find the area of the paper needed to create the flap.**

You can approximate the area of the flap with a triangle, so the area is $A = \frac{1}{2}bh$.

**Step 1: Find the height.**

To find the height, determine the coordinates of the intersection of the diagonals of the envelope, which has vertices at $A(-7, 7)$, $B(7, 7)$, $C(-7, -3)$, and $D(7, -3)$.

$$\left(\frac{x_1 + x_2}{2}, \frac{y_1 + y_2}{2}\right) = \left(\frac{-7 +}{2}, \frac{7 + (\underline{\quad})}{2}\right)$$
$$= (\underline{\quad}, \underline{\quad})$$

The height is the difference in the *y*-coordinates of the intersection of the diagonals and the vertices of the top edge of the envelope.

$h = 7 - 2$ or _____ cm

🧭 **Go Online** You can complete an Extra Example online.

Copyright © McGraw-Hill Education

**Make a Plan** To find the area of the paper needed for the envelope flap, you need to calculate the point of intersection of the diagonals of the envelope. Before solving for the area, analyze the information you are given, develop a plan, and determine the theorems you will need to apply.

💭 **Think About It!**

What assumptions did you make when calculating the area of the paper and the length of washi tape that Jonas needed to make his envelopes?

**Step 2: Find the base.**

The base of the flap is the distance along the top edge of the envelope. You can count the units to determine the base. $b =$ _____ cm

**Step 3: Find the area of the flap.**

$A = \frac{1}{2} bh$  Area Formula

$\phantom{A} = \frac{1}{2} ($_____$)($_____$)$  Substitute.

$\phantom{A} =$ _____ cm² Solve.

**Part B  Find the length of washi tape need to create the border.**

**Step 1: Find the height.**

The length of the envelope is the same as the base of the triangle determined in **Part A**. $\ell =$ _____ cm

**Step 2: Find the width.**

The width of the envelope is the distance between the top edge and the bottom edge of the envelope. $w = 7 - ($_____$) =$ _____ cm

**Step 3: Find the perimeter.**

Because the envelope is a parallelogram, opposite sides are congruent. So, the perimeter is given by $P = 2\ell + 2w$.

$P = 2\ell + 2w$  Perimeter formula

$\phantom{P} = 2(14) + 2(10)$  Substitute.

$\phantom{P} = 48$ cm  Solve.

## Pause and Reflect

Did you struggle with anything in this lesson? If so, how did you deal with it?

Record your observations here

🔵 **Go Online** You can complete an Extra Example online.

# Practice

Go Online You can complete your homework online.

### Example 1

**Use ▱PQRS to find each measure.**

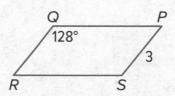

1. m∠R _____ °

2. QR

3. QP

4. m∠S _____ °

5. **DISTANCE** Four friends live at the four corners of a block shaped like a parallelogram. Gracie lives 3 miles away from Keon. How far apart do Teresa and Terek live from each other?

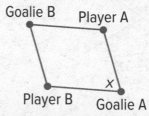

6. **SOCCER** Four soccer players are practicing a drill. Goalie A is facing Player B to receive the ball. Goalie A then turns x° to face Player A in order to pass her the ball. If Goalie B is facing Player A to receive the ball, then what angle measure must Goalie B turn in order to pass the ball to Player B?

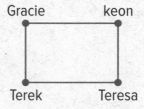

### Example 2

**Write a two-column proof.**

7. **Given:** WXYV and ZYVT are parallelograms

**Prove:** $\overline{WX} \cong \overline{ZY}$

8. **Given:** ▱BDHA, $\overline{CA} \cong \overline{CG}$

**Prove:** $\angle BDH \cong \angle G$

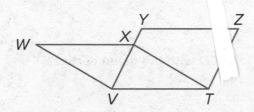

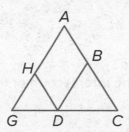

9. **Write a paragraph proof of the theorem *Consecutive angles in a parallelogram are supplementary.***

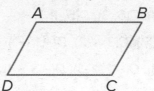

10. **Write a paragraph proof.**

**Given:** ▱PRST and ▱PQVU

**Prove:** $\angle V \cong \angle S$

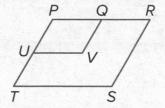

Copyright © McGraw-Hill Education

Example 3

**Find the value of each variable in each parallelogram.**

**11.**

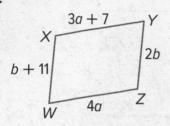

**12.**

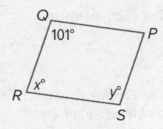

**13.**

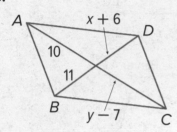

**14.**

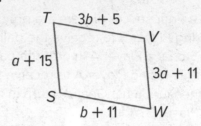

**15.**

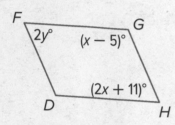

**16.**

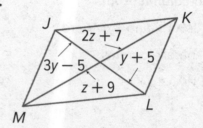

Example 4

**Find the coordinates of the intersection of the diagonals of □ABCD with the given vertices.**

**17.** $A(3, 6)$, $B(5, 8)$, $C(3, -2)$, and $D(1, -4)$

**18.** $A(-4, 3)$, $B(2, 3)$, $C(-1, -2)$, and $D(-7, -2)$

**19.** $A(1, 1)$, $BJ(2, 3)$, $C(6, 3)$, $D(5, 1)$

**20.** $A(-1, 4)$, $B(3, 3)$, $C(3, -2)$, $D(-1, -1)$

**21.** $A(2, 5)$, $B(3, 3)$, $C(-2, -3)$, $D(-3, -1)$

**22.** $A(2, 3)$, $B(1, -2)$, $C(-5, -7)$, $D(-4, -2)$

Mixed Exercises

**ARGUMENTS Write the indicated type of proof.**

**23.** two-column                    (Theorem 7.3)

   **Given:** □PQRS

   **Prove:** $\overline{PQ} \cong \overline{RS}$, $\overline{QR} \cong \overline{SP}$

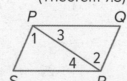

Copyright © McGraw-Hill Education

**24.** paragraph                    (Theorem 7.7)

**Given:** ▱ACDE

**Prove:** $\overline{EC}$ bisects $\overline{AD}$.

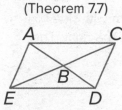

**25.** two-column                    (Theorem 7.5)

**Given:** ▱GKLM

**Prove:** ∠G and ∠K, ∠K and ∠L,

∠L and ∠M, and ∠M and ∠G

are supplementary.

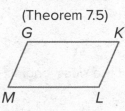

**26.** two-column                    (Theorem 7.8)

**Given:** ▱WXYZ

**Prove:** △WXZ ≅ △YZX

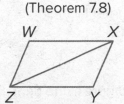

**REASONING Use ▱ABCD to find each measure or value.**

**27.** $x$

**28.** $y$

**29.** $m\angle AFB$

**30.** $m\angle DAC$

**31.** $m\angle ACD$   °

**32.** $m\angle DAB$

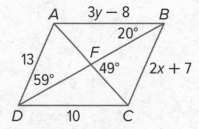

**33. REGULARITY** Use the graph shown.

**a.** Use the Distance Formula to determine if the diagonals of *JKLM* bisect each other. Explain.

**b.** Determine whether the diagonals are congruent. Explain.

**c.** Use slopes to determine if the consecutive sides are perpendicular. Explain.

**34. MODELING** Make a Venn diagram showing the relationship between squares, rectangles, and parallelograms.

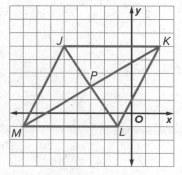

Copyright © McGraw-Hill Education

**35. PERSEVERE** *ABCD* is a parallelogram with side lengths as indicated in the figure at the right. The perimeter of *ABCD* is 22. Find *AB*.

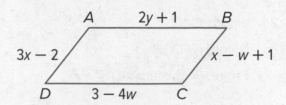

**36. ANALYZE** Explain why parallelograms are *always* quadrilaterals, but quadrilaterals are *sometimes* parallelograms.

**37. WRITE** Summarize the properties of the sides, angles, and diagonals of a parallelogram.

**38. ANALYZE** Provide a counterexample to show that parallelograms are not always congruent if their corresponding sides are congruent.

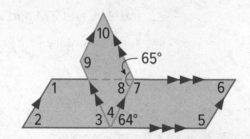

**39. CREATE** Find *m∠1* and *m∠10* in the figure at the right. Explain.

Copyright © McGraw-Hill Education

# Tests for Parallelograms

## Explore Constructing Parallelograms

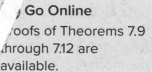 **Online Activity** Use dynamic geometry software to complete the Explore.

> ❓ **INQUIRY** How can you use the properties of parallelograms to construct parallelograms? ✕

## Learn Tests for Parallelograms

If a quadrilateral has each pair of opposite sides parallel, it is a parallelogram by definition. This is not the only test, however, that can be used to determine if a quadrilateral is a parallelogram.

**Theorem 7.9: Conditions for Parallelograms**

If both pairs of opposite sides of a quadrilateral are congruent, then the quadrilateral is a parallelogram.

**Theorem 7.10: Conditions for Parallelograms**

If both pairs of opposite angles of a quadrilateral are congruent, then the quadrilateral is a parallelogram.

**Theorem 7.11: Conditions for Parallelograms**

If the diagonals of a quadrilateral bisect each other, then the quadrilateral is a parallelogram.

**Theorem 7.12: Conditions for Parallelograms**

If one pair of opposite sides of a quadrilateral is both parallel and congruent, then the quadrilateral is a parallelogram.

## Example 1 Identify Parallelograms

**Determine whether the quadrilateral is a parallelogram. Justify your answer.**

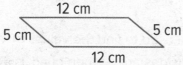

Is the quadrilateral a parallelogram? _____

What theorem can you use to justify your answer?

_____

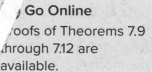 **Go Online** You can complete an Extra Example online.

---

### Go Online

Proofs of Theorems 7.9 through 7.12 are available.

### 💬 Talk About It!

Jude says that by Theorem 7.12, you only need to show that one pair of opposite sides are congruent to show that the quadrilateral is a parallelogram. Do you agree? Justify your answer.

---

**Today's Standards**
G.CO.11, G.GPE.4
MP2, MP3, MP4, MP6

Copyright © McGraw-Hill Education

## 🌐 Example 2  Use Parallelograms and Algebra to Find Values

You can use the conditions of parallelograms along with algebra to find missing values that make a quadrilateral a parallelogram.

SCHOOL SUPPLIES  **The top of the eraser appears to be a parallelogram. Find the values of $x$ and $y$ so that the side of the eraser is a parallelogram.**

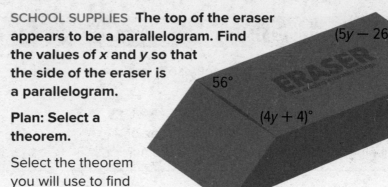

**Plan: Select a theorem.**

Select the theorem you will use to find the values of $x$ and $y$.

**Step 1:** Find $x$ such that $7x = 56$.

$7x = 56$        Opp. angles of a ▱ are ≅.

$x = $ _____        Solve.

**Step 2:** Find $y$ such that $4y + 4 = 5y - 26$.

$4y + 4 = 5y - 26.$        Opp. angles of a ▱ are ≅.

$y = $ _____        Solve.

So, when $x$ is 8 and $y$ is 30, the quadrilateral is a _____.

## Check

MOSAICS  The mosaic pattern of the floor is made up of different tiles.

**Part A**

Find the values of $x$ and $y$ so that the tile is a parallelogram.

$x = $ _____

$y = $ _____

**Part B**

Select the theorem you used to find the values of $x$ and $y$. _____

**A.** If both pairs of opp. sides are ≅, then quad. is a ▱.
**B.** If both pairs of opp. ∠s are ≅, then quad. is a ▱.
**C.** If diag. bisect each other, then quad. is a ▱.
**D.** If one pair of opp. sides is ≅ and ∥, then quad. is a ▱.

🔵 **Go Online** You can complete an Extra Example online.

### Study Tip

**Parallelograms** A quadrilateral needs to pass only one of the five tests to be proven a parallelogram. All of the properties of a parallelogram do not need to be proven.

Copyright © McGraw-Hill Education  Suha Ataoguz/Shutterstock

## Example 3  Parallelograms and Coordinate Geometry

We can use the Distance, Slope, and Midpoint Formulas to determine whether a quadrilateral in the coordinate plane is a parallelogram.

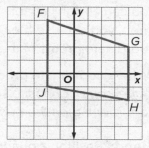

**Determine whether quadrilateral *FGHJ* is a parallelogram. Justify your answer using the Midpoint Formula.**

**Plan: Select a theorem.**

Select the theorem you will use to determine whether quadrilateral *FGHJ* is a parallelogram. _____

**Step 1:** Calculate the midpoint of $\overline{GJ}$.

$$M\left(\frac{x_1+x_2}{2}, \frac{y_1+y_2}{2}\right)$$    Midpoint Formula

$$M\left(\frac{4+(\underline{\phantom{..}})}{2}, \frac{\underline{\phantom{..}}+(-1)}{2}\right)$$    Substitute.

$$M(\underline{\phantom{..}}, \underline{\phantom{..}})$$    Solve.

**Step 2:** Calculate the midpoint of $\overline{FH}$.

$$M\left(\frac{x_1+x_2}{2}, \frac{y_1+y_2}{2}\right)$$    Midpoint Formul

$$M\left(\frac{-2+}{2}, \frac{4+(\underline{\phantom{..}})}{2}\right)$$    Substitute.

$$M(\underline{\phantom{..}}, \underline{\phantom{..}})$$    Solve.

**Step 3:** Determine whether *FGHJ* is a ▱.

If the diagonals of a quadrilateral bisect each oth___ ___en it is a parallelogram. The diagonals of a quadrilateral b___ ___ each other if the midpoints coincide. Because the midpoints ___ ___gonals $\overline{FH}$ and $\overline{GJ}$ _____ have the same coordinates ___ ___ateral *FGHJ* _____ a parallelogram.

## Check

Determine whether quadrilateral *ABCD* is a parallelogram. Justify your answer.

_____

_____

_____

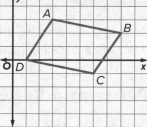

> 🫧 **Think About It!**
>
> Describe another method you could use to determine whether quadrilateral *FGHJ* is a parallelogram.

Copyright © McGraw-Hill Education

🅡 **Go Online** You can complete an Extra Example online.

## Example 4 Parallelograms and Coordinate Proofs

You can assign variable coordinates to the vertices of quadrilaterals. Then, you can use the Distance, Slope, and Midpoint Formulas to write coordinate proofs of theorems.

**Write a coordinate proof for the following statement.**
*If one pair of opposite sides of a quadrilateral is both parallel and congruent, then the quadrilateral is a parallelogram.*

**Step 1:** Position a quadrilateral on the coordinate plane.

Position quadrilateral $ABCD$ on the coordinate plane such that $\overline{AB} \parallel \overline{DC}$ and $\overline{AB} \cong \overline{DC}$.

- Begin by placing the vertex $A$ at the origin.
- Let $\overline{AB}$ have a length of $a$ units. Then $B$ has coordinates $(a, 0)$.
- Because horizontal segments are parallel, position the endpoints of $\overline{DC}$ so that they have the same $y$-coordinate, $c$.
- So that the distance from $D$ to $C$ is also $a$ units, let the $x$-coordinate of $D$ be $b$ and the $x$-coordinate of $C$ be $b + a$.

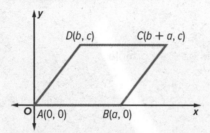

**Step 2:** Use your figure to write a proof.

**Given:** quadrilateral $ABCD$, $\overline{AB} \parallel \overline{DC}$, $\overline{AB} \cong \overline{DC}$

**Prove:** $ABCD$ is a parallelogram.

**Proof:**
By definition, a quadrilateral is a parallelogram if opposite sides are parallel. We are given that $\overline{AB} \parallel \overline{DC}$, so we need to show that $\overline{AD} \parallel \overline{BC}$.

Use the Slope Formula.

slope of $\overline{AD} = \dfrac{c - 0}{b - 0} =$ _____     slope of $\overline{BC} = \dfrac{c - 0}{b + a - a} =$ _____

Because $\overline{AD}$ and $\overline{BC}$ have the same slope, $\overline{AD} \parallel \overline{BC}$. So, quadrilateral $ABCD$ is a parallelogram because opposite sides are parallel.

### Go Online
You may want to complete the construction activities for this lesson.

## Pause and Reflect

Did you struggle with anything in this lesson? If so, how did you deal with it?

Record your observations here

 **Go Online** You can complete an Extra Example online.

Copyright © McGraw-Hill Education

# Practice

⊘ **Go Online** You can complete your homework online.

### Example 1

**Determine whether each quadrilateral is a parallelogram. Justify your answer.**

1.

2.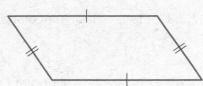

3.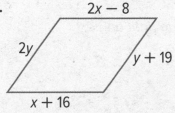

4.

5.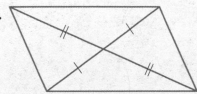

6.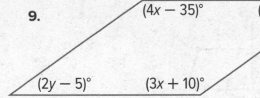

### Example 2

**Find the values of x and y so that each quadrilateral is a parallelogram.**

7.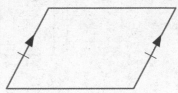
$2x - 8$
$2y$
$y + 19$
$x + 16$

8.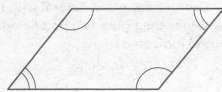
$3x$
$4x - 3$
$y + 5$

9.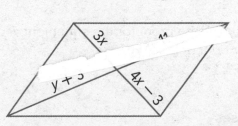
$(4x - 35)°$   $(y + 15)°$
$(2y - 5)°$   $(3x + 10)°$

10.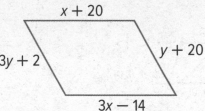
$x + 20$
$3y + 2$
$y + 20$
$3x - 14$

11.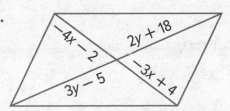
$-4x - 2$   $2y + 18$
$3y - 5$   $-3x + 4$

12.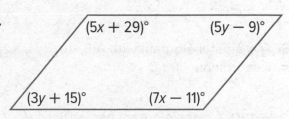
$(5x + 29)°$   $(5y - 9)°$
$(3y + 15)°$   $(7x - 11)°$

Copyright © McGraw-Hill Education

Example 3

13. REGULARITY Determine whether *ABCD* is a parallelogram. Justify your answer.

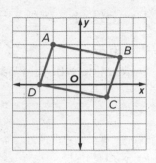

ARGUMENTS Graph each quadrilateral with the given vertices. Determine whether the figure is a parallelogram. Justify your answer with the method indicated.

14. $P(0, 0)$, $Q(3, 4)$, $S(7, 4)$, $Y(4, 0)$; Slope Formula

15. $S(-2, 1)$, $R(1, 3)$, $T(2, 0)$, $Z(-1, -2)$; Distance and Slope Formulas

16. $W(2, 5)$, $R(3, 3)$, $Y(-2, -3)$, $N(-3, 1)$; Midpoint Formula

17. $W(1, -4)$, $X(-4, 2)$, $Y(1, -1)$, and $Z(-2, -3)$; Slope Formula

Example 4

18. Write a coordinate proof for the statement: *If both pairs of opposite sides of a quadrilateral are congruent, then the quadrilateral is a parallelogram.*

19. Write a coordinate proof for the statement: *If a parallelogram has one right angle, it has four right angles.*

Mixed Exercises

Name the missing coordinates for each parallelogram.

20.

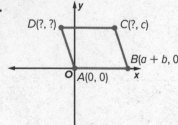

21.

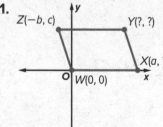

22. *ABCD* is a parallelogram with $A(5, 4)$, $B(-1, -2)$, and $C(8, -2)$. Find one possible set of coordinates for *D*.

23. MODELING A parallelogram has vertices $R(-2, -1)$, $S(2, 1)$, and $T(0, -3)$. Find all possible coordinates for the fourth vertex.

Copyright © McGraw-Hill Education

**24.** If the slope of $\overline{PQ}$ is $\frac{2}{3}$ and the slope of $\overline{QR}$ is $-\frac{1}{2}$, find the slope of $\overline{SR}$ so that *PQRS* is a parallelogram.

**25.** If the slope of $\overline{AB}$ is $\frac{1}{2}$, the slope of $\overline{BC}$ is $-4$, and the slope of $\overline{CD}$ is $\frac{1}{2}$, find the slope of $\overline{DA}$ so that *ABCD* is a parallelogram.

**26. STRUCTURE** The pattern shown in the figure is to consist of congruent parallelograms. How can the designer be certain that the shapes are parallelograms?

**27.** Refer to parallelogram *ABCD*. If *AB* = 8 cm, what is the perimeter of the parallelogram?

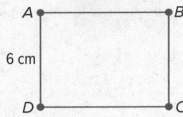

**28. PICTURE FRAME** Aston is making a wooden picture frame in the shape of a parallelogram. He has two pieces of wood that are 3 feet long and two that are 4 feet long.

   **a.** If he connects the pieces of wood at their ends to each other, in what order must he connect them to make a parallelogram?

   **b.** How many different parallelograms could he make with these four lengths of wood?

   **c.** Explain something Aston might do to specify precisely the shape of the parallelogram.

**29. STATE YOUR ASSUMPTION** When a coordinate plane is placed over the Harrisville town map, the four street lamps in the center are located as shown. Do the four lamps form the vertices of a parallelogram? What is an assumption you are making regarding the coordinate plane and the map?

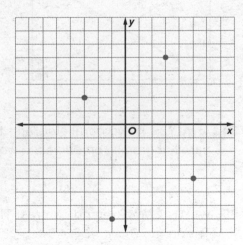

**30. USE TOOLS** Explain how you can use Theorem 7.11 to construct a parallelogram. Then construct a parallelogram using your method.

Copyright © McGraw-Hill Education

**31. BALANCING** Nikia, Madison, Angela, and Shelby are balancing on an "X"-shaped floating object. To balance, they want to stand so they are at the vertices of a parallelogram. In order to achieve this, do all four of them have to be the same distance from the center of the object? Explain.

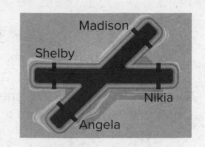

**32. FORMATION** Four jets are flying in formation. Three of the jets are shown in the graph. If the four jets are located at the vertices of a parallelogram, what are the three possible locations of the missing jet?

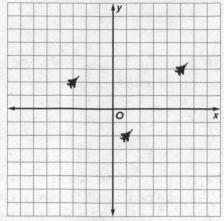

**33. ARGUMENTS** Camila draws the following figure to prove that if the diagonals of a quadrilateral are congruent, then the quadrilateral is a parallelogram. Draw a counterexample to show that Camila is incorrect. What mistake did she make?

**34. ANALYZE** If two parallelograms have four congruent corresponding angles, are the parallelograms *sometimes, always,* or *never* congruent?

**35. WRITE** Compare and contrast Theorem 7.9 and Theorem 7.3.

**36. PERSEVERE** If *ABCD* is a parallelogram and $\overline{AJ} \cong \overline{KC}$, show that quadrilateral *JBKD* is a parallelogram.

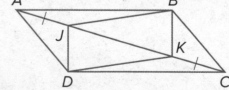

**37. CREATE** The diagonals of a parallelogram meet at the point (0, 1). One vertex of the parallelogram is located at (2, 4), and a second vertex is located at (3, 1). Find the locations of the remaining vertices.

**38. ANALYZE** Write a coordinate proof to prove that the segments joining the midpoints of the sides of any quadrilateral form a parallelogram.

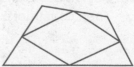

Copyright © McGraw-Hill Education

# Rectangles

## Explore Properties of Rectangles

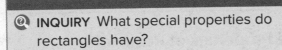 **Online Activity** Use dynamic geometry software to complete the Explore.

---

**? INQUIRY** What special properties do rectangles have?

---

**Today's Standards**
G.CO.11; G.GPE.4
MP1, MP3, MP4

**Today's Vocabulary**
rectangle

## Learn Properties of Rectangles

A rectangle is a parallelogram with four right angles. From this definition, you know that a rectangle has the following properties:

- All four angles are right angles.
- Opposite sides are parallel and congruent.
- Opposite angles are congruent.
- Consecutive angles are supplementary.
- Diagonals bisect each other.

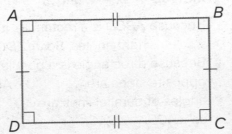

In addition, the diagonals of a rectangle are congruent.

**Theorem 7.13: Diagonals of a Rectangle**

If a parallelogram is a rectangle, then its diagonals are congruent.

**Go Online**
A proof of Theorem 7.13 is available.

**Think About It!**
What does point $G$ represent in the context of the problem?

## 🌐 Example 1 Use Properties of Rectangles

**BASKETBALL Coach is making the basketball team run a new drill along the diagonals of the court, as shown. If BC = 94 feet and FC = 106.5 feet, find DG.**

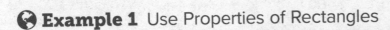

$\overline{FC} \cong$ _____     If a ☐ is a rectangle, diag. are ≅.

$FC =$ _____     Definition of congruence

$BD =$ _____     Substitution

Because *BCDF* is a rectangle, it is a parallelogram. The diagonals of a parallelogram bisect each other, so *DG = BG*.

$DG +$ _____ $= BD$          Segment Addition

$DG + DG = BD$          _____

_____$DG = BD$          Simplify.

$DG = \frac{1}{2}$ _____          Divide each side by 2.

$DG = \frac{1}{2}$ _____ or 53.25 ft          Substitution

Copyright © McGraw-Hill Education

## Check

**FARMING** Jay is framing a barn door with an X-brace as shown. If $RT = 3\frac{9}{16}$, $QR = 7$ feet, and $m\angle RTS = 65°$, find each measure. If a measure is not a whole number, write it as a decimal.

$PS = $ _____ ft          $SQ = $ _____ ft

$m\angle QTR = $ _____°          $m\angle TQR = $ _____°

**Study Tip**

**Right Angles** Recall from Theorem 7.6 that if a parallelogram has one right angle, then it has four right angles.

**🐑 ...ink About It!**

T..re are four ..ngruent right ..riangles formed by the diagonals of a rectangle. How many pairs of congruent triangles are there in all?

## 🌐 **Example 2** Use Properties of Rectangles and Algebra

**ALGEBRA Quadrilateral *ABCD* is a rectangle. If $m\angle BAC = 3x + 3$ and $m\angle ACB = 5x - 1$, find *x*.**

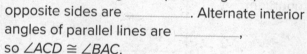

Because *ABCD* is a rectangle, it has _____ right angles. So, $m\angle DCB = 90$. Because a rectangle is a parallelogram, opposite sides are _____. Alternate interior angles of parallel lines are _____, so $\angle ACD \cong \angle BAC$.

| | |
|---|---|
| $m\angle ACD + m\angle ACB = 90$ | Angle Addition |
| $m\angle BAC + \angle ACB = 90$ | _____ |
| _____ $+ 5x - 1 = 90$ | Substitution |
| _____ $x +$ _____ $= 90$ | Add like terms. |
| $8x = $ _____ | Subtract 2 from each side. |
| $x = $ _____ | Divide each side by 8. |

## Check

**ALGEBRA Quadrilateral *JKLM* is a rectangle.**

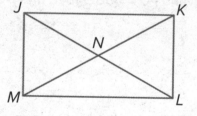

**Part A**

If $MN = 3x + 1$ and $JL = 2x + 9$, find *MK*. Write your answer as a decimal.

$MK = $ _____

**Part B**

If $m\angle JNK = (5x + 2)°$ and $m\angle JNM = (3x - 6)°$, find each measure.

$m\angle JNK = $ _____°

$m\angle JNM = $ _____°

**🡒 Go Online** You can complete an Extra Example online.

Copyright © McGraw-Hill Education   Robert Hamm/123RF

# Learn Prove that Parallelograms are Rectangles

You have learned that the diagonals of parallelograms are congruent. Well, the converse is also true.

> **Theorem 7.14: Diagonals of a Rectangle**
>
> If the diagonals of a parallelogram are congruent, then the parallelogram is a rectangle.

🌐 **Go Online** A proof of Theorem 7.14 is available.

## 🌐 Example 3 Prove Rectangular Relationships

**LANDSCAPING** Jayson is rototilling an area of his lawn to prepare the soil for planting a garden. He marks off the four corners of the planned area and then measures the diagonals with orange tape. If $AB = 50$ feet, $BC = 20$ feet, $CD = 50$ feet, $AD = 20$ feet, $AC = 54$ feet, and $BD = 54$ feet, explain how Jayson can be sure that the planned area is rectangular.

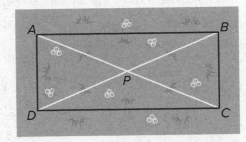

Because $AB = CD$, $BC = AD$, and $AC = BD$, $\overline{AB} \cong \overline{CD}$, $\overline{BC} \cong \overline{AD}$, and $\overline{AC} \cong \overline{BD}$. Because $\overline{AB} \cong \overline{CD}$ and $\overline{BC} \cong \overline{AD}$, $ABCD$ is a _____. Because $\overline{AB}$ and $\overline{BD}$ are congruent diagonals in $ABCD$ is a

_____.

## Check

Complete the proof with the correct statements and reasons.

**Given:** $PQRS$ is a rectangle; $\overline{PT} \cong \overline{ST}$.

**Prove:** $\overline{QT} \cong \overline{RT}$

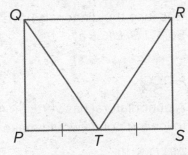

**Proof:**

| Statements | Reasons |
|---|---|
| **1.** $PQRS$ is a rectangle. | **1.** Given |
| **2.** $PQRS$ is a parallelogram. | **2.** Def. of rectangle |
| **3.** _____ | **3.** Opp. Sides of a ▱ are ≅. |
| **4.** _____ | **4.** Def. of rectangle |
| **5.** $\angle S \cong \angle P$ | **5.** All rt $\angle s$ are ≅. |
| **6.** _____ | **6.** SAS |
| **7.** _____ | **7.** CPCTC |

$\angle S$ and $\angle P$ are right angles.    $\overline{RS} \cong \overline{OP}$    $\overline{QT} \cong \overline{RT}$    $\triangle RST \approx \triangle QPT$

🌐 **Go Online** You can complete an Extra Example online.

**Use a Source**

In 1853, the New York State legislature enacted a law to set aside more than 750 acres of land in central Manhattan. This area is now known as Central Park, America's first major landscaped public park, which eventually became one of the most famous and most visited parks in the world. Use available resources to find and use the dimensions of Central Park to prove that it is rectangular.

Copyright © McGraw-Hill Education

## Study Tip

**Rectangles and Parallelograms** All rectangles are parallelograms, but all parallelograms are not necessarily rectangles.

😃 **Think About It!**

Is there another way to show that *GHJK* is a rectangle? If yes, explain.

## 🌐 **Example 4** Rectangles on the Coordinate Plane

You can also use the properties of rectangles to prove that a quadrilateral positioned on a coordinate plane is a rectangle given the coordinates of the vertices.

**COORDINATE GEOMETRY** **Quadrilateral *GHJK* has vertices *G*(−3, 0), *H*(3, 2), *J*(4, −1), and *K*(−2, −3). Determine whether *GHJK* is a rectangle by using the Distance Formula.**

**Step 1:** Determine whether opposite sides are congruent. Use the Distance Formula.

$GH = \sqrt{(-3 - \underline{\hspace{0.5cm}})^2 + (0 - \underline{\hspace{0.5cm}})^2}$ or $\sqrt{\underline{\hspace{0.5cm}}}$

$HJ = \sqrt{(3 - \underline{\hspace{0.5cm}})^2 + [2 - \underline{\hspace{0.5cm}}]^2}$ or $\sqrt{\underline{\hspace{0.5cm}}}$

$KJ = \sqrt{(-2 - \underline{\hspace{0.5cm}})^2 + [-3 - \underline{\hspace{0.5cm}}]^2}$ or $\sqrt{\underline{\hspace{0.5cm}}}$

$GK = \sqrt{[-3 - \underline{\hspace{0.5cm}}]^2 + [0 - \underline{\hspace{0.5cm}}]^2}$ or $\sqrt{\underline{\hspace{0.5cm}}}$

Because opposite sides of the quadrilateral have the same measure, they _____ congruent. So, quadrilateral *GHJK* _____ a parallelogram.

**Step 2:** Determine whether diagonals are congruent. Use the Distance Formula again.

$GJ = \sqrt{(-3 - \underline{\hspace{0.5cm}})^2 + [0 - \underline{\hspace{0.5cm}}]^2}$ or $\sqrt{\underline{\hspace{0.5cm}}}$

$KH = \sqrt{(-2 - \underline{\hspace{0.5cm}})^2 + (-3 - \underline{\hspace{0.5cm}})^2}$ or $\sqrt{\underline{\hspace{0.5cm}}}$

Because the diagonals have the same measure, they are _____.
So, ▱ *GHJK* is a _____.

## Check

A quadrilateral has vertices *A*(2, 6), *B*(3, 7), and *C*(6, 4). Which of the following points would make *ABCD* a rectangle? _____

**A.** *D*(5, 3)

**B.** *D*(5, 2)

**C.** *D*(4, 3)

**D.** *D*(6, 3)

## **Pause and Reflect**

Did you struggle with anything in this lesson? If so, how did you deal with it?

Record your observations here

🌐 **Go Online** You can complete an Extra Example online.

Copyright © McGraw-Hill Education

# Practice

**Go Online** You can complete your homework online.

### Example 1

**FENCING** X-braces are also used to provide support in rectangular fencing.
If $AB = 6$ feet, $AD = 2$ feet, and $m\angle DAE = 65$, find each measure. Round to the nearest tenth, if necessary.

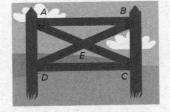

**1.** $BC$

**2.** $DB$

**3.** $m\angle CEB$  °

**4.** $m\angle EDC$  °

**MEASUREMENT** Quadrilateral $ABCD$ is a rectangle. Find each measure if $m\angle 1 = 38°$.

**5.** $m\angle 2$  °

**6.** $m\angle 5$  °

**7.** $m\angle 6$  °

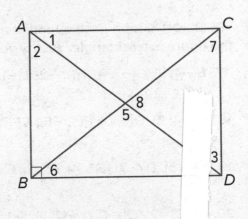

**8. PARKING** The lines of the parking space shown are parallel. How wide is the space (in inches)?  .

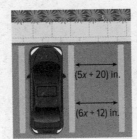

### Example 2

**9. REASONING** Quadrilateral $ABCD$ is a rectangle. If $m\angle ADB = (4x + 8)°$ and $m\angle DBA = (6x + 12)°$, find the value of $x$.

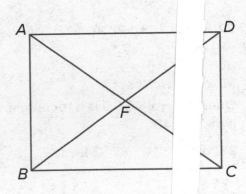

**ALGEBRA** Quadrilateral $EFGH$ is a rectangle.

**10.** If $m\angle FEG = 57°$, find $m\angle GEH$.  °

**11.** If $m\angle HGE = 13°$, find $m\angle FGE$.  °

**12.** If $FK = 32$ feet, find $EG$.

**13.** Find $m\angle HEF + m\angle EFG$.  °

**14.** If $EF = 4x - 6$ and $HG = x + 3$, find $EF$.

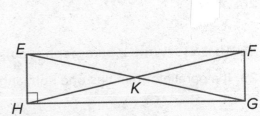

Copyright © McGraw-Hill Education

Example 3

ARGUMENTS  Write a two-column proof.

15. Given: *ABCD* is a rectangle.

Prove: $\triangle ADC \cong \triangle BCD$

16. Given: *QTVW* is a rectangle, $\overline{QR} \cong \overline{ST}$

Prove: $\triangle SWQ \cong \triangle RVT$

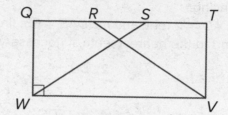

Example 4

PRECISION  Graph each quadrilateral with the given vertices. Determine whether the figure is a rectangle. Justify your answer using the indicated formula.

17. *B*(−4, 3), *G*(−2, 4), *H*(1, −2), *L*(−1, −3); Slope Formula

18. *N*(−4, 5), *O*(6, 0), *P*(3, −6), *Q*(−7, −1); Distance Formula

19. *C*(0, 5), *D*(4, 7), *E*(5, 4), *F*(1, 2); Slope Formula

20. *P*(−3, −2), *Q*(−4, 2), *R*(2, 4), *S*(3, 0); Slope Formula

21. *J*(−6, 3), *K*(0, 6), *L*(2, 2), *M*(−4, −1); Distance Formula

22. *T*(4, 1), *U*(3, −1), *X*(−3, 2), *Y* (−2, 4); Distance Formula

Mixed Exercises

Quadrilateral *ABCD* is a rectangle. Find each measure if $m\angle 2 = 40°$.

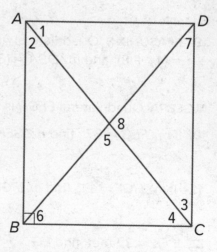

23. $m\angle 1$  °

24. $m\angle 7$  °

25. $m\angle 3$  °

26. $m\angle 5$  °

27. $m\angle 6$  °

28. $m\angle 8$  °

ARGUMENTS  Write a paragraph proof of each statement.

29. If a parallelogram has one right angle, then it is a rectangle.

30. If a quadrilateral has four right angles, then it is a rectangle.

Copyright © McGraw-Hill Education

**31. LANDSCAPING** Huntington Park officials approved a rectangular plot of land for a Japanese Zen garden. Is it sufficient to know that opposite sides of the garden plot are congruent and parallel to determine that the garden plot is rectangular? Explain.

**32.** Name a property that is true for a rectangle and not always true for a parallelogram.

**33. USE TOOLS** Construct a rectangle using the construction for congruent segments and the construction for a line perpendicular to another line through a point on the line. Justify each step of the construction.

**34. SIGNS** The sign is attached to the front of Jackie's lemonade stand. Based on the dimensions given, can Jackie be sure that the sign is a rectangle: Explain your reasoning.

**ALGEBRA** **Quadrilateral WXYZ is a rectangle.**

**35.** If $XW = 3$, $WZ = 4$, and $XZ = b$, find $YW$.

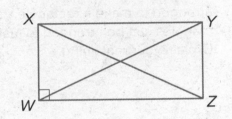

**36.** If $XZ = 2c$ and $ZY = 6$, and $XY = 8$, find $WY$.

**37. FRAMES** Jalen makes the rectangular frame shown. In order to make sure that it is a rectangle, Jalen measures the distances $BD$ and $AC$. How should these two distances compare if the frame is a rectangle?

**38. BOOKSHELVES** A bookshelf consists of two vertical planks with five horizontal shelves. Are each of the four sections for books rectangles? Explain.

**39. LANDSCAPING** A landscaper is marking off the corners of a rectangular plot of land. Three of the corners are in the place as shown. What are the coordinates of the fourth corner?

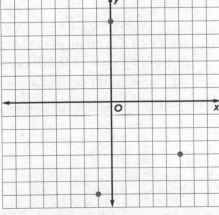

Copyright © McGraw-Hill Education

**40. STRUCTURE** Veronica made the pattern shown out of 7 rectangles with four equal sides. The side length of each rectangle is written inside the rectangle.

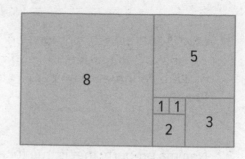

   **a.** How many rectangles can be formed using the lines in this figure?

   **b.** If Veronica wanted to extend her pattern by adding another rectangle with 4 equal sides to make a larger rectangle, what are the possible side lengths of rectangles that she can add?

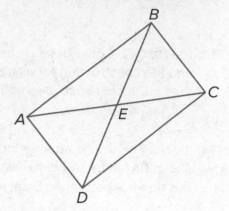

**41. PERSEVERE** In rectangle $ABCD$, $m\angle EAB = (4x + 6)°$, $m\angle DEC = (10 - 11y)°$, and $m\angle EBC = 60°$. Find the values of $x$ and $y$.

**42. FIND THE ERROR** Parker says that any two congruent acute triangles can be arranged to make a rectangle. Tamika says that only two congruent right triangles can be arranged to make a rectangle. Is either of them correct? Explain your reasoning.

**43. WRITE** Why are all rectangles parallelograms, but all parallelograms are not rectangles? Explain.

**44. CREATE** Write the equations of four lines having intersections that form the vertices of a rectangle. Very your answer using coordinate geometry.

**45. ANALYZE** David argues that to prove a quadrilateral is a rectangle, it is sufficient to prove that its diagonals are congruent. Do you agree? If so, explain why. If not, explain and draw a counterexample.

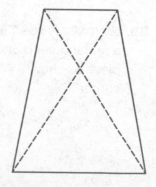

Copyright © McGraw-Hill Education

# Rhombi and Squares

## Explore Properties of Rhombi and Squares

**Today's Standards**
G.CO.11, G.GPE.4
MP1, MP7

**Online Activity** Use dynamic geometry software to complete the Explore.

> ⊘ **INQUIRY** What special properties do rhombi and squares have?

**Today's Vocabulary**
rhombus
square

## Learn Properties of Rhombi and Squares

A **rhombus** is a parallelogram with all four sides congruent. All of the properties of a parallelogram hold true for a rhombus, in addition to the following two theorems.

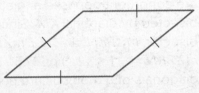

**Theorem 7.15: Diagonals of a Rhombus**

If a parallelogram is a rhombus, then its diagonals are perpendicular.

**Theorem 7.16: Diagonals of a Rhombus**

If a parallelogram is a rhombus, then each diagonal bisects a pair of opposite angles.

**Go Online**
Proofs of theorems 7.15 and square are available.

A **square** is a parallelogram with all four sides and all four angles congruent. All of the properties of parallelograms, rectangles, and rhombi apply to squares. For example, the diagonals of a square bisect each other (parallelogram), are congruent (rectangle), and are perpendicular (rhombus).

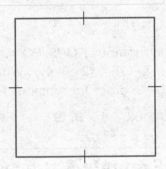

**Think About It!**

How are the definition of a rhombus and the definition of congruence used to justify the first and second steps?

## ⊙ Example 1 Use the Definition of Rhombus

**ALGEBRA** If $LM = 2x - 9$ and $KN = x + 15$ in rhombus *KLMN*, find the value of *x*.

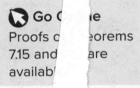

| | |
|---|---|
| $\overline{LM} \cong$ _____ | Definition of rhombus |
| $LM =$ _____ | Definition of congruence |
| $2x - 9 =$ _____ | Substitution |
| $x - 9 = 15$ | Subtract _____ from each side. |
| $x =$ _____ | Add 9 to each side. |

⊙ **Go Online** You can complete an Extra Example online.

Copyright © McGraw-Hill Education

📢 **Talk About It!**

Compare all of the properties of the following quadrilaterals: parallelograms, rectangles, rhombi, and squares.

## Check

**ALGEBRA** Quadrilateral *WXYZ* is a rhombus. If *AY* = 14, *ZY* = 22, and *m∠WYZ* = 35°, find each measure.

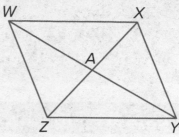

*WY* = _____    *m∠XYZ* = _____°    *m∠WXZ* = _____°

---

## Example 2  Use Diagonals of a Rhombus

**The diagonals of rhombus *KLMN* intersect at *P*. If *m∠LMN* = 75°, find *m∠KNP*.**

Because we know that *KLMN* is a rhombus, we can use the definition of a rhombus to say that ∠*LKN* and ∠_____ are congruent opposite angles that are bisected by diagonal $\overline{KM}$. Since $\overline{KM}$ is a bisector, *m∠PKN* = _____ *m∠LKN*. So *m∠PKN* = $\frac{1}{2}$(_____°) or 37.5°. Because the diagonals of a rhombus are perpendicular, *m∠KPN* = _____° by the definition of perpendicular lines.

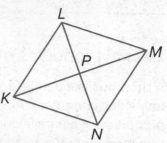

| | |
|---|---|
| *m∠PKN* + *m∠KPN* + *m∠*_____ = 180 | Triangle Sum Theorem |
| 37.5 + 90 + *m∠KNP* = 180 | _____ |
| _____ + *m∠KNP* = 180 | Substitution |
| *m∠KNP* = _____° | Subtract 127.5 from each side. |

## Check

In rhombus *PQRS*, *PQ* = 4*X* + 3, *QR* = 41, and *m∠PQT* = (2*x* + 4*y*)°. What must the value of *y* be in order for rhombus *PQRS* to be a square?

**A.** 6.5    **B.** 9.5    **C.** 45    **D.** 90

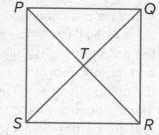

---

## Example 3  Use the Definition of a Square

**EFGH is a square. If FJ = 19, find FH.**

Because *EFGH* is square, it is both a parallelogram and a rectangle. Therefore, we know that its diagonals bisect each other and are congruent.

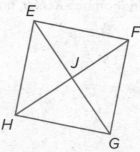

| | |
|---|---|
| $\overline{FJ} \cong \overline{JH}$ | Definition of a square |
| *FJ* = _____ | Definition of congruence |
| _____ = *JH* | Substitution |
| *FJ* + *JH* = _____ | Definition of bisector |
| 19 + _____ = *FH* | Substitution |
| _____ = *FH* | Simplify. |

🗲 **Go Online** You can complete an Extra Example online.

Copyright © McGraw-Hill Education

# Learn Properties of Rhombi and Squares

If a parallelogram meets certain conditions, you can conclude that it is a rhombus or a square.

> **Theorem 7.17: Conditions for Rhombi and Squares**
>
> If the diagonals of a parallelogram are perpendicular, then the parallelogram is a rhombus.

> **Theorem 7.18: Conditions for Rhombi and Squares**
>
> If one diagonal of a parallelogram bisects a pair of opposite angles, then the parallelogram is a rhombus.

> **Theorem 7.19: Conditions for Rhombi and Squares**
>
> If two consecutive sides of a parallelogram are congruent, then the parallelogram is a rhombus.

> **Theorem 7.20: Conditions for Rhombi and Squares**
>
> If a quadrilateral is both a rectangle and a rhombus, then it is a square.

**Go Online** Proofs of Theorems 7.17 through 7.20 are available.

## Example 4 Use Conditions for Rhombi and Squares

You can use the properties of rhombi and squares to write proofs.

**Write a paragraph proof.**

**Given:** $TUVW$ is a parallelogram.

$\triangle TSW \cong \triangle TSU$

**Prove:** $TUVW$ is a rhombus.

**Paragraph Proof:**

Since it is given that $\triangle TSW \cong \triangle TSU$, it must be true that _____ $\cong \overline{UT}$. Since _____ and $\overline{UT}$ are congruent consecutive sides of the given parallelogram, we can prove that $TUVW$ is a rhombus by using Theorem _____.

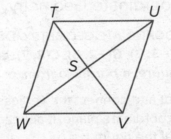

## 🌐 Example 5 Use Properties of a Rhombus

**GARMENT DESIGN** **Ananya is designing a sweater using an argyle pattern. All four sides of quadrilateral *ABCD* are 2 inches long. How can Ananya be sure that the argyle pattern is a square?**

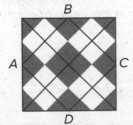

**Make a Plan**

A square has all of the properties of a parallelogram, a rhombus, and a rectangle. To prove that quadrilateral $ABCD$ is a square, prove that it is a parallelogram, a rhombus, and a rectangle.

*(continued on the next page)*

**Go Online** You can complete an Extra Example online.

**Go Online** You may want to complete the Concept Check to check your understanding.

**Math History Minute**

**Robert Ammann (1946–1994)** was a programmer who considered himself an amateur mathematician. Although he did not study mathematics in college, Ammann discovered new ways to tile a plane by using quadrilaterals including rhombi. One of the tilings, the Ammann-Beenker tiling, is named for him.

**Study Tip**

**Common Misconceptions** The conditions for rhombi and squares only apply if you already know that a quadrilateral is a parallelogram.

Copyright © McGraw-Hill Education

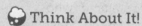

🍥 **Think About It!**

How could Ananya show
that the pattern is a
square in a different way?

**Study Tip**

**Assumptions** Although
it would be difficult to
measure a perfect
square on a piece of
fabric due to its
stretching and shrinking
qualities, we assume
that the properties of a
square will hold true, for
example, the congruence
of its diagonals.

🍥 **Think About It!**

What other way could
you use to determine
whether a quadrilateral
is a rhombus?

**Study Tip**

**Make a Graph** When
analyzing a figure using
coordinate geometry,
graph the figure to help
formulate a conjecture
and also to help check the
reasonableness of your
answer. Use the same
scale on the x- and y-axes
so the representation is as
accurate as possible. Be
sure to choose a window
that will allow you to see
all of the vertices.

**Rhombus**

Draw and label quadrilateral *ABCD* to
model the argyle pattern. Each side
of quadrilateral *ABCD* measures
2 inches. Because _____ pairs of
opposite sides are congruent, *ABCD*
is a _____.

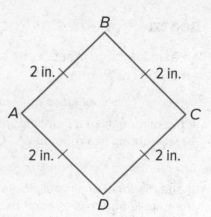

**Rectangle**

Because _____ sides of □*ABCD*
are congruent, it is a rhombus.

**Show Diagonals**

If the diagonals of a parallelogram are
congruent, then the parallelogram is a
_____. So, if Ananya measures
the length of each diagonal and finds
that they are equal, then *ABCD* is a
_____.

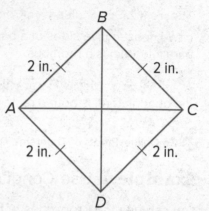

## 🌐 **Example 6** Classify Parallelograms Using Coordinate Geometry

**COORDINATE GEOMETRY** Determine whether □*ABCD* with vertices
*A*(−3, 2), *B*(−2, 6), *C*(2, 7), and *D*(1, 3) is a *rhombus*, a *rectangle*, a
*square*, a *parallelogram*, or *none*. List all that apply. Explain.

Plot and connect the vertices on
a coordinate plane. It appears
that the figure is a _____.

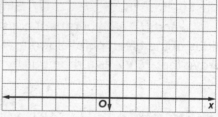

*ABCD* appears to be a
parallelogram, but is it also a
rhombus? To check if figure can
be further classified, compare the
slopes of the diagonals to determine
whether they are perpendicular.

slope of $\overline{AC} = \dfrac{7}{-3} = \dfrac{5}{}$ or _____

slope of $\overline{BD} = \dfrac{3 - }{- 2 - } = \dfrac{-3}{}$ or _____

Since the product of the slopes of the diagonals is _____, the diagonals
are _____, so □*ABCD* is a rhombus.

ℝ **Go Online** You can complete an Extra Example online.

Copyright © McGraw-Hill Education

# Practice

🔵 **Go Online** You can complete your homework online.

### Examples 1 and 2

**Quadrilateral *ABCD* is a rhombus. Find each value or measure.**

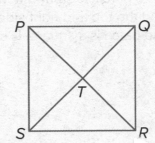

1. If $m\angle ABD = 60°$, find $m\angle BDC$. °

2. If $AE = 8$, find $AC$.

3. If $AB = 26$ and $BD = 20$, find $AE$.

4. Find $m\angle CEB$. °

5. If $m\angle CBD = 58°$, find $m\angle ACB$. °

6. If $AE = 3x - 1$ and $AC = 16$, find $x$.

7. If $m\angle CDB = 6y°$ and $m\angle ACB = (2y + 10)°$, find the value of $y$.

8. If $AD = 2x + 4$ and $CD = 4x - 4$, find the value of $x$.

### Example 3

9. *PQRS* is a square. If $PR = 42$, find $TR$.

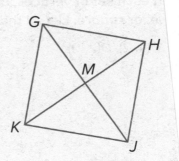

10. *GHJK* is a square. If $KM = 26.5$, find $KH$.

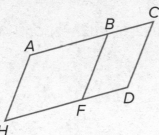

### Example 4

**Write a two-column proof.**

11. **Given:** *ACDH* and *BCDF* are parallelograms, $\overline{BF} \cong \overline{AB}$
    **Prove:** *ABFH* is a rhombus.

12. **Given:** *QRST* is a parallelogram $\overline{?R} \cong \overline{QS}$
    $m\angle QPR = 90$
    **Prove:** *QRST* is a square.

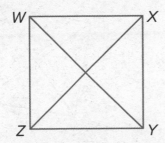

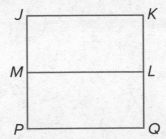

13. **Given:** $\overline{WZ} \parallel \overline{XY}$, $\overline{WX} \parallel \overline{ZY}$, $\overline{WX} \cong \overline{ZY}$
    **Prove:** *WXYZ* is a rhombus.

14. **Given:** *JKQP* is a square, $\overline{ML}$ bisects $\overline{JP}$ and $\overline{KQ}$
    **Prove:** *JKLM* is a parallelogram.

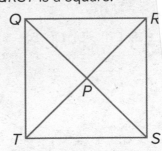

Copyright © McGraw-Hill Education

## Example 5

**15.** PRECISION  Jorge is using this box garden to plant his vegetables this year. What does Jorge need to know to ensure that the box garden is a square? Explain.

4 ft

4 ft

**16.** PRECISION  Ingrid is designing a quilt with patches like the one shown. The patch is a parallelogram with all four angles having the same measure and the top and right sides having the same measure. Ingrid says that the patch is a square. Is she correct? Explain.

## Example 6

REGULARITY  **Given each set of vertices, determine whether □*ABCD* is a** *rhombus, rectangle,* **or** *square.* **List all that apply. Explain.**

**17.** $A(0, 2)$, $B(2, 4)$, $C(4, 2)$, $D(2, 0)$

**18.** $A(-2, -1)$, $B(0, 2)$, $C(2, -1)$, $D(0, -4)$

**19.** $A(-6, -1)$, $B(4, -6)$, $C(2, 5)$, $D(-8, 10)$

**20.** $A(2, -4)$, $B(-6, -8)$, $C(-10, 2)$, $D(-2, 6)$

**21.** $A(1, 3)$, $B(7, -3)$, $C(1, -9)$, $D(-5, -3)$

**22.** $A(-9, 1)$, $B(2, 3)$, $C(12, -2)$, $D(1, -4)$

## Mixed Exercises

***BCDF* is a square with** $BC = 2x + 9$, $FD = 55$, **and** $GD = x + 8$**. Find each measure.**

**23.** *BC*

**24.** *CD*

**25.** *GD*

**26.** *BD*

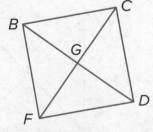

Copyright © McGraw-Hill Education   Andrey Simonenko/123RF

27. **REASONING** In rhombus $ABCD$, $\overline{AB} = 2x + 3$, $\overline{BC} = 21$, and $m\angle ABF = (x + 2y)°$. What must the value of $y$ be in order for rhombus $ABCD$ to be a square?

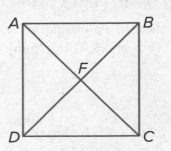

**PRYZ is a rhombus. If RK = 5, RY = 13 and $m\angle YRZ = 67°$, find each measure.**

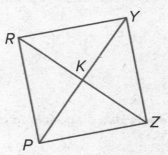

28. $KY$

29. $m\angle MNP$ °

30. $m\angle APQ$ °

31. $PM$

**WXYZ is a square. If WT = 3, find each measure.**

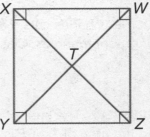

32. $ZX$

33. $XY$

34. $WTZ$ °

35. $m\angle WYX$ °

36. **USE TOOLS (ESTIMATION)** The figure is an example of a tessellation. Estimate the type and number of shapes in the figure. Use a ruler or protractor to measure the shapes and then name the quadrilaterals used to form the figure. Compare this to your estimation.

37. **USE TOOLS (ESTIMATION)** The figure is an example of a quilt pattern. Estimate the type and number of shapes in the figure. Use a ruler or protractor to measure the shapes and then name the quadrilaterals used to form the figure. Compare this to your estimation.

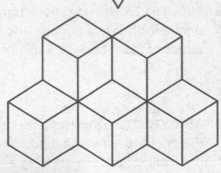

**Classify each quadrilateral.**

38.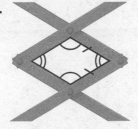

39.

40.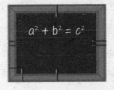

Copyright © McGraw-Hill Education

**Arguments** Write a two-column proof.

**41. Given:** Parallelogram *RSTU*. $\overline{RS} \cong \overline{ST}$
**Prove:** *RSTU* is a rhombus.

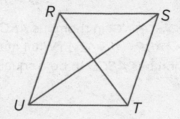

**42. Given:** *RSTU* is a parallelogram. $\overline{RX} \cong \overline{TX} \cong \overline{SX} \cong \overline{UX}$
**Prove:** *RSTU* is a rectangle.

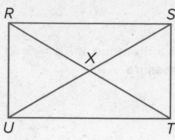

**Use Tools** Use diagonals to construct each figure. Justify each construction.

**43.** rhombus

**44.** square

**45. SLICING** Charles cuts a rhombus along both diagonals He ends up with four congruent triangles. Classify these triangles as *acute*, *obtuse*, or *right*.

**46. PERSEVERE** The area of square *ABCD* is 36 square units and the area of $\triangle EBF$ is 20 square units. If $\overline{EB} \perp \overline{BF}$ and $\overline{AE} = 2$, find the length of $\overline{CF}$.

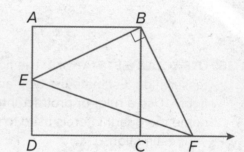

**47. WRITE** Compare all of the properties of the following quadrilaterals: parallelograms, rectangles, rhombi, and squares.

**48. FIND THE ERROR** In parallelogram *PQRS*, $\overline{PR} \cong \overline{QS}$. Lola thinks that the parallelogram is a square, and Xavier thinks that it is a rhombus. Is either of them correct. Explain your reasoning.

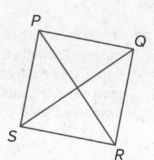

**49. ANALYZE** Determine whether the statement is *true* or *false*. Then write the converse, inverse, and contrapositive of the statement and determine the truth value of each. Explain your reasoning.
    *If a quadrilateral is a square, then it is a rectangle.*

**50. CREATE** Find the vertices of a square with diagonals that are contained in the graphs of $y = x$ and $y = -x + 6$. Justify your reasoning.

Copyright © McGraw-Hill Education

# Trapezoids and Kites

## Explore Properties of Trapezoids and Kites

**Today's Standards**
G.GPE.4

MP1, MP2, MP3, MP4, MP5, MP6

**Today's Vocabulary**
trapezoid
bases of a trapezoid
legs of a trapezoid
base angles of a trapezoid
isosceles trapezoid
midsegment of a trapezoid
kite

**Online Activity** Use dynamic geometry software to complete the Explore.

> ❓ **INQUIRY** What special properties do trapezoids and kites have?

## Learn Trapezoids

A **trapezoid** is a quadrilateral with exactly one pair of parallel sides. The parallel sides are called **bases of a trapezoid**. The nonparallel sides are called **legs of a trapezoid**. A **base angle of a trapezoid** is formed by a base and one of the legs of the trapezoid. In trapezoid *ABCD*, ∠A and ∠B are one pair of base angles, and ∠C and ∠D are the other pair. If the legs of a trapezoid are congruent, then it is an **isosceles trapezoid**.

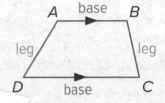

---

**Theorem 7.21: Isosceles Trapezoids**

If a trapezoid is isosceles, then each pair of base angles is congruent.

---

**Theorem 7.22: Isosceles Trapezoids**

If a trapezoid has one pair of congruent base angles, then it is an isosceles trapezoid.

---

**Theorem 7.23: Isosceles Trapezoids**

If a trapezoid is isosceles if and only if its diagonals are congruent.

---

**Go Online** Proofs of Theorems 7.21 through 7.23 are available.

## 🌐 Example 1 Use Properties of Isosceles Trapezoids

MUSIC **The body of the guitar shown is a trapezoidal prism. The front face of the guitar is an isosceles trapezoid. $AB = 3x - 2$, $CD = 3x + 9$, $AD = 4x + 5$, and $BC = 5x - 6$.**

*(continued on the next page)*

Copyright © McGraw-Hill Education

## Part A

**Prove $x = 11$.**

| Statements | Reasons |
|---|---|
| 1. *ABCD* is an isosceles trapezoid. | 1. Given |
| 2. _____ | 2. Def. of isosceles trapezoid |
| 3. $AD = BC$ | 3. Def. of congruent _____ |
| 4. _____ = _____ | 4. Substitution |
| 5. $5 = x - 6$ | 5. _____ Prop. of Equality |
| 6. _____ $= x$ | 6. _____ Prop. of Equality |
| 7. $x = 11$ | 7. _____ Prop. of Equality |

## Part B

**Find $m\angle A$ if $m\angle C = 72°$.**

Since *ABCD* is an isosceles trapezoid, $\angle C$ and $\angle D$ are congruent base angles. So, $m\angle C = m\angle\_\_\_ = \_\_\_$.

Since *ABCD* is a trapezoid, $\overline{AB} \parallel \overline{CD}$.

$$m\angle A + m\angle\_\_\_ = 180 \qquad \text{Triangle Sum Theorem}$$
$$m\angle A + \_\_\_ = 180 \qquad \text{Substitution}$$
$$m\angle A = \_\_\_ \qquad \text{Subtract 72 from each side.}$$

## Part C

**Find the perimeter of the front face of the guitar in centimeters.**

$$P = AB + BC + CD + AD \qquad \text{Perimeter of trapezoid } ABCD$$
$$= 3x - 2 + 5x - 6 + 3x + 9 + 4x + 5 \qquad \text{Substitution}$$
$$= _____ \qquad \text{Combine like terms.}$$
$$= 15(\_\_\_) + 6 \qquad x = 11$$
$$= \_\_\_ \qquad \text{Simplify.}$$

So, the perimeter of the front face of the guitar is 171 centimeters.

## Example 2 Isosceles Trapezoids and Coordinate Geometry

**Quadrilateral *QRST* has vertices $Q(-8, -4)$, $R(0, 8)$, $S(6, 8)$, and $T(-6, -10)$. Show that *QRST* is a trapezoid, and determine whether *QRST* is an isosceles trapezoid.**

**Step 1 Graph quadrilateral *QRST*.**

Graph and connect the vertices of *QRST*.

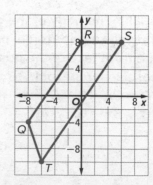

*(continued on the next page)*

Copyright © McGraw-Hill Education

**Step 2 Compare the slopes of the opposite sides.**

Use the Slope Formula to compare the slopes of opposite sides $\overline{QR}$ and $\overline{ST}$ and opposite sides $\overline{QT}$ and $\overline{RS}$. A quadrilateral is a trapezoid if exactly _____ pair of opposite sides is _____.

Opposite sides $\overline{QR}$ and $\overline{ST}$:          Opposite sides $\overline{QT}$ and $\overline{RS}$:

slope of $\overline{QR} = \dfrac{8 - (\quad)}{0 - (\quad)} = \dfrac{12}{\quad}$ or $^3$    slope of $\overline{QT} = \dfrac{-10 - (\quad)}{-6 - (\quad)} = \dfrac{-6}{\quad}$ or ___

slope of $\overline{ST} = \dfrac{-10 -}{-6 -} = \dfrac{-18}{\quad}$ or $^3$    slope of $\overline{RS} = \dfrac{8 -}{6 -} = \dfrac{0}{\quad}$ or ___

Since the slopes of $\overline{QR}$ and $\overline{ST}$ are equal, $\overline{QR}$ _____ $\overline{ST}$.

Since the slopes of $\overline{QT}$ and $\overline{RS}$ are not equal $\overline{QT}$ _____ $\overline{RS}$. Since quadrilateral QRST has only one pair of opposite sides that are parallel, quadrilateral QRST _____ is a trapezoid.

**Step 3 Compare the lengths of the legs.**

Use the Distance Formula to compare the lengths of the legs $\overline{QT}$ and $\overline{RS}$. A trapezoid is isosceles if its legs are congruent.

$QT = \sqrt{[-6 - (\quad)]^2 + [-10 - (\quad)]^2}$ or $\sqrt{\quad}$

$RS = \sqrt{(6 - \quad)^2 + (8 - \quad)^2} = \sqrt{\quad}$ or 6

Since $QT \neq RS$, legs $\overline{QT}$ and $\overline{RS}$ are *not* _____. Therefore, trapezoid QRST _____ isosceles.

## Learn Midsegments of Trapezoids

The **midsegment of a trapezoid** is the segment that connects the midpoints of the legs of the trapezoid.

---

**Theorem 7.24: Trapezoid Midsegment Theorem**

The midsegment of a trapezoid is parallel to each base and its length is one half the sum of the lengths of the bases.

---

🔗 **Go Online** A proof of Theorem 7.24 is available.

## Example 3 Midsegments of Trapezoids

**In the figure, $\overline{UR}$ is the midsegment of trapezoid PQST. Find UR.**

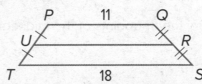

By the Trapezoid Midsegment Theorem, UR is equal to one half the sum of PQ and TS.

$\underline{\quad} = \dfrac{1}{2}(PQ + TS)$     Trapezoid Midsegment Theorem

$= \dfrac{1}{2}(11 + \underline{\quad})$     Substitution.

$= \underline{\quad}$     Simplify.

🔗 **Go Online** You can complete an Extra Example online.

Copyright © McGraw-Hill Education

💭 **Think About It!**

What other method could you have used to show that trapezoid QRST is not isosceles?

**Study Tip**

**Midsegment** _he midsegment _ _a trapezoid ca_ _lso be called a *med_ _n*.

## Example 4 Find Missing Values in Trapezoids

**In the figure, $\overline{RN}$ is the midsegment of trapezoid *LMPQ*. What is the value of *x*?**

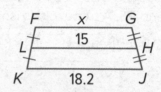

You can use the Trapezoid Midsegment Theorem to write an equation and find the value of *x*.

$RN = \frac{1}{2}(LM + QP)$    Trapezoid Midsegment Theorem

_____ $= \frac{1}{2}($ ____ $+$ ____ $)$    Substitution

_____ $= x + 16.7$    Multiply each side by ____.

_____ $= x$    Subtract 16.7 from each side.

## Check

In the figure, $\overline{LH}$ is the midsegment of trapezoid *FGJK*. What is the value of *x*?

$x =$ ____

---

## Example 5 Midsegments and Coordinate Geometry

**In trapezoid *ABCD*, $\overline{AD} \parallel \overline{BC}$. Find the endpoints of the midsegment.**

You can use the Midpoint Formula to find the midpoints of $\overline{AB}$ and $\overline{DC}$. These midpoints are the endpoints of the midsegment of trapezoid *ABCD*.

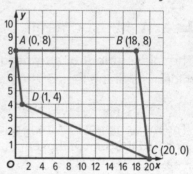

midpoint of $\overline{AB} = \left(\frac{0+18}{2}, \frac{8+8}{2}\right) = (9, \underline{\quad})$

midpoint of $\overline{DC} = \left(\frac{1+20}{2}, \frac{4+0}{2}\right) = (\underline{\quad}, 2)$

So, the endpoints of the midsegments are (9, 8) and (10.5, 2).

## Check

In trapezoid *PQRT*, $\overline{PQ} \parallel \overline{TR}$. Find the endpoints of the midsegment.

_____

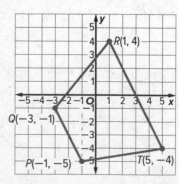

🄝 **Go Online** You can complete an Extra Example online.

---

🍪 **Think About It!**

If the parallel sides of a trapezoid are contained by the lines $y = x + 4$ and $y = x - 8$, what equation represents the line contained by the midsegment?

Copyright © McGraw-Hill Education

## Learn Kites

A **kite** is a quadrilateral with exactly two distinct pairs of adjacent congruent sides. Unlike a parallelogram, the opposite sides of a kite are not congruent or parallel.

> **Theorem 7.25: Kites**
>
> If a quadrilateral is a kite, then its diagonals are perpendicular.

> **Theorem 7.26: Kites**
>
> If a quadrilateral is a kite, then exactly one pair of opposite angles is congruent.

**Go Online** Proofs of Theorems 7.25 and 7.26 are available.

## Example 6 Find Angle Measures in Kites

**If *KLMN* is a kite, find *m∠N*.**

Since a kite can only have one pair of opposite congruent angles and ∠K ≇ ∠M, then ∠N ≅ ∠L. So, m∠N = m∠L. You can write and solve an equation to find m∠N.

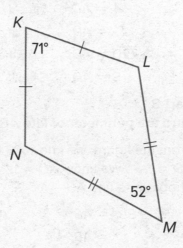

$m∠K + m∠L + m∠M + m∠N = 360$    Polygon Interior Angles Sum Theorem

_____ + $m∠N + 52 + m∠N = 360$    Substitution

$2m∠N +$ ___ $= 360$    Simplify.

$2m∠N =$ ___    Subtract.

$m∠N =$ ___    Divide each side by 2.

## Check

If *FGHJ* is a kite, find *m∠F*.

$m∠F =$ ___

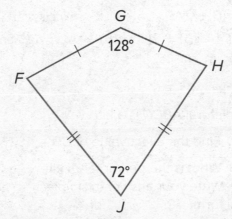

**Go Online** You can complete an Extra Example online.

### Talk About It!

If the congruent sides of a kite are marked, how can you identify which angles are congruent?

Copyright © McGraw-Hill Education

## Example 7 Find Lengths in Kites

**Quadrilateral *ABCD* is a kite.**

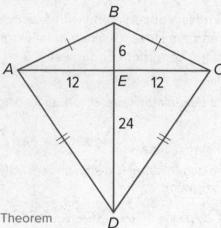

**Part A**
**Find *AD*.**

Since the diagonals of a kite are perpendicular, they divide *ABCD* into four _____ triangles. You can use the Pythagorean Theorem to find *AD*, the length of the _____ of right △*AED*.

$$AE^2 + ED^2 = AD^2 \qquad \text{Pythagorean Theorem}$$
$$\underline{\quad}^2 + \underline{\quad}^2 = AD^2 \qquad \text{Substitution}$$
$$\underline{\quad} + \underline{\quad} = AD^2 \qquad \text{Simplify.}$$
$$\underline{\quad} = AD^2 \qquad \text{Simplify.}$$
$$\sqrt{720} = AD \qquad \text{Take the square root of each side.}$$
$$\underline{\quad}\sqrt{5} = AD \qquad \text{Simplify.}$$

**Part B**
**Find the perimeter of kite *ABCD*.**

From the figure, we know $\overline{AB} \cong$ ___ and $\overline{AD} \cong$ ___. So, $AB =$ ___ and $AD =$ ___. We know $AD = 12\sqrt{5}$. So, we can use the _____ Theorem to find *AB*.

$$AE^2 + EB^2 = AB^2 \qquad \text{Pythagorean Theorem}$$
$$\underline{\quad}^2 + \underline{\quad}^2 = AB^2 \qquad \text{Substitution}$$
$$\underline{\quad} + \underline{\quad} = AB^2 \qquad \text{Simplify.}$$
$$\underline{\quad} = AB^2 \qquad \text{Simplify.}$$
$$\sqrt{180} = AB \qquad \text{Take the square root of each side.}$$
$$\underline{\quad}\sqrt{5} = AB \qquad \text{Simplify.}$$

Use the values of *AB* and *AD* to find the perimeter of kite *ABCD*.

$$P = AB + BC + CD + AD \qquad \text{Perimeter of kite}$$
$$= AB + AB + AD + AD \qquad AB = BC \text{ and } AD = CD$$
$$= 2AB + 2AD \qquad \text{Simplify.}$$
$$= 2(\underline{\quad}\sqrt{5}) + 2(\underline{\quad}\sqrt{5}) \qquad AB = 6\sqrt{5} \text{ and } AD = 12\sqrt{5}$$
$$= 36\sqrt{5} \qquad \text{Simplify.}$$

### Check

Quadrilateral *ABCD* is a kite.

**Part A** Find the exact value of *CD*. _____

**Part B** What is the perimeter of kite *ABCD*. Write your answer in simplest radical form. _____ or _____

 **Go Online** You can complete an Extra Example online.

---

**Think About It!**

How can you find the area of kite *ABCD*? Justify your argument.

**Go Online**
to practice what you've learned in Lessons 7-2 through 7-6.

Copyright © McGraw-Hill Education

# Practice

🐦 **Go Online** You can complete your homework online.

### Example 1

1. **SIGNS** The medical sign shown is a trapezoidal prism. The front face of the sign is an isosceles trapezoid. $WX = 2x - 2$, $YZ = 2x + 6$, $WZ = 4x + 5$, $XY = 5x - 3$.

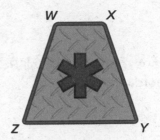

**Part A** Find the value of $x$.

**Part B** Find $m\angle Z$ if $m\angle W = 106°$.

**Part C** Find the perimeter of the front face of the sign.

**Find each measure.**

2. $m\angle T$

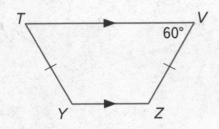

3. $m\angle Y$

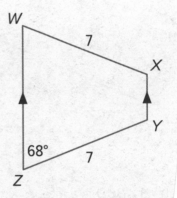

### Example 2

4. **COORDINATE GEOMETRY** *RSTU* is a quadrilateral with vertices $R(-3, -3)$, $S(5, 1)$, $T(1, -2)$, $U(-4, -9)$.

   a. Verify that *RSTU* is a trapezoid.

   b. Determine whether *RSTU* is an isosceles trapezoid. Explain.

### Example 3

**For trapezoid *HJKL*, *T* and *S* are midpoints of the legs.**

5. If $HJ = 14$ and $LK = 42$, find $TS$.

6. If $LK = 19$ and $TS = 15$, find $HJ$.

7. If $HJ = 7$ and $TS = 10$, find $LK$.

8. If $KL = 17$ and $JH = 9$, find $ST$.

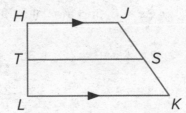

Copyright © McGraw-Hill Education

### Example 4

9. In the figure, $\overline{HD}$ is the midsegment of trapezoid *BCFG*. What is the value of *x*?

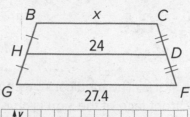

### Example 5

10. In trapezoid *ABCD*, $\overline{AD} \parallel \overline{BC}$, and $\overline{FG}$ is a midsegment. Find the *x*-coordinate of point *G*.

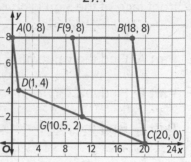

### Example 6

**Find each measure in the kites.**

11. $m\angle Q$

12. $m\angle D$

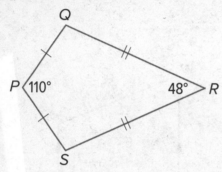
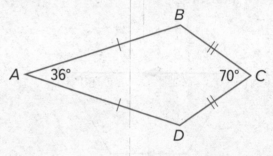

### Example 7

**REASONING**

13. Quadrilateral *ABCD* is a kite.

**Part A** Find *BC*.

**Part B** Find the perimeter of kite *ABCD*. Express as a decimal; round to the nearest tenth.

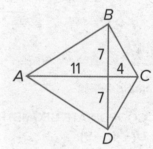

14. Quadrilateral *HRSE* is a kite.

**Part A** Find *RH*. Express in radical form.

**Part B** Find the perimeter of kite *HRSE*. Express as a decimal; round to the nearest tenth.

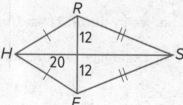

### Mixed Exercises

***ABCD* is a trapezoid.**

15. If $AC = 3x - 7$ and $BD = 2x + 8$, find the value of *x* so that *ABCD* is isosceles.

16. If $m\angle ABC = (4x + 11)°$ and $m\angle DAB = (2x + 33)°$, find the value of *x* so that *ABCD* is isosceles.

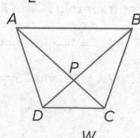

***WXYZ* is a kite.**

17. If $m\angle WXY = 120°$, $m\angle WZY = 4x°$, and $m\angle ZWX = 10x°$, find $m\angle ZYX$.

18. If $m\angle WXY = (13x + 24)°$, $m\angle WZY = 35°$, and $m\angle ZWX = (13x + 14)°$, find $m\angle ZYX$.

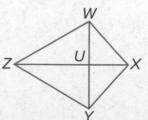

Copyright © McGraw-Hill Education

**19. MODELING** A set of stairs leading to the entrance of a building is designed in the shape of an isosceles trapezoid with the longer base at the bottom of the stairs and the shorter base at the top. If the bottom of the stairs is 21 feet wide and the top is 14 feet wide, find the width of the stairs halfway to the top.

**20. DESK TOPS** A carpenter needs to replace several trapezoid-shaped desktops in a classroom. The carpenter knows the lengths of both bases of the desktop. What other measurements, if any, does the carpenter need?

**21. USE A SOURCE** Go online to research the world's biggest kite.

    **a.** What type of quadrilateral does it represent?

    **b.** Find the perimeter of the kite.

    **c.** Find the area of the kite.

**Determine whether each statement is *always*, *sometimes*, or *never* true. Explain.**

**22.** The opposite angles of a trapezoid are supplementary.

**23.** One pair of opposite sides are parallel in a kite.

**24.** A square is a rhombus.

**25.** A rectangle is a square.

**26.** A parallelogram is a rectangle.

**For trapezoid *QRST*, *M* and *P* are midpoints of the legs.**

**27.** If $QR = 16$, $PM = 12$, and $TS = 4x$, find the value of $x$.

**28.** If $TS = 2x$, $PM = 20$, and $QR = 6x$, find the value of $x$.

**29.** If $PM = 2x$, $QR = 3x$, and $TS = 10$, find $PM$.

**30.** If $TS = 2x + 2$, $QR = 5x + 3$, and $PM = 13$, find $TS$.

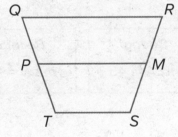

**31. AIRPORTS** A simplified drawing of the reef runway complex at Honolulu International Airport is shown below. How many trapezoids are there in this image?

32. **PERSPECTIVE** Artists use different techniques to make things appear to be 3-dimensional when drawing in two dimensions. Kevin drew the walls of a room. In real life, all of the walls are rectangles. In what shape did he draw the side walls to make them appear 3-dimensional?

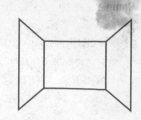

33. **CREATE** A kite manufacturer is experimenting with different designs. The designer wants to modify a current design layout.

   a. A current kite design is represented in the coordinate plane with vertices at $A(4, 20)$, $B(20, 34)$, $C(36, 20)$, and $D(20, 0)$. The designer wants to modify the design by shortening the length of the kite. Draw the kite design in the coordinate plane and determine which point should be moved to modify the kite. What are the new coordinates if the kite is to be in the shape of a parallelogram?

   b. Prove that the new kite design is in fact in the shape of a parallelogram.

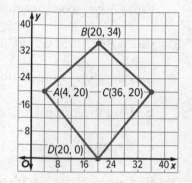

34. **WHICH ONE DOESN'T BELONG** The following are characteristics that describe several quadrilaterals.

   *Opposite sides are parallel.*

   *Diagonals are perpendicular.*

   *All sides are congruent.*

All three characteristics describe all but which of the following quadrilateral?

   **A.** square          **B.** rhombus          **C.** rectangle

35. **FIND THE ERROR** Bedagi and Belinda are trying to determine $m\angle A$ in kite $ABCD$ shown. Is either of them correct? Explain.

| **Bedagi** | **Belinda** |
|:---:|:---:|
| $m\angle A = 45$ | $m\angle A = 115$ |

36. **PERSEVERE** If the parallel sides of a trapezoid are contained by the lines $y = x + 4$ and $y = x - 8$, what equation represents the line contained by the midsegment?

37. **WRITE** Describe the properties of a quadrilateral must possess in order for the quadrilateral to be classified as a trapezoid, an isosceles trapezoid, or a kite. Compare the properties of all three quadrilaterals.

38. **ANALYZE** Is it *sometimes*, *always*, or *never* true that a square is also a kite? Explain.

**Essential Question**

**What are the different types of quadrilaterals, and how can their characteristics be used to model real-world situations?**

## Module Summary

### Lesson 7-1

Angles of Polygons

- The sum of the interior angle measures of an $n$-sided convex polygon is $(n - 2) \cdot 180°$.

- The sum of the exterior angle measures of a convex polygon, one angle at each vertex, is 360°.

### Lessons 7-2 and 7-3

Parallelograms

- A parallelogram is a quadrilateral with both pairs of opposite sides parallel.

- In a parallelogram, opposite sides and opposite angles are congruent.

- If the diagonals of a quadrilateral bisect each other, then the quadrilateral is a parallelogram.

- If one pair of opposite sides of a quadrilateral is both parallel and congruent, then the quadrilateral is a parallelogram.

### Lesson 7-4

Rectangles

A rectangle has the following properties:

- All four angles are right angles.
- Opposite sides are parallel and congruent.
- Opposite angles are congruent.
- Consecutive angles are supplementary.
- Diagonals bisect each other.

### Lessons 7-4 and 7-5

Rectangles, Rhombi, and Squares

- A rhombus is a special type of parallelogram with all four sides congruent.

- A square is a special type of parallelogram with all four sides and all four angles congruent.

- If a quadrilateral is both a rectangle and a rhombus, then it is a square.

### Lesson 7-6

Trapezoids and Kites

- A trapezoid is a quadrilateral with exactly one pair of parallel sides.

- If a trapezoid is isosceles, then each pair of base angles is congruent.

- The midsegment of a trapezoid is the segment that connects the midpoints of the legs of the trapezoid.

- The midsegment of a trapezoid is parallel to each base and its length is one half the sum of the lengths of the bases.

- A kite is a quadrilateral with exactly two distinct pairs of adjacent congruent sides.

### Study Organizer

 Foldables

Use your Foldable to review this module. Working with a partner can be helpful. Ask for clarification of concepts as needed.

# Test Practice

**1. GRIDDED RESPONSE** What is the value of x in degrees? (Lesson 7-1)

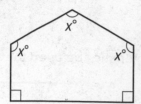

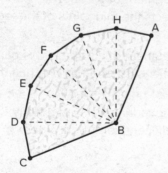

**2. MULTIPLE CHOICE** A paper fan is made by folding the pattern shown in the diagram.

Angles A and C measure 80° and angle B measures 135°. If the remaining angles are congruent to each other, what is the measure of each angle? (Lesson 7-1)

(A) 135°  (B) 143°  (C) 157°  (D) 173°

**3. GRIDDED RESPONSE** Find the measure of each interior angle in a regular 10-sided polygon in degrees. (Lesson 7-1)

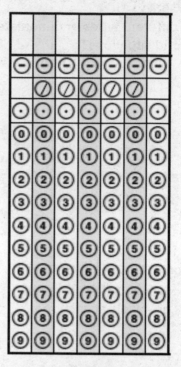

**4. OPEN RESPONSE** Quadrilateral PQRS is a parallelogram. If m∠P = 72°, then find m∠Q and m∠R. (Lesson 3-3)

**5. OPEN RESPONSE** A repeating tile design is made from a rhombus and four congruent parallelograms. (Lesson 7-2)

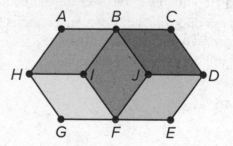

If m∠IBJ = 54°, find each angle measure.

m∠BIF = ___°

m∠JBC = ___°

m∠BJD = ___°

**6. TABLE ITEM** Identify whether each quadrilateral can be proven to be a parallelogram. (Lesson 7-3)

**A.**

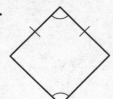

**B.**

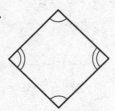

**C.**

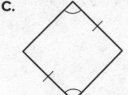

**D.**

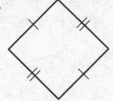

| Quadrilateral | Parallelogram? | |
|---|---|---|
| | **Yes** | **No** |
| A | | |
| B | | |
| C | | |
| D | | |

**7. OPEN RESPONSE** Physicists can use a diagram with parallelograms to determine the result of two forces acting on an object.

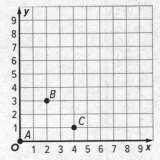

If $\overline{AB}$ and $\overline{AC}$ and are two sides of a diagram, at which coordinates in Quadrant I should point $D$ be placed so that $ABDC$ is a parallelogram? (Lesson 7-3)

**8. MULTIPLE CHOICE** Which measurements will ensure that $PQRS$ is a parallelogram? (Lesson 7-3)

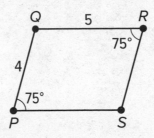

Ⓐ $PS = 5$ or $m\angle Q = 105°$

Ⓑ $PS = 4$ or $m\angle Q = 105°$

Ⓒ $PS = 5$ or $m\angle Q = 75°$

Ⓓ $PS = 4$ or $m\angle Q = 75°$

**9. OPEN RESPONSE** Given rectangle WXYZ, if $m\angle XZY = 27°$, then find $m\angle WYX$ and $m\angle WVZ$. (Lesson 7-4)

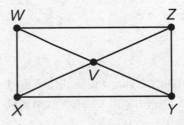

**10. MULTIPLE CHOICE** A carpenter builds a frame from two 6-foot long boards and two 8 foot long boards. Given these side lengths, how can the carpenter ensure that the frame is a rectangle? (Lesson 7-4)

Ⓐ If the diagonals are congruent, the frame must be a rectangle.

Ⓑ If the opposite sides are congruent, the frame must be a rectangle.

Ⓒ If the opposite angles are congruent, the frame must be a rectangle.

Ⓓ If the diagonals are perpendicular, the frame must be a rectangle.

**11. MULTIPLE CHOICE** In parallelogram $ABCD$, $AB = 2x$, $AC = 3x - 2$, and $AD = x + 2$. If the length of segment $BD$ is 10, what value of $x$ will ensure that $ABCD$ is a rectangle? (Lesson 7-4)

(A) 2

(B) 4

(C) 5

(D) 8

**12. TABLE ITEM** If four bars of equal length are joined at their endpoints, which shape(s) can be created? (Lesson 7-5)

| Shape | Can be created? | |
|---|---|---|
| | Yes | No |
| Kite | | |
| Parallelogram | | |
| Rectangle | | |
| Rhombus | | |
| Square | | |
| Trapezoid | | |

**13. MULTIPLE CHOICE** Given kite JKLM and $JN = 12$, find the length of segment $KM$. (Lesson 7-5)

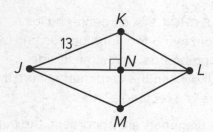

(A) $KM = 5$

(C) $KM = 13$

(B) $KM = 12$

(D) $KM = 10$

**14. MULTI-SELECT** If $ABCD$ is a rhombus that is not a square, select all of the true statements. (Lesson 7-5)

(A) $\overline{AB} \cong \overline{CD}$

(B) $\overline{AC} \perp \overline{BD}$

(C) $\angle A \cong \angle C$

(D) $\overline{AC} \cong \overline{BD}$

**15. GRIDDED RESPONSE** Find the measure of angle B, in degrees. (Lesson 7-6)

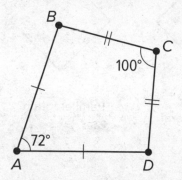

**16. MULTIPLE CHOICE** Trapezoid $ABCD$ has vertices $A(0, 0)$, $B(2, 5)$, $C(3, 5)$, and $D(8, 0)$. What is the length of its midsegment? (Lesson 7-6)

(A) 4     (B) 4.5     (C) 5     (D) 5.5

## e Essential Question

What does it mean for objects to be similar, and how is similarity useful for modeling in the real world?

G.CO.2 G.SRT.1, G.SRT.2, G.SRT.3, G.SRT.4, G.SRT.5, G.GPE.5, G.CO.10, G.CO.12
**Mathematical Practices:** MP1, MP2, MP3, MP4, MP5, MP6, MP7, MP8

## What will you learn?

Place a checkmark (✓) in each row that corresponds with how much you already know about each topic **before** starting this module.

KEY

🖓 — I don't know.  👍 — I've heard of it.  👍 — I know it!

| | Before | | | After | | |
|---|---|---|---|---|---|---|
| | 🖓 | 👍 | 👍 | 🖓 | 👍 | 👍 |
| draw and analyze dilated figures using tools or functions | | | | | | |
| solve problems using the definition of similar polygons | | | | | | |
| solve problems involving identifying the corresponding parts of similar polygons | | | | | | |
| solve problems involving identifying similar polygons based on corresponding sides and angles | | | | | | |
| solve problems using the AA Theorem of triangle similarity | | | | | | |
| solve problems involving parts of similar triangles | | | | | | |
| solve problems using the SSS and SAS Theorems of triangle similarity | | | | | | |
| prove geometric theorems using triangle similarity | | | | | | |
| use the Converse of the Triangle Proportionality Theorem to determine if lines are parallel | | | | | | |
| solve problems and prove theorems using the Triangle Midsegment Theorem and its corollaries | | | | | | |

📖 **Foldables** Make this Foldable to help you organize your notes about similarity. Begin with four sheets of notebook paper.

1. **Fold** the four sheets of paper in half.

2. **Cut** along the top fold of the papers. Staple along the side to form a book.

3. **Cut** the right sides of each paper to create a tab for each lesson.

4. **Label** each tab with a lesson number as shown.

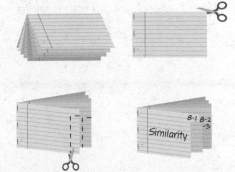

# What Vocabulary Will You Learn?

Check the box next to each vocabulary term that you may already know.

- ☐ center of dilation
- ☐ dilation
- ☐ enlargement
- ☐ midsegment
- ☐ nonrigid transformation

- ☐ reduction
- ☐ scale factor of dilation
- ☐ similar polygons
- ☐ similar triangles
- ☐ similarity transformation

## Are You Ready?

Complete the Quick Review to see if you are ready to start this module.
Then complete the Quick Check.

### Quick Review

**Example 1**

**Simplify the fraction.**

$\dfrac{6}{27}$

$= \dfrac{6 \div 3}{27 \div 3}$   Divide the numerator and denominator by the GCF.

$= \dfrac{2}{9}$   Simplify.

**Example 2**

**Use the scale drawing to find the actual base length and height of the triangle.**

Multiply the base length in the scale drawing by 12.
actual base length =
$4 \times 12 = 48$ cm
Multiply the height in the scale drawing by 12.
actual height = $9 \times 12 = 108$ cm

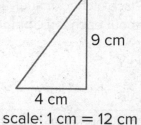

9 cm

4 cm

scale: 1 cm = 12 cm

### Quick Check

**Simplify each fraction.**

1. $\dfrac{4}{16}$

2. $\dfrac{8}{24}$

3. $\dfrac{15}{25}$

4. $\dfrac{12}{18}$

5. $\dfrac{36}{45}$

6. $\dfrac{10}{12}$

**Use the scale drawing to find the actual length and width of the rectangle.**

3 in.

5 in.

scale: 1 in. = 8 in.

7. actual length

8. actual width

### How Did You Do?

Which exercises did you answer correctly in the Quick Check? Shade those Exercise numbers below.

① ② ③ ④ ⑤ ⑥ ⑦ ⑧

# Dilations

## Explore Verifying the Properties of Dilations

▶ **Online Activity** Use dynamic geometry software to complete the Explore.

> ⓠ **INQUIRY** What special properties do dilations have?

## Learn Dilations

Recall that a *rigid transformation* is an operation that maps an original figure, the *preimage*, onto a new figure called the *image*.

A **nonrigid transformation** is a transformation that changes the dimensions of a given figure. A **dilation** is a nonrigid transformation that enlarges or reduces a geometric figure. When a figure is enlarged or reduced, the sides of the image are proportional to the sides of the original figure.

The **center of dilation** is the center point from which dilations are performed. The **scale factor of a dilation** is the ratio of a length on an image to a corresponding length on the preimage.

The value of *k* determines whether the dilation is an enlargement or a reduction.

---

**Key Concept • Types of Dilations**

**Enlargement** A dilation with a scale factor greater than 1 produces an **enlargement**, or an image that is larger than the original figure.

**Reduction** A dilation with a scale factor between 0 and 1 produces a **reduction**, or an image that is smaller than the original figure.

---

## Example 1 Identify a Dilation and Find its Scale Factor

**Determine whether the dilation from △ABC to △DEF is an *enlargement* or a *reduction*. Then find the scale factor of the dilation.**

△*DEF* is _____ than △*ABC*, so the dilation is a reduction. The scale factor is equal to the side length of △*DEF* divided by the corresponding side length of △*ABC*. So, the scale factor is $\frac{2}{6}$ or ____.

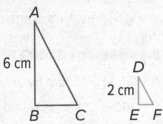

### Today's Standards
G.CO.2, G.SRT.1
MP1, MP5, MP6

### Today's Vocabulary
nonrigid transformation
dilation
center of dilation
scale factor of dilation
enlargement
reduction

▶ **Go Online**
You may want to complete the Concept Check to check your understanding.

💬 **Talk About It!**
Does your answer seem reasonable? Explain.

**Study Tip**

**Multiple Representations** The scale factor of a dilation can be represented as a fraction, a decimal, or as a percent. For example, a scale factor of $\frac{2}{5}$ can also be written as 0.4 or as 40%.

## Check

Consider the dilation from *WXYZ* to *JKLM*.

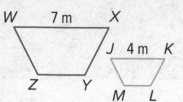

### Part A

The dilation of *WXWXZ* to *JKLM* is _____.

### Part B

What is the scale factor of the dilation? Write your answer as a fraction in simplest form. _____

---

## 🌐 Example 2  Find and Use a Scale Factor

**SCHOOL SPIRIT**
**Jalal is printing a banner for his school's wheelchair tennis team based on the design shown. By what percent should Jalal enlarge his design so that the dimensions of the banner are 5 times that of the original design? What will be the dimensions of the banner?**

Jalal wants to create a dilated image of his banner design using a commercial printer. The scale factor of his enlargement is 5. Written as a percent, the scale factor is (5 × 100)% or _____ %.

Now we can find the dimensions of the enlarged image using the scale factor.

width: 6 in. × 500% = _____ in.    length: 14 in. × 500% = _____ in.

The banner will be 30 inches by 70 inches.

**Study Tip**

**Units of Measure** Sometimes when a figure is reduced or enlarged, the unit of measure should change to better fit the context of the problem. In Example 2, the dimensions of the banner is approximately 2.5 feet by 5.8 feet.

## Check

**PORTRAITS** Natalia wants to print an enlarged family portrait for her mother. By what percent should Natalia enlarge the portrait so that the dimensions of its image are 2.5 times that of the original? What will be the dimensions of the enlarged portrait? _____

**A.** 125%; 22.5 cm by 30 cm

**B.** 125%; 45 cm by 60 cm

**C.** 250%; 22.5 cm by 30 cm

**D.** 250%; 45 cm by 60 cm

🧭 **Go Online** You can complete an Extra Example online.

# Learn Dilations on the Coordinate Plane

On the coordinate plane, a dilation is a function in which the coordinates of the vertices of a figure are multiplied by the same ratio $k$.

| Key Concept • Dilations on the Coordinate Plane | |
|---|---|
| **Words** | To dilate a figure by a scale factor of $k$ with respect to the center of dilation $(0, 0)$, multiply the $x$- and $y$-coordinate of each vertex by $k$. |
| **Symbols** | $(x, y) \longrightarrow (kx, ky)$ |
| **Example** | The image of $\triangle PQR$ dilated by $k = 3$ is $\triangle P'Q'R'$. |

**Think About It!**

What is the relationship between $PQ$ and $P'Q'$? How does this relate to the scale factor?

**Study Tip**

**Center of Dilation**
Unless otherwise stated, all dilations on the coordinate plane use the origin as their center of dilation.

# Example 3 Dilate a Figure

$\triangle TRS$ has vertices $T(-4, -5)$, $R(0, 6)$, and $S(4, 3)$. Find the coordinates of the vertices of $\triangle T'R'S'$ after a dilation of $\triangle TRS$ by a scale factor of $\frac{1}{2}$.

Because the scale factor is $\frac{1}{2}$, the coordinates of the vertices of $\triangle T'R'S'$ should be half of the value of the coordinates of the vertices of $\triangle TRS$.

Complete the calculations for the dilation when $k = \frac{1}{2}$.

$(x, y)$                              $(x, y) \longrightarrow (kx, ky)$

$T(-4, -5)$                 $T'\left(\frac{1}{2}(-4), \frac{1}{2}(-5)\right)$ or $T'(-2, -2.5)$

$R(0, 6)$                    $R'\left(\frac{1}{2}(\underline{\phantom{xx}}), \frac{1}{2}(\underline{\phantom{xx}})\right)$ or $R'(0, 3)$

$S(4, 3)$                    $S'\left(\frac{1}{2}(\underline{\phantom{xx}}), \frac{1}{2}(\underline{\phantom{xx}})\right)$ or $S'(2, 1.5)$

## Check

$\triangle XYZ$ has vertices $X(3, -4)$, $Y(6, 5)$, and $Z(8, -2)$. Dilate $\triangle XYZ$ by a scale factor of 4. Find the coordinates of the vertices of $\triangle X'Y'Z'$.

$X'(\underline{\phantom{xx}}, \underline{\phantom{xx}})$        $Y'(\underline{\phantom{xx}}, \underline{\phantom{xx}})$        $Z'(\underline{\phantom{xx}}, \underline{\phantom{xx}})$

🅑 **Go Online** You can complete an Extra Example online.

## Example 4 Find the Scale Factor of Dilation

$\triangle A'B'C'$ is the image of $\triangle ABC$ after a dilation. Find the scale factor of the dilation.

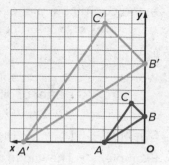

To find the scale factor of the dilation, you must compare the lengths of corresponding sides in $\triangle ABC$ and $\triangle A'B'C'$.

**Step 1 Identify two corresponding sides and their endpoints.**

$\overline{AB}$ and $\overline{A'B'}$ are corresponding sides. The endpoints of $\overline{AB}$ are $A(\underline{\quad}, \underline{\quad})$ and $B(\underline{\quad}, \underline{\quad})$. The endpoints of $\overline{A'B'}$ are $A'(\underline{\quad}, \underline{\quad})$ and $B'(\underline{\quad}, \underline{\quad})$.

**Step 2 Find the lengths of the corresponding sides.**

Use the Distance Formula to find $AB$ and $A'B'$.

$A(-3, 0)$ and $B(0, 2)$                    $A'(-9, 0)$ and $B'(0, 6)$

$AB = \sqrt{[0 - (\underline{\quad})]^2 + (2 - \underline{\quad})^2}$          $A'B' = \sqrt{[0 - (\underline{\quad})]^2 + (6 - \underline{\quad})^2}$

$\phantom{AB} = \sqrt{\underline{\quad}^2 + 2^2}$          $\phantom{A'B'} = \sqrt{9^2 + \underline{\quad}^2}$

$\phantom{AB} = \sqrt{\underline{\quad}}$          $\phantom{A'B'} = \sqrt{\underline{\quad}}$

**Step 3 Calculate the scale factor.**

To find the scale factor, find the ratio of $A'B'$ to $AB$.

$AB = \sqrt{13}$ and $A'B' = \sqrt{117}$

$\dfrac{A'B'}{AB} = \dfrac{\sqrt{117}}{\sqrt{13}}$

$\phantom{\dfrac{A'B'}{AB}} = \dfrac{\sqrt{\underline{\quad}}}{\sqrt{13}}$

$\phantom{\dfrac{A'B'}{AB}} = \sqrt{\underline{\quad}}$

$\phantom{\dfrac{A'B'}{AB}} = \underline{\quad}$

So, the scale factor of the dilation is 3.

**Go Online** You can complete an Extra Example online.

# Practice

🔵 **Go Online** You can complete your homework online.

### Example 1

**Determine whether the dilation from the figure on the left to the figure on the right is an *enlargement* or a *reduction*. Then find the scale factor of the dilation.**

**1.**

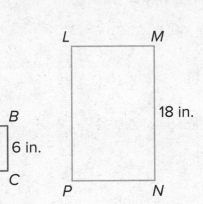

**2.**

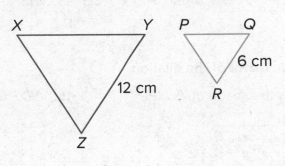

**3.**

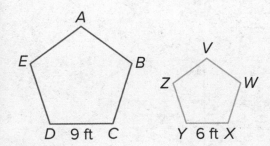

**4.**

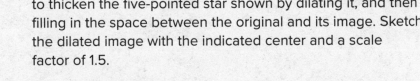
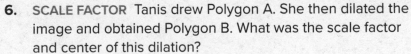
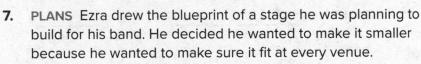

### Example 2

**5.** **DILATIONS** Cara is making images for a poster. She wants to thicken the five-pointed star shown by dilating it, and then filling in the space between the original and its image. Sketch the dilated image with the indicated center and a scale factor of 1.5.

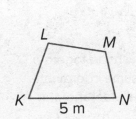

**6.** **SCALE FACTOR** Tanis drew Polygon A. She then dilated the image and obtained Polygon B. What was the scale factor and center of this dilation?

**7.** **PLANS** Ezra drew the blueprint of a stage he was planning to build for his band. He decided he wanted to make it smaller because he wanted to make sure it fit at every venue.

**a.** Sketch the image of Ezra's stage after a dilation with scale factor 0.5.

**b.** The perimeter of the image is 26 units. What is the perimeter of the original figure?

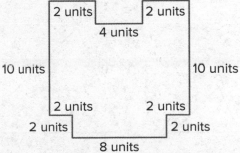

**Example 3**

**Find the image of each polygon with the given vertices after a dilation centered at the origin with the given scale factor.**

8. $J(-8, 0)$, $K(-4, 4)$, $L(-2, 0)$, $k = 0.5$

9. $S(0, 0)$, $T(-4, 0)$, $V(-8, -8)$, $k = 1.25$

10. $A(9, 9)$, $B(3, 3)$, $C(6, 0)$, $k = \frac{1}{3}$

11. $D(4, 4)$, $F(0, 0)$, $G(8, 0)$, $k = 0.75$

12. $M(-2, 0)$, $P(0, 2)$, $Q(2, 0)$, $R(0, -2)$, $k = 2.5$

13. $W(2, 2)$, $X(2, 0)$, $Y(0, 1)$, $Z(1, 2)$, $k = 3$

**Example 4**

**Find the scale factor of the dilation.**

14. $\triangle J'K'P'$ is the image of $\triangle JKP$.

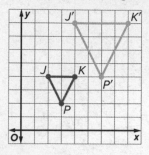

15. $\triangle D'F'G'$ is the image of $\triangle DFG$.

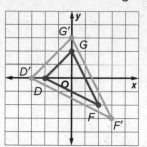

16. **SCALE FACTORS** Tyrone drew a shape together with one of its dilations on the same coordinate plane as shown. What is the scale factor of the dilation?

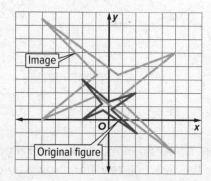

**Mixed Exercises**

**Determine whether the dilation from figure A to A' is an *enlargement* or a *reduction*. Then find the scale factor of the dilation.**

17.

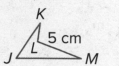
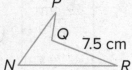

5 cm  7.5 cm

18.

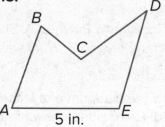

5 in.

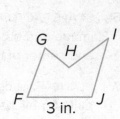

3 in.

19.

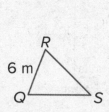

6 m  14 m

20.

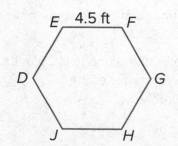

4.5 ft

2.5 ft

**Find the image of each polygon with the given vertices after a dilation centered at the origin with the given scale factor.**

**21.** $F(-10, 4)$, $G(-4, 4)$, $H(-4, -8)$, $k = 0.25$

**22.** $X(2, -1)$, $Y(-6, 4)$, $Z(-2, -5)$, $k = \frac{5}{4}$

**23.** $M(4, 6)$, $N(-6, 2)$, $P(0, -8)$, $k = \frac{3}{4}$

**24.** $R(-2, 6)$, $S(0, -1)$, $T(-5, 3)$, $k = 1.5$

**Find the scale factor of the dilation.**

**25.** $A'B'C'D'$ is the image of $ABCD$.

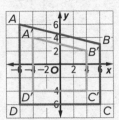

**26.** $\triangle P'Q'R'$ is the image of $\triangle PGR$.

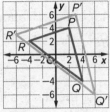

**27.** Determine whether the dilation from Figure $N$ to $N'$ is an *enlargement* or a *reduction*. Find the scale factor of the dilation.

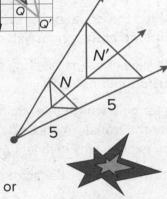

**28.** CENTERS Elena superimposed the image of the dilation of a figure on its original figure as shown. Identify the center of this dilation. Explain how you found it.

**29.** Determine if the dilation from $A$ to $B$ is an *enlargement* or *reduction*. Then find the scale factor of the dilation.

**30.** $\triangle ABC$ has vertices $A(2, 2)$, $B(3, 4)$, and $C(5, 2)$. What are the coordinates of point $C$ of the image of the triangle after a dilation centered at the origin with a scale factor of 2.5?

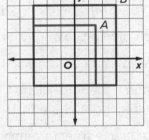

**31.** USE TOOLS Use a ruler to draw the image of the figure under a dilation with center $M$ and the scale factor of $\frac{1}{5}$.

**32.** Davion is using a coordinate plane to experiment with quadrilaterals, as shown in the figure.

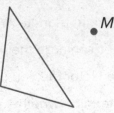

**a.** PRECISION Davion creates $M'N'P'Q'$ by enlarging $MNPQ$ with a dilation with scale factor 2. Then he creates $M''N''P''Q''$ by dilating $M'N'P'Q'$ with a scale factor of $\frac{1}{3}$. The center of dilation for each dilation is the origin. Draw and label the final image, $M''N''P''Q''$.

**b.** REASONING Can Davion map $MNPQ$ directly to $M''N''P''Q''$ with a single transformation? If so, what transformation should he use?

**c.** STATE YOUR ASSUMPTION In general, what can you say about a dilation with scale factor $k_1$ that is followed by a dilation with scale factor $k_2$? State a conjecture and then provide an argument to justify it.

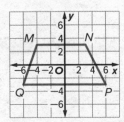

**33.** The point $P'$ is the image of point $P(a, b)$ under a dilation centered at the origin with scale factor $k \neq 1$.

   **a.** REGULARITY Assuming that point $P$ does not lie on the $y$-axis, what is the slope of $\overleftrightarrow{PP'}$? Explain how you know.

   **b.** STRUCTURE In **part a**, why is it important that $P$ does not lie on the $y$-axis?

**34.** A city planner is designing the streets in a new shopping district. She has already planned Palm Street, as shown on the coordinate plane.

   **a.** REASONING The city planner will apply a dilation to Palm Street to create Cedar Street. Describe how the two streets will be related.

   **b.** STRUCTURE The city planner uses a dilation with scale factor 2 and center of dilation at the origin to create Cedar Street. Draw Cedar Street on the coordinate plane and write its equation

**35.** PERSEVERE Find the equation for the dilated image of the line $y = 4x - 2$ if the dilation is centered at the origin with a scale factor of 1.5.

**36.** WRITE Are parallel lines (parallelism) and collinear points (collinearity) preserved under all transformations? Explain.

**37.** ANALYZE Determine whether invariant points are *sometimes*, *always*, or *never* maintained for the transformations described below. If so, describe the invariant point(s). If not, explain why invariant points are not possible.

   **a.** dilation of $ABCD$ with scale factor of 1

   **b.** rotation of $\overline{AB}$ 74° about $B$

   **c.** reflection of $\triangle MNP$ in the $x$-axis

   **d.** translation of $PQRS$ along $(7, 3)$

   **e.** dilation of $\triangle XYZ$ centered at the origin with scale factor 2

**38.** CREATE Graph a triangle. Dilate the triangle so that its area is four times the area of the original triangle. State the scale factor and center of your dilation.

**39.** WRITE Can you use transformations to create congruent figures? Explain.

**40.** CREATE Draw right triangle $ABC$ and point $P$ not on the triangle.

   **a.** Use a ruler to enlarge the triangle under a dilation with center $P$.

   **b.** What scale factor did you use?

**41.** ANALYZE Determine whether each statement is *sometimes*, *always*, or *never* true. Explain.

   **a.** If $c$ is a real number, then a dilation centered at the origin maps the line $y = cx$ to itself.

   **b.** If $k > 1$, then a dilation with scale factor $k$ maps $\overline{AB}$ to a segment that is congruent to $\overline{AB}$.

# Similar Polygons

## Explore Similarity in Polygons

**Online Activity** Use graphing technology to complete the Explore.

@ **INQUIRY** How can you identify whether two polygons are similar?

## Learn Similar Polygons

A dilation is a type of similarity transformation. A **similarity transformation** occurs when a figure and its image have the same shape. Two figures are **similar polygons** if one can be obtained from the other by a dilation or a dilation with one or more rigid transformations. When two polygons are similar, their angles and sides are related.

Two polygons are similar if and only if their corresponding angles are congruent and their corresponding side lengths are proportional.

Like congruence, similarity is reflexive, symmetric, and transitive.

**Theorem 8.1: Properties of Similarity**

**Reflexive Property of Similarity**
$\triangle ABC \sim \triangle ABC$

**Symmetric Property of Similarity**
If $\triangle ABC \sim \triangle DEF$, then $\triangle DEF \sim \triangle ABC$.

**Transitive Property of Similarity**
If $\triangle ABC \sim \triangle DEF$ and $\triangle DEF \sim \triangle XYZ$, then $\triangle ABC \sim \triangle XYZ$.

**Go Online** A proof of Theorem 8.1 is available.

## Example 1 Use a Similarity Statement

**ABCD ~ PQRS. Complete the given angles and side lengths to list all pairs of congruent angles, and write a proportion that relates the corresponding sides.**

Use the similarity statement.

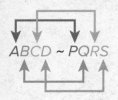

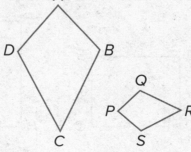

Congruent angles: $\angle A \cong$ _____ , $B \cong$ _____ , $C \cong$ _____ , $D \cong$ _____

Proportion: $\dfrac{AB}{} = \dfrac{BC}{} = \dfrac{}{RS} = \dfrac{}{}$

**Today's Standards**
..SRT.2
MP3, MP4, MP6

**Today's Vocabulary**
similarity transformation
similar polygons

**Talk About It!**
If two polygons are congruent, are they also similar? Justify your argument.

## Check

NPQR ~ UVST

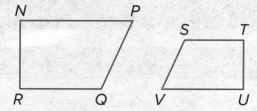

**Part A** List all pairs of congruent angles.

∠N ≅ _____ , P ≅ _____ , Q ≅ _____ , R ≅ _____

**Part B** Complete the proportions that relate the corresponding sides.

$\frac{UV}{NP} = \frac{VS}{\phantom{xx}} = \frac{\phantom{xx}}{QR} = $ ——

---

## Example 2 Identify Similar Polygons

**Determine whether △NQP is similar to △RST. If so, find the scale factor. Explain your reasoning.**

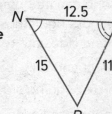
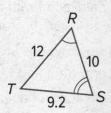

🫐 **Think About It!**

What transformations can be used to create △NQP from △RST?

**Step 1 Compare the corresponding angles.**

Since ∠N ≅ ∠_____ and ∠Q ≅ ∠_____ , by the _____ , ∠P ≅ U∠_____. So, the corresponding angles are congruent.

**Step 2 Compare the corresponding sides.**

$\frac{NQ}{RS} = \frac{12.5}{10}$ or $\frac{5}{4}$      $\frac{QP}{ST} = \frac{11.5}{9.2}$ or $\frac{5}{4}$      $\frac{PN}{TR} = \frac{15}{12}$ or $\frac{5}{4}$

Because the corresponding angles are congruent and the corresponding sides are proportional, △NQP ~ △_____. So, the triangles are similar with a scale factor from △_____ to △NQP of $\frac{5}{4}$.

## Check

Is quadrilateral *ABCD* similar to quadrilateral *EFGH*? If so, find the scale factor from □*EFGH* to □*ABCD*. _____

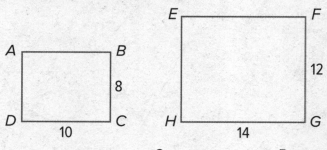

**A.** no      **B.** yes; $\frac{2}{3}$      **C.** yes; $\frac{5}{7}$      **D.** yes; $\frac{4}{5}$

🔵 **Go Online** You can complete an Extra Example online.

## Example 3 Use Similar Figures to Find Missing Measures

**In the diagram, *WXYZ ~ PQRS*. Find the value of *y*.**

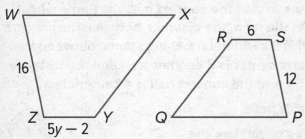

Use the corresponding side lengths to write a proportion.

$\dfrac{WZ}{PS} = \dfrac{YZ}{RS}$      Similarity proportion

$\dfrac{16}{12} = \dfrac{5y-2}{6}$      $WZ = 16, PS = 12, YZ = 5y - 2, RS = 6$

$16(6) = 12(5y - 2)$      Cross Products Property

$96 = \underline{\quad} y - \underline{\quad}$      Multiply.

$\underline{\quad} = 60y$      Add 24 to each side.

$\underline{\quad} = y$      Divide each side by 60.

## Check

In the diagram, $\triangle JLM \sim \triangle QST$. Find the value of *x*. Write your answer as a decimal, if necessary.

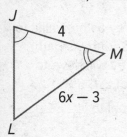 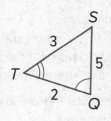

$x = \underline{\quad}$

 **Go Online** You can complete an Extra Example online.

## Learn Perimeters of Similar Polygons

In similar polygons, the ratio of any two corresponding lengths is equal to the scale factor between them. So, you can write a proportion to relate them. This leads to the following theorem about the perimeters of two similar polygons.

> **Theorem 8.2: Perimeters of Similar Polygons**
>
> If two polygons are similar, then their perimeters are proportional in the same ratio as the scale factor between them.

**Go Online** A proof of Theorem 8.2 is available.

**Problem-Solving Tip**

**Redraw Diagrams**
When solving problems that use similar figures, sometimes it is difficult to identify corresponding parts. Redraw the given diagram so the similar figures have the same orientation. This will allow you easily compare corresponding parts and set up a similarity proportion.

**Think About It!**

Will any two regular polygons with the same number of sides be similar? Justify your argument.

### 🌐 Example 4  Use Similar Polygons to Find Perimeter

**STATE FAIR**  **Geoffrey plans on going to the state fair this summer. He has downloaded a map of the fairgrounds that shows all of the attractions. Geoffrey plans to visit the concert hall, the Ferris wheel, and the sports center. On the map, the distance between the concert hall and the Ferris wheel is 9 centimeters, the distance between the Ferris wheel and the sports center is 8 centimeters, and the distance between the sports center and the concert hall is 4 centimeters.**

**Part A  Describe Geoffrey's path.**

Geoffrey wants to visit the concert hall, the Ferris wheel, and the sports center before going to any other attractions. If Geoffrey returns to the concert hall after visiting the sports center, what polygon can be used to model Geoffrey's path between the three attractions?

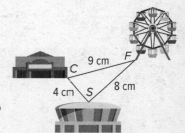

Geoffrey's path can be modeled by a _____.

On the map, draw line segments between the three attractions. Then label the points that represent the three attractions. Label the concert hall as $C$, the sports center as $S$, and the Ferris wheel as $F$.

**Part B  Find the total distance.**

**If the actual distance between the concert hall and the sports center is 20 meters, how far will Geoffrey have to travel to visit all three attractions?**

$\triangle CSF$ will be similar to the triangle formed by Geoffrey as he walks to the three attractions. Let's call the figure formed by Geoffrey's path $\triangle C'S'F'$. So, $\triangle CSF \sim \triangle C'S'F'$. To find how far Geoffrey will have to travel to visit all three attractions, find the perimeter of $\triangle C'S'F'$.

Because 1 meter = 100 centimeters, 20 meters is equal to _____ centimeters. So, $C'S' = 2000$ centimeters.

The scale factor of $\triangle CSF$ to $\triangle C'S'F'$ $\dfrac{C'S'}{CS} = \dfrac{2000}{4}$ or _____

The perimeter of $\triangle CSF$ is $4 +$ _____ $+ 9$ or _____.

Use the perimeter of $\triangle CSF$ and the scale factor to write a proportion. Let $w$ represent the perimeter of $\triangle C'S'F'$.

$$\dfrac{1}{500} = \dfrac{\text{perimeter of } \triangle CSF}{\text{perimeter of } \triangle C'S'F'}$$

$$\dfrac{1}{500} = \dfrac{\phantom{xx}}{w}$$

$$w = \underline{\phantom{xxx}} \text{ (21)}$$

$$w = \underline{\phantom{xxxxx}}$$

So, the perimeter of $\triangle C'S'F'$ is 10,500 centimeters or 105 meters.

🌐 **Go Online**  You can complete an Extra Example online.

---

### 🤔 Think About It!

What assumption did you make while solving this problem?

---

### Study Tip

**Units of Measure**
When finding a scale factor, the measurements being compared must have the same unit of measure. If they do not have the same unit of measure, you will need to convert one of the measurements.

# Practice

Go Online You can complete your homework online.

**Example 1**

**List all pairs of congruent angles, and write a proportion that relates the corresponding sides for each pair of similar polygons.**

**1.** *ABCD ~ WXYZ*

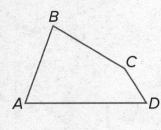

**2.** *MNPQ ~ RSTU*

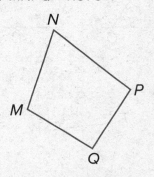

**3.** △*FGH ~* △*JKL*

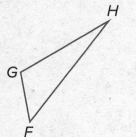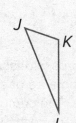

**4.** △*DEF ~* △*VWX*

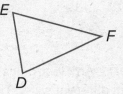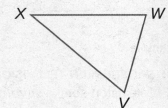

**Example 2**

**Determine whether each pair of figures is similar. If so, find the scale factor. Explain your reasoning.**

**5.**

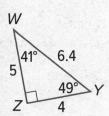

W  41°  6.4
5          Y
  49°
Z    4

K
8.2
M  42°    7.4
   11    48°  L

**6.**

X
W          Y
Z

**7.**

G    8    H

10

M  2  N
    4
L    P    F        J

**8.**

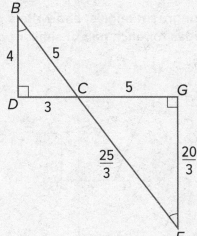

B
4    5
D    3    C    5    G
25/3              20/3
              F

Example 3

**Each pair of polygons is similar. Find the value of x.**

9.

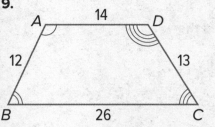
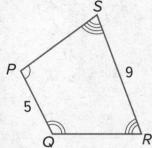

10.

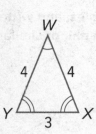
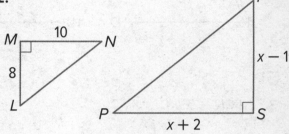

11.

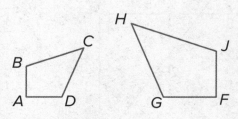

12.

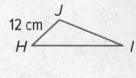

Example 4

13. **WIDESCREEN TELEVISIONS** An electronics company manufactures widescreen television sets in several different sizes. The rectangular viewing area of each television size is similar to the viewing areas of the other sizes. The company's 42-inch widescreen television has a viewing area perimeter of approximately 144.4 inches. What is the approximate viewing area perimeter of the company's 46-inch widescreen television?

14. **PERIMETER** In a rectangle, the ratio of the length to the width is 5:2, and its perimeter is 126 centimeters. Find the width of the rectangle.

15. **TRIANGLES** If $\triangle ABC \sim \triangle HIJ$, find the perimeter of $\triangle HIJ$.

16. **RECTANGLES** A rectangle has a perimeter of 14 inches. A similar rectangle has a perimeter of 10 inches. If the length of the larger rectangle is 4 inches, what is the length of the smaller rectangle? Round to the nearest tenth.

C
32 cm        60 cm
A _____ B
        80 cm

J
12 cm
H _____ I

**Mixed Exercises**

**List all pairs of congruent angles, and write a proportion that relates the corresponding sides for each pair of similar polygons.**

17. *ABCD ~ FGHJ*

18. $\triangle MNP \sim \triangle QRP$

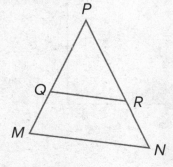

H
C
B
A    D

J
G        F

P
Q        R
M
        N

**ARGUMENTS Determine whether each pair of figures is similar. If so, find the scale factor. Explain your reasoning.**

**19.**

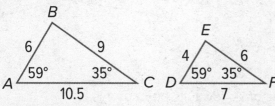

**20.**

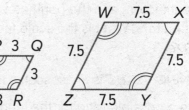

**Each pair of polygons is similar. Find the value of x.**

**21.**

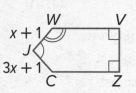
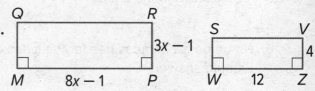

**22.**

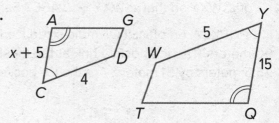

**23.**

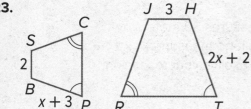

**24.**

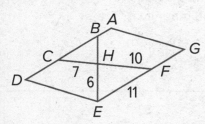
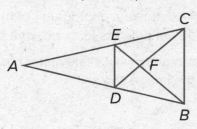

**Find the perimeter of the given triangle.**

**25.** △CBH, if △CBH ~ △FEH, ADEG is a parallelogram, CH = 7, FH = 10, FE = 11, and EH = 6

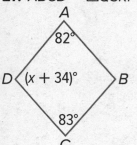

**26.** △DEF, if △DEF ~ △CBF, perimeter of △CBF = 27, DF = 6, FC = 8

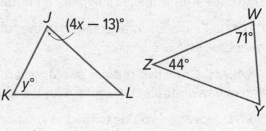

**Find the value of each variable.**

**27.** ABCD ~ △QSRP

**28.** △JKL ~ △WYZ

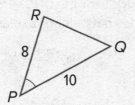

**29. STRUCTURE** Are any two of the three triangles similar? If so, write the appropriate similarity statement.

**30. REASONING** If △*ABC* ~ △*DEC*, find *x* and the scale factor from △*DEC* to △*ABC*.

**31.** Rectangle *ABCD* ~ rectangle *EFGH*, the perimeter of *ABCD* is 54 centimeters and the perimeter of *EFGH* is 36 centimeters. What is the scale factor from *EFGH* to *ABCD*?

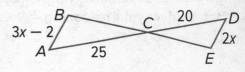

**32. REASONING** △*ABC* is an isosceles triangle.

   **a.** Write a possible ratio for the lengths of the sides of △*ABC* if its perimeter is 42 inches. Sample answer:

   **b.** Name possible measures for the sides of △*ABC* using your answer to **part a**. Sample answer:

   **c.** If △*WXY* is also isosceles and has a perimeter of 28 and △*ABC* has sides with the measures you gave in **part b**, what must be the measure of the sides of △*WXY* so that △*WXY* ~ △*ABC*? Sample answer:

**33. ICE HOCKEY** An official Olympic-sized ice hockey rink measures 30 meters by 60 meters. The ice hockey rink at the local community college measures 25.5 meters by 51 meters. Are the ice hockey rinks similar? Explain your reasoning.

**34. BIOLOGY** A paramecium is a small single-cell organism. The magnified paramecium shown is actually one tenth of a millimeter long.

   **a.** If you want to make a photograph of the original paramecium so that its image is 1 centimeter long, by what scale factor should you magnify it?

   **b.** If you want to make a photograph of the original paramecium so that its image is 15 centimeters long, by what scale factor should you magnify it?

   **c.** By approximately what scale factor has the paramecium been enlarged to make the image shown?

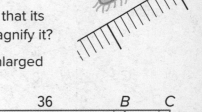

**35. PERSEVERE** For what value(s) of *x* is *BEFA* ~ *EDCB*?

**36. ANALYZE** Recall that an *equivalence relation* is any relationship that satisfies the Reflective, Symmetric, and Transitive Properties. Is similarity an equivalence relation? Explain.

**37. CREATE** Find a counterexample of the following statement.
*All rectangles are similar.*

**38. ANALYZE** Draw two regular pentagons of different sizes. Are the pentagons similar? Will any two regular polygons with the same number of sides be similar? Explain.

**39. WRITE** How can you describe the relationship between two figures?

**40. WHICH ONE DOESN'T BELONG?** Polygon *ABCD* is similar to polygon *PQRS*. Which proportion does not belong? Explain.

   **A.** $\dfrac{AC}{AD} = \dfrac{PR}{PS}$      **B.** $\dfrac{BC}{CD} = \dfrac{QR}{RS}$      **C.** $\dfrac{AB}{BD} = \dfrac{PQ}{QS}$      **D.** $\dfrac{CD}{AB} = \dfrac{PQ}{RS}$

# Similar Triangles: AA Similarity

## Explore Similarity Transformations and Triangles

⊙ **Online Activity** Use dynamic geometry software to complete the Explore.

> ⊚ **INQUIRY** How can you determine whether two triangles are similar? ✕

## Explore Conditions that Prove Triangles Similar

⊙ **Online Activity** Use dynamic geometry software to complete the Explore.

> ⊚ **INQUIRY** What shortcut can be used to determine whether two triangles are similar? ✕

## Learn Similar Triangles: AA Similarity

In **similar triangles**, all of the corresponding angles are congruent and all of the corresponding sides are proportional. However, you don't need to show that all of the criteria are met to show that two triangles are similar. Angle-Angle Similarity is one of several shortcuts.

**Postulate 8.1: Angle-Angle (AA) Similarity**

If two angles of one triangle are congruent to two angles of another triangle, then the triangles are similar.

## Example 1 Use the AA Similarity Postulate

**Determine whether the triangles are similar. Explain your reasoning.**

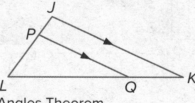

$\angle L \cong \angle L$ by the _____ Property of Congruence.

$\angle LPQ \cong \angle$_____ by the Corresponding Angles Theorem. By AA Similarity, $\triangle KLJ \sim \triangle QLP$.

## Check

**Part A**
True or False: $\triangle ABC \sim \triangle FGD$ _____

**Part B**
Explain your reasoning. _____

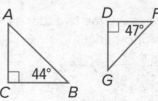

_____

⊙ **Go Online** You can complete an Extra Example online.

**Today's Standards**
G.SRT.2, G.SRT.3
MP2, MP3, MP5

**Today's Vocabulary**
similar triangles

**Math History Minute**

Italian mathematician **Margherita Beloch (1879-1976)** was particularly interested in a field of study called photogrammetry, which is the science of making measurements from photographs. Photogrammetry involves using scales and similarity to locate points. She is also known for her contributions to paper folding, entitling one of her papers "On the Method of Paper Folding for the Resolution of Geometric Problems."

## 🌐 **Example 2** Parts of Similar Triangles

**HANDBALL Demarco is teaching Taye how to play handball. Taye prefers to return the ball on a serve when it bounces to a height of 42 inches. When Demarco serves the ball, where should he aim for the ball to hit the front wall to ensure that it bounces at the short line and up to Taye standing 4 feet from the back of the court? Assume that the angles formed by the path of the bounding handball are congruent.**

**Draw a Diagram/Create and Describe the Model**

- The ball should bounce at the short line, 25 feet from the front wall.

- The ball should bounce to Taye, who is standing 11 feet from the short line.

- Taye will hit the ball when it bounces to a height of _____.

- Find the point on the front wall $x$ where the ball bounces.

- You can model the path of the handball using two triangles.

- You are given that the angles formed by the bouncing handball are congruent. The front wall forms a _____° angle with the ground, and the height to which the ball bounces forms a 90° angle with the ground. Therefore, the two triangles are similar by the _____.

**Watch Out!**

**Units** Remember to convert inches to feet when you solve for $x$ in the example.

🙂 **Think About It!**

What assumptions did you make when creating your model for the path of the handball?

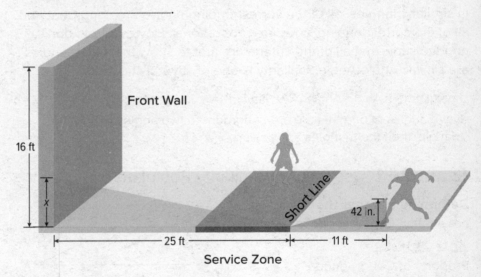

16 ft

$x$

Front Wall

Short Line

42 in.

25 ft

11 ft

Service Zone

**Solve**

Because the two triangles are similar, the corresponding sides are proportional.

$$\frac{11 \text{ ft}}{25 \text{ ft}} = \frac{\underline{\quad} \text{ in.}}{x}$$     AA Similarity Postulate

$$\frac{11 \text{ ft}}{25 \text{ ft}} = \frac{\underline{\quad} \text{ ft}}{x}$$     12 inches = 1 foot

$$x = \underline{\quad} \text{ ft}$$     Solve.

🌐 **Go Online** You can complete an Extra Example online.

Name _____ Period _____ Date _____

# Practice

Go Online You can complete your homework online.

## Example 1

**Determine whether each pair of triangles is similar. If so, write a similarity statement. If not, what would be sufficient to prove the triangles similar? Explain your reasoning.**

**1.**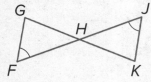

**2.**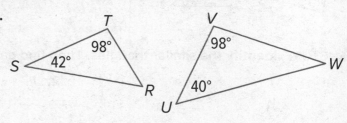

**3.**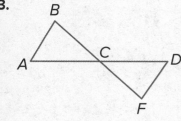

**4.**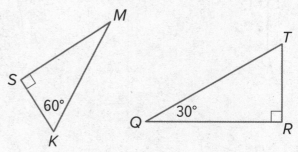

## Example 2

**5.** **CHAIRS** A local furniture store sells two versions of the same chair: one for adults, and one for children. Find the value of $x$ such that the chairs are similar.

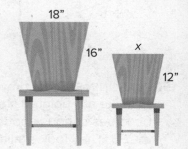

**6.** Two right triangles are similar. One has legs that are 3 and 5 inches long. The corresponding legs of the other triangle are 2.1 and $x$ inches long, respectively. Find the value of $x$.

**7.** **SHADOWS** A cell phone tower casts a shadow 10 feet long at the same time that a 75-foot building casts a shadow 15 feet long. How tall is the cell phone tower?

**8.** **LIGHTHOUSE** Maya wants to know how far she is standing from a lighthouse. The end of Maya's shadow coincides with the end of the lighthouse's shadow. She carefully made the drawing shown.
   **a.** What is the distance from the lighthouse to the end of the lighthouse's shadow, $x$?
   **b.** What is the distance from Maya to the lighthouse, $y$?

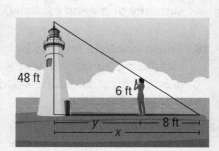

**Mixed Exercises**

**ARGUMENTS** Determine whether each pair of triangles is similar. Justify your answer.

**9.**

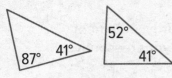

**10.**

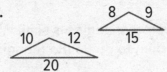

**STRUCTURE** Identify the similar triangles. Then find each measure.

**11.** *AC*

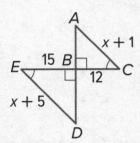

**12.** *JL*

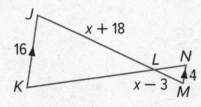

**13.** *EH*

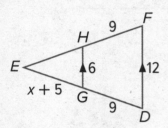

**14.** *VT*

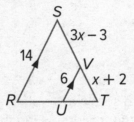

**15. REASONING** Olivia draws a regular pentagon and starts connecting its vertices to make a 5-pointed star. After drawing three of the lines in the star, she becomes curious about two triangles that appear in the figure, △*ABC* and △*CEB*. They look similar to her. Prove that this is the case.

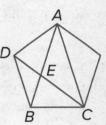

**16. ANALZYE** Write as many triangle similarity statements as possible for the figure shown. How do you know that these triangles are similar?

**17. PERSEVERE** In the figure, $\overline{KM} \perp \overline{JL}$ and $\overline{JK} \perp \overline{KL}$. Is △*JKL* ~ △*JMK*? Provide a proof to demonstrate their similarity or give an explanation of why they are not similar.

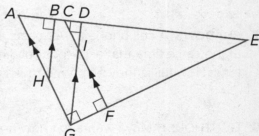

**18.** Pedro wants to find the height of a building. On a sunny afternoon, he stands near the building and finds that his shadow is 3 feet long, while the building's shadow is 84 feet long. Pedro is 5 feet 6 inches tall.

   **a. WRITE** Explain why you can conclude that the two triangles in the figure are similar.

   **b. ANALZYE** Explain how you can use similar triangles to find the height of the building.

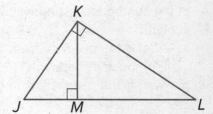

# Similar Triangles: SSS and SAS Similarity

**Today's Standards**
G.SRT.2, G.SRT.5
MP3, MP6, MP7

## Explore Similarity Criteria: SSS, SAS, and SSA

**Online Activity** Use dynamic geometry software to complete the Explore.

> **INQUIRY** What shortcuts can be used to determine whether two triangles are similar?

## Learn Similar Triangles: SSS and SAS Similarity

You can use the AA Similarity Postulate to prove the following two theorems.

**Theorem 8.3: Side-Side-Side (SSS) Similarity**

If two corresponding side lengths of two triangles are proportional, then the triangles are similar.

**Theorem 8.4: Side-Angle-Side (SAS) Similarity**

If the lengths of two sides of one triangle are proportional to the lengths of two corresponding sides of another triangle and the included angles are congruent, then the triangles are similar.

**Go Online** Proofs of Theorems 8.3 and 8.4 are available.

## Example 1 Use the SSS and SAS Similarity Theorems

**Determine whether the triangles are similar. Explain your reasoning.**

$\dfrac{JL}{QM} = \underline{\quad}$ or $\underline{\quad}$

$\dfrac{LK}{MP} = \underline{\quad}$ or $\underline{\quad}$   $\dfrac{JK}{QP} = \underline{\quad}$ or $\underline{\quad}$

By the $\underline{\quad}$ Similarity Theorem, $\triangle JKL \sim \triangle QMP$.

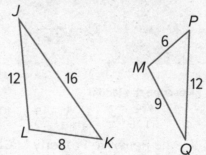

**Study Tip**

**Corresponding Sides**
To determine which sides of two triangles correspond, begin by comparing the longest sides, then the next longest sides, and finish by comparing the shortest sides.

## Example 2 Parts of Similar Triangles

**Find QN and PO. Justify your answer.**

**Step 1 Show that $\triangle NQM \sim \triangle OPM$.**

Because $\dfrac{MP}{MQ} = \dfrac{8}{5}$ and $\dfrac{MO}{MN} = \dfrac{9\frac{3}{5}}{6}$ or $\dfrac{8}{5}$, these two sides of $\triangle NQM$ and $\triangle OPM$ are proportional. By the Reflexive Property of Congruence, $\angle M \cong \angle M$. So, by the SAS Similarity Theorem,

**Go Online** You can complete an Extra Example online.

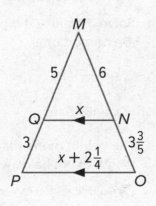

**Go Online** An alternate method is available for this example.

**Step 2** Find **QN** and **PO**.

| | |
|---|---|
| $\dfrac{MP}{MQ} = \dfrac{}{QN}$ | Definition of Similar Polygons |
| $\dfrac{8}{\phantom{xx}} = \dfrac{}{x}$ | Substitution |
| $8x = 5\underline{\phantom{xxxxx}}$ | Cross Multiply. |
| $x = \dfrac{\phantom{x}}{\phantom{x}}$ | Solve for $x$. |
| $QN = \underline{\phantom{xx}}$ | Substitution |
| $PO = 3\frac{3}{4} + 2\frac{1}{4}$ or $6$ | Solve. |

### 🌐 Example 3 Use Similar Triangles to Solve Problems

ARCHITECTURAL DESIGN **Julia is designing an A-frame house. The entire house will be 40 feet tall and the base of the house will be 60 feet long. She will build a second-floor balcony around the outside of the house, 15 feet above the ground. The left side of the house will be 50 feet long and the balcony will intersect the side 18.75 feet from the bottom. The height of the house bisects the base of the house and the balcony. Calculate the total length of the balcony.**

**Draw a Diagram**          **Create a Model**

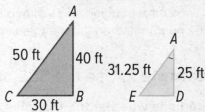

**Describe the Model**

$\dfrac{AC}{AE} = \dfrac{}{31.25}$ or $\underline{\phantom{xxxx}}$; $\dfrac{AB}{AD} = \dfrac{40}{}$ or $\underline{\phantom{xxxx}}$

By the Reflexive Property of Congruence, $\angle A \cong \angle\underline{\phantom{xxx}}$. Therefore, the two triangles are similar by the $\underline{\phantom{xxxx}}$ Similarity Theorem.

**Solve** Because the two triangles are similar, the corresponding sides are proportional.

$\dfrac{50 \text{ ft}}{31.25 \text{ ft}} = \dfrac{30 \text{ ft}}{ED}$          SAS Similarity Theorem

$ED = 18.75$ ft.          Solve.

The height of the triangle bisects the length of the balcony, so the total length of the balcony is 2(18.75) or 37.5 feet.

🌐 **Go Online** You can complete an Extra Example online.

---

💭 **Think About It!**
What assumptions did you make when creating your model for the side of the house?

🖱 **Go Online** to learn how to prove the slope criteria in Expand 8-4.

🖱 **Go Online** to practice what you've learned in Lessons 8-3 through 8-4.

# Practice

🡒 **Go Online** You can complete your homework online.

## Example 1

**Determine whether each pair of triangles is similar. If so, write a similarity statement. If not, what would be sufficient to prove the triangles similar? Explain your reasoning.**

1.
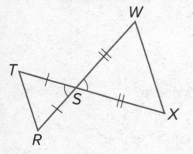

2.
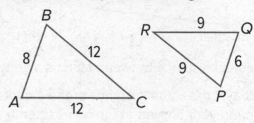

3.
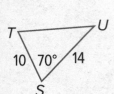

4.
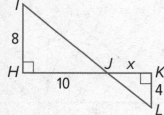

## Example 2

**Identify the similar triangles. Then find each measure.**

5.
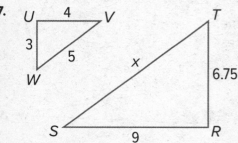

6.
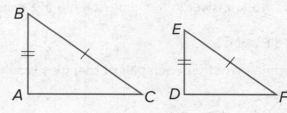

7.

8.
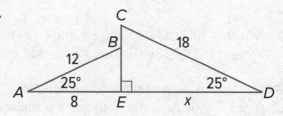

## Example 3

9. **ROOFS** The outline of a roof includes similar triangles. Find the value of x such that triangles *DEF* and *FBC* in the outline of the roof are similar.

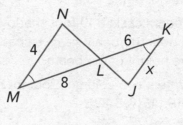

10. **SHADOWS** A radio tower casts an 8-foot-long shadow at the same time that a vertical yardstick casts a shadow one half inch long. How tall is the radio tower?

11. **BOATING** The two sailboats shown are participating in a regatta. If the sails are similar, what is the value of x?

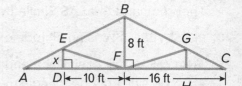
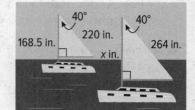

**12. MOUNTAIN PEAKS** Marcus and Skye want to know how far a mountain peak is from their houses. They measure the angles between the line of sight to the peak and to each other's houses and carefully make the drawing shown.

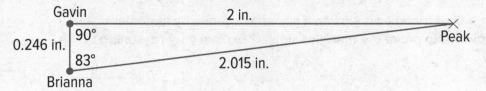

The actual distance between Marcus and Skye's houses is $1\frac{1}{2}$ miles.

**a.** What is the actual distance of the mountain peak from Marcus's house? Round your answer to the nearest tenth of a mile.

**b.** What is the actual distance of the mountain peak from Skye's house? Round your answer to the nearest tenth of a mile.

**Mixed Exercises**

**Determine whether each pair of triangles is similar. Justify your answer.**

**13.**

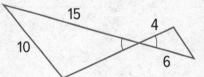

**14.**

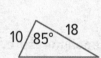

**15. REASONING** Mia drew the figure shown in art class. She said △*STU* and △*SQR* look similar. Prove that this is the case.

**16. STRUCTURE** A pole casts a 70-foot shadow. At the same time, a 3-foot mailbox near the pole casts a shadow 3 feet 6 inches long. Find the height of the pole.

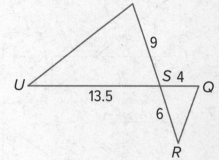

**17. STRUCTURE** A beam casts a 19-foot shadow. At the same time, a 6-foot person near the beam casts a shadow 4 feet 9 inches long. What is the height of the beam?

**18. ARGUMENTS** Write a two-column proof for the following.
Given: *AD* = 3*AB*; *AE* = 3*AC*
Prove: △*ABC* ~ △*ADE*

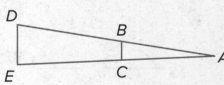

**19. WRITE** Compare and contrast the AA Similarity Postulate, the SSS Similarity Theorem, and the SAS Similarity Theorem.

**20. PERSEVERE** $\overline{YW}$ is an altitude of △*XYZ*. Find *YW*.

**21. ANALYZE** A pair of similar triangles has angles that measure 50°, 85°, and 45°. The sides of one triangle measure 3, 3.25, and 4.23 units, and the sides of the second triangle measure *x* − 0.46, *x*, and *x* + 1.81 units. Find the value of *x*. Round to thte nearest integer.

**22. CREATE** Draw a triangle that is similar to △*ABC* shown. Explain how you know that it is similar.

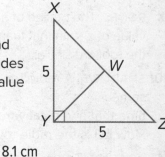

# Triangle Proportionality

## Explore Proportions in Triangles

▶ **Online Activity** Use dynamic geometry software to complete the Explore.

> ⊘ **INQUIRY** How do the midsegments of a triangle compare to its sides?

## Learn Triangle Proportionality

When a triangle contains a line that is parallel to one of its sides, the two triangles formed can be proven similar using the Angle-Angle Similarity Postulate. Since the triangles are similar, their sides are proportional.

**Theorem 8.7: Triangle Proportionality Theorem**

If a line is parallel to one side of a triangle and intersects the other two sides, then it divides the sides into segments of proportional lengths.

**Theorem 8.8: Converse of Triangle Proportionality Theorem**

If a line is parallel to one side of a triangle and intersects the other two sides, then it divides the sides into segments of proportional lengths.

## Example 1 Use Triangle Proportions to Find the Length of a Side

In $\triangle BCD$, $\overline{PQ} \parallel \overline{CD}$. If $QD = 14.5$, $BP = 9$, and $PC = 15$, find $BQ$.

$\dfrac{BP}{PC} = \dfrac{\phantom{BQ}}{QD}$    Triangle Proportionality Theorem

$\underline{\phantom{xx}} = \dfrac{BQ}{14.5}$    Substitution

$15 \cdot BQ = \underline{\phantom{xx}}(\underline{\phantom{xx}})$    Cross Products Property

$15BQ = \underline{\phantom{xxxx}}$    Multiply 9 and 14.5.

$BQ = \underline{\phantom{xxx}}$    Divide each side by 15.

## Check

In $\triangle FGH$, $\overline{FH} \parallel \overline{BC}$. If $GB = 14.4$, $BF = 4.8$, and $GH = 21$.

**Part A** What is a good estimate for the measure of $\overline{CH}$? _____

**A.** 2    **B.** 5    **C.** 10    **D.** 20

**Part B** Find $GC$ and $CH$.

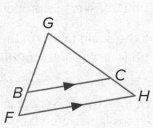

$GC = $ _____      $CH = $ _____

**Today's Standards**
G.SRT.4, G.CO.10, G. CO.12
MP1, MP7, MP8

**Today's Vocabulary**
midsegment of a triangle

▶ **Go Online**
Proofs of Theorems 8.7 and 8.8 are available.

💭 **Think About It!**
If $BP = 9$, $PC = 15$, and $PQ = 7.5$, could you use $\dfrac{9}{15} = \dfrac{7.5}{CD}$ to find $CD$? Justify your reasoning.

## Example 2 Use Triangle Proportions to Determine if Lines are Parallel

**In △WXY, YL = 5, LX = 20, and** $\overline{JX}$ **is four times as long as** $\overline{WJ}$. **Is** $\overline{JL} \parallel \overline{WY}$?

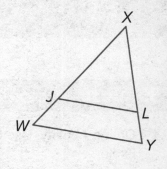

Using the Converse of the Triangle Proportionality Theorem, in order to show that $\overline{JL} \parallel \overline{WY}$, we must show that $\frac{WJ}{JX} = \frac{YL}{LX}$. Find and simplify each ratio. Since $\overline{JX}$ is four times as long as $\overline{WJ}$, you can represent their lengths with x and 4x, respectively.

$\frac{WJ}{JX} = \frac{x}{4x}$ or $\frac{1}{4}$          $\frac{YL}{LX} = \frac{5}{20}$ or $\frac{1}{4}$

Since $\frac{1}{4} = \frac{1}{4}$, the sides are proportional. Therefore, $\overline{JL} \parallel \overline{WY}$.

## Check

**In △PQR, PK = 34, KQ = 20, and PJ = 1.7JR.**

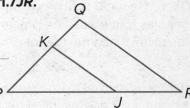

**Part A** Is $\overline{QR} \parallel \overline{KJ}$? _____

**Part B** Select the correct statement. _____

**A.** $\frac{PK}{KQ} = \frac{PJ}{JR} = \frac{10}{17}$          **B.** $QR = 1.7KJ$

**C.** $\frac{PK}{KQ} = \frac{PJ}{JR} = \frac{17}{10}$          **D.** $\frac{PK}{KQ} \neq \frac{PJ}{JR}$

## Learn Midsegments and Parallel Lines

A **midsegment of a triangle** is a segment that connects the midpoints of the legs of the triangle. Every triangle has three midsegments. The midsegments of △ABC are $\overline{RP}$, $\overline{PQ}$, and $\overline{RQ}$.

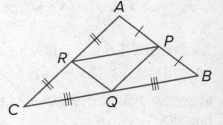

One special case of the Triangle Proportionality Theorem is the Triangle Midsegment Theorem.

**Theorem 8.9: Triangle Midsegment Theorem**

A midsegment of a triangle is parallel to one side of the triangle, and its length is one half the length of that side.

Another special case of the Triangle Proportionality Theorem involves three or more parallel lines cut by two transversals.

**Corollary 8.1: Proportional Parts of Parallel Lines**

If three or more parallel lines intersect two transversals, then they cut off the transversals proportionality.

If the scale factor of the proportional segments is 1, they separate the transversals into congruent parts.

**Corollary 8.2: Congruent Parts of Parallel Lines**

If three or more parallel lines cut off congruent segments on one transversal, then they cut off congruent segments on every transversal.

### ◆ Go Online
Proofs of Theorem 8.9, Corollary 8.1, and Corollary 8.2 are available.

# Example 3 Use the Triangle Midsegment Theorem

**In the figure, $\overline{EF}$ and $\overline{DE}$ are midsegments of $\triangle ABC$. Find each measure.**

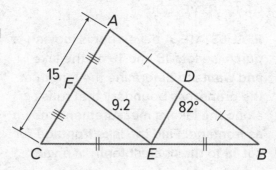

## DE

$DE = \frac{1}{2}$ _____          Triangle Midsegment Theorem

$DE = \frac{1}{2}$ _____          Substitution

$DE =$ _____          Simplify.

## DB

_____ $= \frac{1}{2} AB$          Triangle Midsegment Theorem

_____ $= \frac{1}{2} AB$          Substitution

_____ $= AB$          Simplify.

$DB = \frac{1}{2}$ _____          Definition of midpoint

$DB = \frac{1}{2}$ _____          Substitution

$DB =$ _____          Simplify.

## m∠FED

$\angle FED \cong \angle EDB$          Alternate Interior Angles Theorem

$m\angle FED = m\angle$ _____          Definition of congruence

$m\angle FED =$ _____          Substitution

## Check

**In the figure, $\overline{FG}$, $\overline{GH}$, and $\overline{FH}$ are midsegments of $\triangle ABC$. Find the measure of each of the segments.**

$AC =$ _____

$AB =$ _____

$CB =$ _____

$CG =$ _____

$AH =$ _____

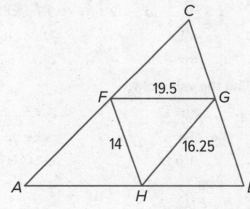

---

**Study Tip**

**Midsegments** The triangle Midsegment Theorem is similar to the Trapezoid Midsegment Theorem, which states that the midsegment of a trapezoid is parallel to the bases and its length is one half the sum of the measures of the bases.

---

🧠 **Think About It!**

If $BC = 24$, what is $DF$?

_____

---

### 🌐 **Example 4** Use Proportional Segments of Transversals

**REAL ESTATE** A developer is looking to purchase lots 18 and 19 on the lake and wants to determine the length of the property's boundary that runs along the lake, a measurement known as frontage. Find the lake frontage for Lot 18 to the nearest tenth of a yard.

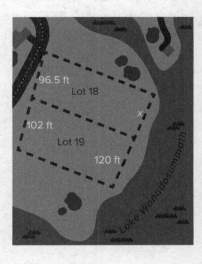

96.5 ft
Lot 18
x
102 ft
Lot 19
120 ft
Lake Wanadosumnath

By Corollary 8.1, since the three boundaries are approximately parallel, the segments formed by the front and back property lines are divided into proportional parts. Let $x$ represent the missing length.

| | |
|---|---|
| $\dfrac{102}{x} = \dfrac{96.5}{x}$ | Corollary 8.1 |
| _____ $\cdot x =$ _____ $\cdot 120$ | Cross Products Property |
| $102x =$ _____ | Simplify. |
| $x =$ _____ | Divide each side by 102. |

### Example 5  Use Congruent Segments of Transversals

**ALGEBRA  Find the values of $x$ and $y$.**

Since $\overline{AJ} \parallel \overline{BK} \parallel \overline{CL}$ and $\overline{JK} \cong \overline{KL}$, then $\overline{AB} \cong \overline{BC}$.

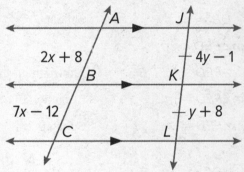

$2x + 8$  A    J
B  $4y - 1$  K
$7x - 12$  C    L  $y + 8$

**Find the value of $x$.**

| | |
|---|---|
| $AB = BC$ | Definition of congruence |
| $2x + 8 =$ _____ | Substitution |
| $8 =$ _____ $x - 12$ | Subtract $2x$ from each side. |
| _____ $= 5x$ | Add 12 to each side. |
| _____ $= x$ | Divide each side by 5. |

**Find the value of $y$.**

| | |
|---|---|
| $JK = KL$ | Definition of congruence |
| $4y - 1 =$ _____ | Substitution |
| _____ $y - 1 = 8$ | Subtract $y$ from each side. |
| $3y =$ _____ | Add 1 to each side. |
| $y =$ _____ | Divide each side by 3. |

🌐 **Go Online** You can complete an Extra Example online.

### 🐾 Go Online
You may want to complete the construction activities for this lesson.

# Practice

**Go Online** You can complete your homework online.

### Example 1

**Use the figure at the right.**

1. If $AB = 6$, $BC = 4$, and $AE = 9$, find $ED$.

2. If $AB = 12$, $AC = 16$, and $ED = 5$, find $AE$.

3. If $AC = 14$, $BC = 8$, and $AD = 21$, find $ED$.

4. If $AD = 27$, $AB = 8$, and $AE = 12$, find $BC$.

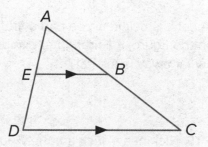

### Example 2

**Determine whether $\overline{NR} \parallel \overline{PQ}$. Justify your answer.**

5. $PM = 18$, $PN = 6$, $QM = 24$, and $RM = 16$

6. $NM = 7.5$, $PM = 24$, $QR = 27.5$, and $QM = 40$

7. $PN = 8$, $NM = 2$, and $RM = \frac{1}{2}\,QR$

8. $QM = 31$, $RM = 21$, and $PM = 4PN$

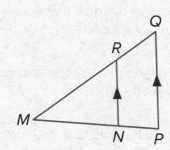

### Example 3

$\overline{VX}$, $\overline{VZ}$, and $\overline{ZR}$ are midsegments of $\triangle UWY$. Find the value of $x$.

9.

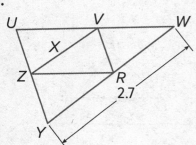

10.

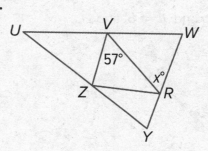

11.

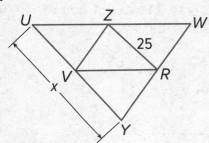

12.

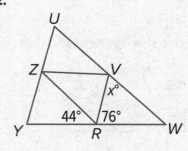

### Example 4

13. **STREETS** In the diagram, Cay Street and Bay Street are parallel. Find $x$.

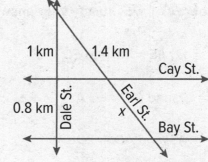

*not drawn to scale*

**14. JUNGLE GYMS** Prassad is building a two-story jungle gym according to the plans shown. Find the value of *x*.

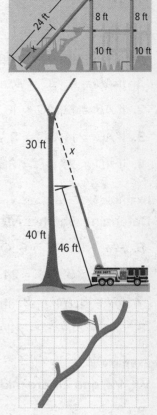

24 ft
*x*
8 ft    8 ft
10 ft    10 ft

**15. FIREMEN** A cat is stuck in a tree and a firefighter is trying to rescue it. Based on the figure, if a firefighter climbs to the top of the ladder, how far away is the cat?

30 ft    *x*
40 ft
46 ft

**16. EQUAL PARTS** Nick has a stick that he would like to divide into 9 equal parts. He places it on a piece of grid paper as shown. The grid paper is ruled so that vertical and horizontal lines are equally spaced.

a. Explain how he can use the grid paper to help him find where he needs to cut the stick.

b. Suppose Nick wants to divide his stick into 5 equal parts utilizing the grid paper. What can he do?

**Mixed Exercises**

**17.** If $JK = 7$, $KH = 21$, and $JL = 6$, find $LI$.

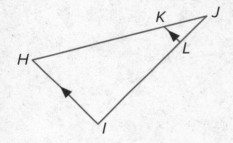

**18.** If $RU = 8$, $US = 14$, $TV = x - 1$, and $VS = 17.5$, find the value of $x$ and $TV$.

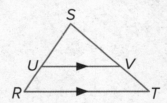

**For Exercises 19 and 20, use the figure at the right.**

**19.** If $BE = 4$, $BD = 5$, and $DA = 2$, find $BC$.

**20.** If $BD = 6$, $BE = 9$, and $BC = 15$, find $AD$.

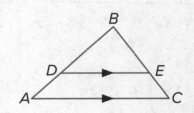

**Determine whether $\overline{BC} \parallel \overline{DE}$. Justify your answer.**

**21.** $AD = 15$, $DB = 12$, $AE = 10$, and $EC = 8$

**22.** $BD = 9$, $BA = 27$, and $CE = \frac{1}{3}EA$

**23.** $AE = 30$, $AC = 45$, and $AD = 2DB$

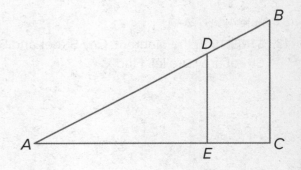

**$\overline{JH}$ is a midsegment of △KLM. Find the value of x.**

**24.**

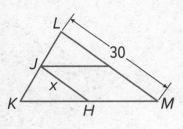

**25.**

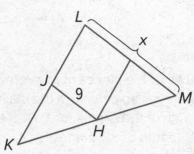

**26.**

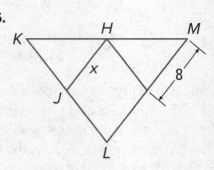

**27.** In △ABC, $\overline{DE}$ is parallel to $\overline{AC}$ and $DE = 10$. Find the length of $\overline{AC}$ if $\overline{DE}$ is the midsegment of △ABC.

**Find the values of x and y.**

**28.**

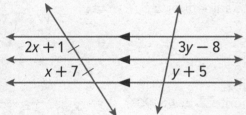

**29.**

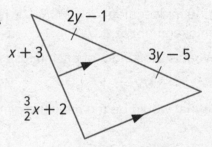

**30.** CARPENTRY Jake is fixing an A-frame. He wants to add a horizontal support beam halfway up and parallel to the ground. How long should this beam be?

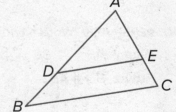

**31.** Follow these steps to prove the Converse of the Triangle Proportionality Theorem.

**Given:** $\frac{AD}{DB} = \frac{AE}{EC}$

**Prove:** $\overline{DE} \parallel \overline{BC}$

   **a.** REASONING Explain how you can use the given proportion to show that $\frac{AB}{AD} = \frac{AC}{EC}$.

   **b.** ARGUMENTS Explain how to use the proportion $\frac{AB}{AD} = \frac{AC}{EC}$ to complete the proof.

**32.** REGULARITY In the figure, $\overline{DE} \parallel \overline{BC}$, $BD = 12$, $EC = 10$, and $AE = 15$. Explain how to find the length of $\overline{AD}$.

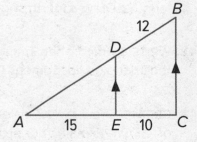

**USE TOOLS** Use a compass and straightedge to locate a point that partitions the given line segment in the given ratio. Label the required point *X*.

**33.** 1:2

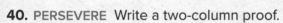

R    S

**34.** 3:1

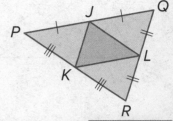

M    N

**35.** MODELING  The divider bars between the pieces of colored glass in a stained glass window are called *cames*. In the stained window at the right, the total length of the cames for △*PQR* is 78 centimeters. What is the total length of the cames for △*JKL*? Give an argument to support your answer.

**36.** MODELING  A crew is laying out a shuffleboard court using the plan shown at the right. A member of the crew wants to know the lengths of $\overline{AB}$, $\overline{BD}$, and $\overline{DF}$ to the nearest tenth of a foot. Explain how the crew member can calculate these lengths. Justify your steps.

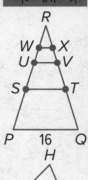

**37.** In △*PQR*, the length of $\overline{PQ}$ is 16 units. A series of midsegments are drawn such that $\overline{ST}$ is the midsegment of △*PQR*, $\overline{UV}$ is the midsegment of △*STR*, and $\overline{WX}$ is the midsegment of △*UVR*.

  **a.** PRECISION  What is the length of each midsegment?

  ST =          UV =          WX =

  **b.** STRUCTURE  What would be the measure of midsegment $\overline{YZ}$ of △*WXR*?

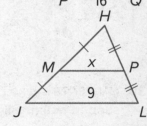

**38.** FIND THE ERROR  Jacob and Sebastian are finding the value of *x* in △*JHL*. Jacob says that *MP* is one half of *JL*, so *x* is 4.5. Sebastian says that *JL* is one half of *MP*, so *x* is 18. Is either of them correct? Explain.

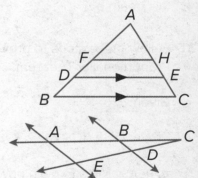

**39.** ANALYZE  In △*ABC*, *AF* = *FB* and *AH* = *HC*. If *D* is $\frac{3}{4}$ of the way from *A* to *B* and *E* is $\frac{3}{4}$ of the way from *A* to *C*, is *DE always*, *sometimes*, or *never* $\frac{3}{4}$ of *BC*? Explain.

**40.** PERSEVERE  Write a two-column proof.

  **Given:** *AB* = 4, *BC* = 4, and *CD* = *DE*

  **Prove:** $\overline{BD} \parallel \overline{AE}$

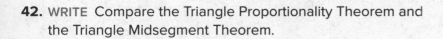

**41.** CREATE  Draw three segments *a*, *b*, and *c*, of all different lengths. Draw a fourth segment, *d*, such that $\frac{a}{b} = \frac{c}{d}$.

**42.** WRITE  Compare the Triangle Proportionality Theorem and the Triangle Midsegment Theorem.

**43.** CREATE  Draw △*XYZ* inside △*PQR* with half the perimeter of △*PQR*. Explain your process and why it works.

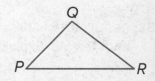

# Parts of Similar Triangles

**Today's Standards**
G.SRT.4
MP1, MP4, MP6, MP8

## Explore Special Segments in Triangles

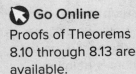 **Online Activity** Use dynamic geometry software to complete the Explore.

> ✕
>
> ⓠ **INQUIRY** What relationships exist among special segments in triangles?

## Learn Parts of Similar Triangles

**Theorem 8.10: Parts of Similar Triangles**

If two triangles are similar, the lengths of the corresponding altitudes are proportional to the lengths of corresponding sides.

**Theorem 8.11: Parts of Similar Triangles**

If two triangles are similar, the lengths of corresponding angle bisectors are proportional to the lengths of the corresponding sides.

**Theorem 8.12: Parts of Similar Triangles**

If two triangles are similar, the lengths of corresponding medians are proportional to the lengths of corresponding sides.

**Theorem 8.13: Triangle Angle Bisector**

An angle bisector in a triangle separates the opposite side into two segments that are proportional to the lengths of the other two sides.

**Go Online**
Proofs of Theorems 8.10 through 8.13 are available.

## Example 1 Use Special Segments in Similar Triangles

**In the figure, $\triangle MNP \sim \triangle XYZ$. Find the value of $x$.**

$\overline{MF}$ and $\overline{XG}$ are corresponding medians, and $\overline{NP}$ and $\overline{YZ}$ are corresponding sides of similar triangles $MNP$ and $XYZ$. Because $\overline{MF}$ and $\overline{XG}$ are the medians of triangles $MNP$ and $XYZ$, respectively, $NF = FP$ and $YG$ and $GZ$.

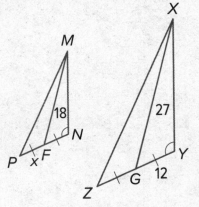

$$\frac{NP}{\phantom{XG}} = \frac{}{XG} \qquad \sim \triangle s \text{ have corr. Medians proportional to corr. sides}$$

$$\frac{2x}{\phantom{27}} = \frac{}{27} \qquad \text{Substitution}$$

$$\underline{\phantom{x}} \, x = \underline{\phantom{x}} \qquad \text{Cross Products Property}$$

$$x = \underline{\phantom{x}} \qquad \text{Solve for } x.$$

**Go Online**
An alternate method is available for this example.

**Study Tip**

**Assumptions** In this example, we assume that the sensor on the camera is pointing straight, creating altitudes in triangles *ABC* and *DCF*.

**Watch Out!**

Not all altitudes separate a triangle proportionally? The altitudes of an obtuse triangle are outside of the triangle and don't separate the triangle at all.

🔵 **Go Online**
You can complete an Extra Example online.

## 🌐 **Example 2** Use Similar Triangles to Solve Problems

PHOTOGRAPHY **A digital camera projects an image through its lens and onto its sensor, where it is converted into a digital image. The distance between the camera's lens and its sensor is known as the focal length and is adjusted depending on the size of the object being photographed and its distance from the camera lens. Ms. Elgin sets her camera up 3 meters away from her subject, who is 1.6 meters tall. If the sensor on her camera is 4.8 mm tall, what is the optimal focal length?**

**Step 1** Determine congruent angles.
Since $\overline{AB} \parallel \overline{DF}$, $\angle BAC \cong \angle$_____ and $\angle CBA \cong \angle$_____ by the Alternate Interior Angles Theorem.

**Step 2** Determine whether the triangles are similar.
$\triangle ABC \sim \triangle$_____ by AA Similarity.

**Step 3** Write a proportion, and solve for *x*.

$\dfrac{LC}{\quad} = \dfrac{}{AB}$     ~ △s have corr. Medians proportional to corr. sides

$\dfrac{3}{\quad} = \dfrac{}{0.0048}$     Substitution

$3 \cdot 0.0048 = 1.6 \cdot x$     Cross Products Property

_____ $=$ _____$x$     Simplify.

_____ $= x$     Divide each side by 1.6.

So, the estimated focal length is 0.009 meters or 9 millimeters.

---

## **Example 3** Use the Triangle Angle Bisector Theorem

**Find the value of *x*.**
Since $\overline{FJ}$ is an angle bisector of $\triangle FGH$, you can use the _____ Bisector Theorem to write a proportion.

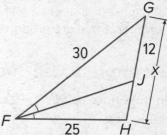

$\dfrac{GJ}{\quad} = \dfrac{}{FH}$     Triangle Angle Bisector Theorem Medians

$\dfrac{12}{\quad} = \dfrac{}{25}$     Substitution

$12 \cdot$ _____ $= 30($_____$)$     Cross Products Property

_____ $=$ _____$x - 360$     Simplify.

_____ $= 30x$     Add 360 to each side.

_____ $= x$     Divide each side by 30.

🔵 **Go Online** You can complete an Extra Example online.

# Practice

⟳ **Go Online** You can complete your homework online.

### Example 1

**Find the value of x.**

**1.**
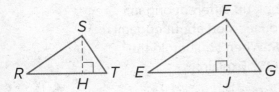

**2.**
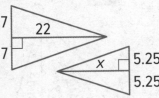

**3.**
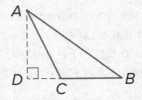

**4.**
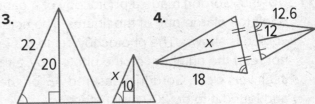

### Example 2

**5.** If △RST ~ △EFG, $\overline{SH}$ is an altitude of △RST, $\overline{FJ}$ is an altitude of △EFG, ST = 6, SH = 5, and FJ = 7, find FG.

**6.** If △ABC ~ △MNP, $\overline{AD}$ is an altitude of △ABC, $\overline{MQ}$ is an altitude of △MNP, AB = 24, AD = 14, and MQ = 10.5, find MN.

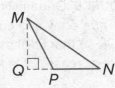

**7. TENTS** Jana went camping and stayed in a tent shaped like a triangle. In a photo of the tent, the base of the tent is 6 inches and the altitude is 5 inches. The actual base was 12 feet long. What was the height of the actual tent?

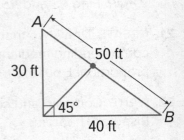

### Example 3

**Find the value of each variable.**

**8.**
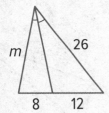

**9.**
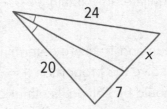

**Mixed Exercises**

**For Exercises 11 and 12, △ABC ~ △DEF.**

**10.** Find the length of $\overline{XC}$ to the nearest tenth.

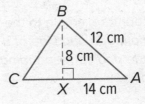
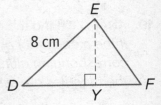

**11.** Find the length of $\overline{EY}$ to the nearest tenth.

**12. PLAYGROUND** The playground at the park in Hank's neighborhood has a large right triangle painted in the ground. Hank starts at the right angle corner and walks toward the opposite side along an angle bisector and stops when he gets to the hypotenuse. How much farther from Hank is point B versus point A?

13. **FLAG POLES** A flag pole attached to the side of a building is supported with a network of strings as shown in the figure. The rigging is done so that $AE = EF$, $AC = CD$, and $AB = BC$. What is the ratio of $CF$ to $BE$?

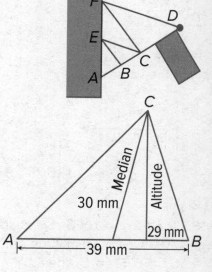

14. **COPIES** Gordon made a photocopy of a page from his geometry book to enlarge one of the figures. The actual figure that he copied is shown. The photocopy came out poorly. Gordon could not read the numbers on the photocopy, although the triangle itself was clear. Gordon measured $\overline{AB}$ on the enlarged triangle and found it to be 200 millimeters.

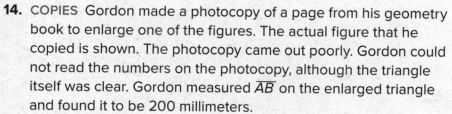

a. What is the length of the drawn altitude of the enlarged triangle? Round your answer to the nearest millimeter.

b. What is the length of the drawn median of the enlarged triangle? Round your answer to the nearest millimeter.

15. **USE A SOURCE** Origami is the art of paper folding. Look up different origami images. Then draw and cut out a $\triangle PQR$. Use what you learned about origami to fold $\triangle PQR$ so that $\overline{QR}$ lies on top of $\overline{PR}$ and $PR > QR$. When you unfold the triangle, the crease $\overline{SR}$ is formed. Is $PS$ greater than $QS$? Explain.

16. **MODELING** Amani uses copper wire to make earrings. She bends a piece of wire 45 mm long to make $\triangle ABC$. Then she adds a piece of wire to make the angle bisector $\overline{BD}$. The lengths of $\overline{AD}$ and $\overline{DC}$ are as shown. Explain how Amani can find the lengths of $\overline{AB}$ and $\overline{BC}$ without measuring.

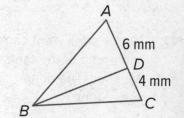

17. **REASONING** If the angle bisector of a triangle divides the side opposite it into two congruent segments, can you conclude that the triangle is equilateral? Explain.

18. **FIND THE ERROR** Chun and Traci are determining the value of $x$ in the figure. Chun says to find $x$, solve the proportion $\frac{5}{8} = \frac{15}{x}$, but Traci says to find $x$, the proportion $\frac{5}{x} = \frac{8}{15}$ should be solved. Is either of them correct? Explain.

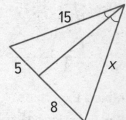

19. **ANALYZE** Find the counterexample to the following statement. Explain.
*If the measure of an altitude and side of a triangle are proportional to the corresponding altitude and corresponding side of another triangle, then the triangles are similar.*

20. **PERSEVERE** The perimeter of $\triangle PQR$ is 94 units. $\overline{QS}$ bisects $\angle PQR$. Find $PS$ and $RS$.

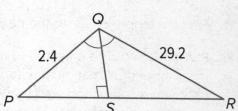

21. **CREATE** Draw two triangles so that the measures of corresponding medians and a corresponding side are proportional, but the triangles are not similar.

22. **WRITE** Compare and contrast Theorem 8.9 and the Triangle Angle Bisector Theorem.

##  Essential Question

What does it mean for objects to be similar, and how is similarity useful for modeling in the real world?

## Module Summary

### Lessons 8-1 and 8-2

#### Dilations and Similar Polygons

- A dilation is a nonrigid transformation that enlarges or reduces a geometric figure.

- When a figure is enlarged or reduced, the sides of the image are proportional to the sides of the original figure.

- To dilate a figure by a scale factor of $k$ with respect to (0, 0), multiply the $x$- and $y$-coordinate of each vertex by $k$.

- Two figures are similar if one can be obtained from the other by a dilation with one or more rigid transformations.

### Lessons 8-3 and 8-4

#### Criteria for Similar Triangles

- Angle-Angle Similarity: If two angles of one triangle are congruent to two angles of another triangle, then the triangles are similar.

- Side-Side-Side (SSS) Similarity: If the corresponding side lengths of two triangles are proportional, then the triangles are similar.

- Side-Angle-Side (SAS) Similarity: If the lengths of two sides of one triangle are proportional to the lengths of two corresponding sides of another triangle and the included angles are congruent, then the triangles are similar.

### Lesson 8-5

#### Triangle Proportionality

- If a line is parallel to one side of a triangle and intersects the other two sides, then it divides the sides into segments of proportional lengths.

- If three or more parallel lines intersect two transversals, then they cut off the transversals proportionality.

### Lesson 8-6

#### Parts of Similar Triangles

- If two triangles are similar, the lengths of the corresponding altitudes are proportional to the lengths of corresponding sides.

- If two triangles are similar, the lengths of corresponding angle bisectors are proportional to the lengths of the corresponding sides.

### Study Organizer

####  Foldables

Use your Foldable to review this module. Working with a partner can be helpful. Ask for clarification of concepts as needed.

# Test Practice

1. **MULTIPLE CHOICE** Which kind of dilation is the transformation from trapezoid *EFGH* to trapezoid *JKLM*? (Lesson 8-1)

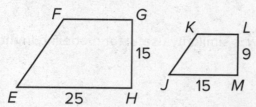

Ⓐ Enlargement with a scale factor of $\frac{5}{3}$

Ⓑ Enlargement with a scale factor of $\frac{3}{5}$

Ⓒ Reduction with a scale factor of $\frac{5}{3}$

Ⓓ Reduction with a scale factor of $\frac{3}{5}$

2. **TABLE ITEM** Match each effect on a figure with its corresponding dilation scale factor value. (Lesson 8-1)

| | Scale Factor | | | |
|---|---|---|---|---|
| | < −1 | −1 to 0 | 0 to 1 | > 1 |
| reduction | | | | |
| rotation and reduction | | | | |
| enlargement | | | | |
| rotation and enlargement | | | | |

3. **OPEN RESPONSE** The vertices of △*ABC* are *A*(5, 4), *B*(10, 8), and *C*(20, 0). A dilation centered at the origin with a scale factor of $\frac{4}{5}$ maps △*ABC* onto △*A″B′C′*. What are the coordinates of the vertices of △*A′B′C″*?

(Lesson 8-1)

4. **GRIDDED RESPONSE** An image viewed on a computer screen has a height of 2 inches and a width of 3 inches. If the image is enlarged 150%, what is the area of the enlargement in square inches? (Lesson 8-1)

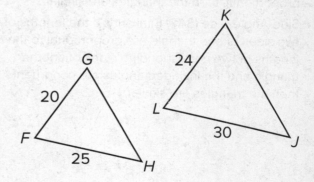

5. **OPEN RESPONSE** Which angle corresponds to ∠*A*? (Lesson 8-2)

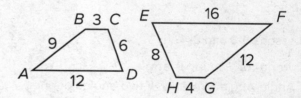

6. **OPEN RESPONSE** If a similarity transformation maps △*FGH* onto △*LKJ*, what is the similarity ratio? (Lesson 8-2)

**7. GRIDDED RESPONSE** On a blueprint, a rectangular kitchen has a length of 4 inches and a width of 3 inches. If the length of the kitchen is 6 feet, what is the perimeter of the kitchen in feet? (Lesson 8-2)

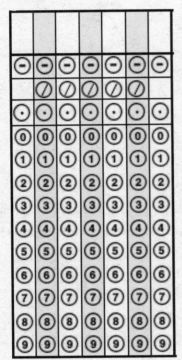

**8. MULTIPLE CHOICE** Given: $\angle B \cong \angle E$

Prove: $\triangle ADE \cong \triangle CDB$

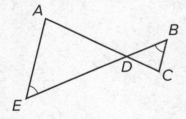

Complete the paragraph proof.

It is given that $\angle B \cong \angle E$. $\angle ADE \cong \angle CDB$ by the ___. Therefore, $\triangle ADE$ is similar to $\triangle CDB$ by the ___. (Lesson 8-3)

(A) Corresponding Angles Postulate, Angle-Angle Similarity Postulate

(B) Corresponding Angles Postulate, Third Angles Theorem

(C) Vertical Angles Theorem, Angle-Angle Similarity Postulate

(D) Vertical Angles Theorem, Third Angles Theorem

**9. OPEN RESPONSE** In $\triangle ABC$, $m\angle A = 44°$ and $m\angle B = 56°$. In $\triangle DEF$, $m\angle D = 44°$ and $m\angle F = 80°$. Is $\triangle ABC$ similar to $\triangle DEF$? Justify your answer. (Lesson 8-3)

**10. MULTIPLE CHOICE** What can be used to prove that triangle $ABC$ is similar to triangle $DBE$? (Lesson 8-4)

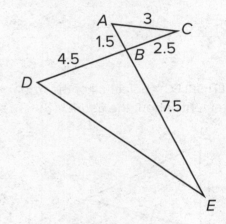

(A) AA Similarity Postulate

(B) SAS Similarity Theorem

(C) SSS Similarity Theorem

(D) AAS Similarity Theorem

**11. OPEN RESPONSE** What proves that $\triangle CDE$ is not similar to $\triangle FGH$? (Lesson 8-4)

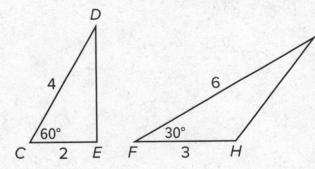

**12. TABLE ITEM** If $\overline{BD} \parallel \overline{AE}$ and $AC = 4$, identify the lengths of the other sides. (Lesson 8-5)

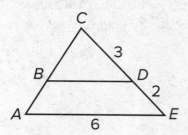

| Sides | Lengths | | | |
|---|---|---|---|---|
| | **1.6** | **2.0** | **2.4** | **3.6** |
| $\overline{AB}$ | | | | |
| $\overline{BD}$ | | | | |
| $\overline{BC}$ | | | | |

**13. MULTI-SELECT** Select all of the true statements about the figure. (Lesson 8-5)

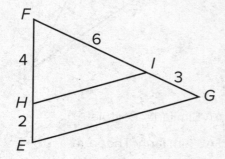

Ⓐ $\overline{HI} \parallel \overline{EG}$

Ⓑ $\dfrac{FH}{HE} = \dfrac{FI}{IG}$

Ⓒ $\dfrac{FH}{HE} = \dfrac{HI}{EG}$

Ⓓ $\dfrac{FH}{GI} = \dfrac{FI}{EH}$

**14. OPEN RESPONSE** Suppose $AD = 7$. What is the length of $\overline{AB}$? Write the answer as a fraction or decimal. (Lesson 8-6)

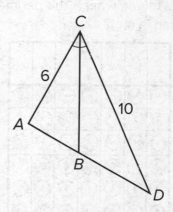

**15. MULTIPLE CHOICE** If $\triangle ABC \sim \triangle GHE$ which of these proportions must be true?

(Lesson 8-6)

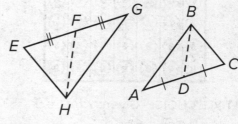

Ⓐ $\dfrac{FH}{BD} \parallel \dfrac{GH}{AB}$

Ⓑ $\dfrac{EH}{EF} = \dfrac{AB}{AD}$

Ⓒ $\dfrac{EF}{FH} = \dfrac{BD}{AD}$

Ⓓ $\dfrac{FH}{AC} = \dfrac{BD}{EG}$

**16. OPEN RESPONSE** If lines $k$, $m$, and $n$ are parallel, what is the value of $x$? (Lesson 8-6)

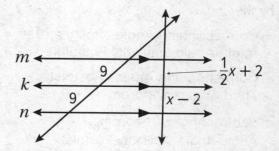

# Right Triangles and Trigonometry

## e Essential Question
How are right triangle relationships useful in solving real-world problems?

G.SRT.4, G.SRT.5, G.SRT.6, G.SRT.7, G.SRT.8, G.SRT.9, G.SRT.10, G.SRT.11
**Mathmatical Practices:** MP1, MP2, MP3, MP4, MP5, MP6, MP7, MP8

## What will you learn?

Place a checkmark (✓) in each row that corresponds with how much you already know about each topic **before** starting this module.

KEY

👎 — I don't know.     👊 — I've heard of it.     👍 — I know it!

| | Before | | | After | | |
|---|---|---|---|---|---|---|
| | 👎 | 👊 | 👍 | 👎 | 👊 | 👍 |
| solve problems using geometric mean and relationships between parts of a right triangle when an altitude is drawn to the hypotenuse | | | | | | |
| solve problems using the Pythagorean Theorem and its converse | | | | | | |
| graph points and find distances using the Distance Formula in three dimensions | | | | | | |
| solve problems using the properties of 45° −45° −90° and 30° −60° −90° right triangles | | | | | | |
| solve problems using the trigonometric ratios for acute angles | | | | | | |
| solve problems using the inverse trigonometric ratios for acute angles | | | | | | |
| derive and use a formula for the area of a triangle using trigonometry | | | | | | |
| solve problems using the Law of Sines and Law of Cosines | | | | | | |
| determine whether three given measures of a triangle define 0, 1, or 2 triangles using the Law of Sines | | | | | | |

📖 **Foldables** Make this Foldable to help you organize your notes about right triangles and trigonometry. Begin with four sheets of notebook paper and one sheet of construction paper.

1. **Stack** the notebook paper on the construction paper.

2. **Fold** the paper diagonally to form a triangle and cut off the excess.

3. **Open** the paper and staple the inside fold to form a booklet.

4. **Label** each tab with a lesson number and title.

1

2

3

4

Right Triangles and Trigonometry

# What Vocabulary Will You Learn?

Check the box next to each vocabulary term that you may already know.

- ☐ ambiguous case
- ☐ angle of depression
- ☐ angle of elevation
- ☐ cosine
- ☐ geometric mean

- ☐ inverse cosine
- ☐ inverse sine
- ☐ inverse tangent
- ☐ Law of Sines
- ☐ Law of Cosines

- ☐ ordered triple
- ☐ sine
- ☐ tangent
- ☐ trigonometry
- ☐ trigonometric ratio

## Are You Ready?

Complete the Quick Review to see if you are ready to start this module.
Then complete the Quick Check.

### Quick Review

**Example 1**

**The two triangles are similar. Solve for x.**

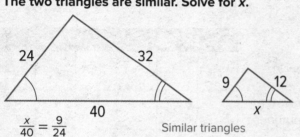

$$\frac{x}{40} = \frac{9}{24}$$     Similar triangles
$$24x = 360$$     Cross products
$$x = 15$$     Simplify.

**Example 2**

**Find $\sqrt{144}$.**

What number multiplied by itself equals 144?

$12 \cdot 12 = 144$

So, $\sqrt{144} = 12$.

### Quick Check

**Solve for x.**

**1.** $\triangle PQR \sim \triangle JKL$

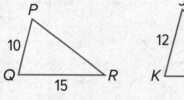
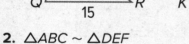

**2.** $\triangle ABC \sim \triangle DEF$

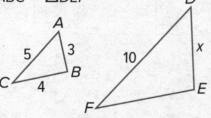

**Find each square root.**

**3.** $\sqrt{100}$      **4.** $\sqrt{64}$

**5.** $\sqrt{196}$      **6.** $\sqrt{81}$

**7.** $\sqrt{289}$      **8.** $\sqrt{625}$

### How Did You Do?

Which exercises did you answer correctly in the Quick Check? Shade those exercise numbers below.

① ② ③ ④ ⑤ ⑥ ⑦ ⑧

# Geometric Mean

## Explore Discovering Relationships in Right Triangles

**Today's Standards**
G.SRT.4, G.SRT.5
MP2, MP3, MP4

**Today's Vocabulary**
geometric mean

**◆ Online Activity** Use graphing technology to complete the Explore.

> **⊘ INQUIRY** In a right triangle, what relationship exists when an altitude is drawn from the vertex of the right angle to the hypotenuse?

## Learn Geometric Mean

The **geometric mean** of a set of numbers is the $n$th root, where $n$ is the number of elements in a set of numbers, of the product of the numbers. So the geometric mean of two numbers is the positive square root of their product.

### Key Concept • Geometric Mean

**Words** The geometric mean of two positive numbers $a$ and $b$ is $x$ such that $\frac{a}{x} = \frac{x}{b}$. So, $x^2 = ab$ and $x = \sqrt{ab}$.

The geometric mean can be found in the parts of a right triangle. An altitude drawn from the vertex of the right angle to the hypotenuse of a right triangle forms two additional right triangles. These three right triangles share a special relationship.

### Theorem 9.1: Altitudes in Right Triangles

If the altitude is drawn to the hypotenuse of a right triangle, then the two triangles formed are similar to the original triangle and to each other.

### Theorem 9.2: Geometric Mean (Altitude) Theorem

The altitude drawn to the hypotenuse of a right triangle separates the hypotenuse into two segments. The length of this altitude is the geometric mean between the lengths of these two segments.

**Example**
If $\overline{AB}$ is the altitude to hypotenuse of right $\triangle ABC$, then $\frac{x}{h} = \frac{h}{y}$, or $h = \sqrt{xy}$.

### Theorem 9.3: Geometric Mean (Leg) Theorem

The altitude drawn to the hypotenuse of a right triangle separates the hypotenuse into two segments. The length of a leg of this triangle is the geometric mean between the length of the hypotenuse and the segment of the hypotenuse adjacent to that leg.

**Example**
If $\overline{CD}$ is the altitude to hypotenuse $\overline{AB}$ of right $\triangle ABC$, then $\frac{c}{a} = \frac{a}{y}$, or $a = \sqrt{cy}$.

> **Study Tip**
>
> **Altitudes** Remember, the altitude of a triangle is a segment from a vertex to the line containing the opposite side and perpendicular to the line containing that side.

## Example 1 Find a Geometric Mean

**Find the geometric mean between 5 and 45.**

$$x = \sqrt{ab}$$    Definition of geometric mean

$$= \sqrt{5 \cdot 45}$$    $a = 5$ and $b =$ _____

$$= \sqrt{225}$$    Multiply.

$$= \underline{\hspace{1cm}}$$    Simplify.

The geometric mean between 5 and 45 is _____.

## Check

Find the geometric mean between 12 and 15. Write your answer in simplest radical form, if necessary. _____

---

## Example 2 Identify Similar Right Triangles

**Write a similarity statement identifying the three similar right triangles in the figure.**

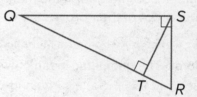

The diagram below shows the reorientation of the three triangles so that their corresponding angles and sides are in the same position as the original triangle.

**Complete the labels in the diagram.**

 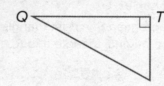 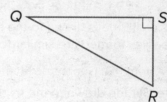

So by Theorem 9.1, $\triangle STR \sim \triangle$ _____ $\sim \triangle$ _____.

## Check

Select the similarity statement that identifies the three similar right triangles in the figure. _____

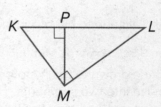

**A.** $\triangle KML \sim \triangle MPK \sim \triangle LPM$

**B.** $\triangle KPM \sim \triangle MPL \sim \triangle LMK$

**C.** $\triangle LKM \sim \triangle LMP \sim \triangle KMP$

**D.** $\triangle KML \sim \triangle KPM \sim \triangle MPL$

**Go Online** You can complete an Extra Example online.

## Example 3 Use Geometric Mean with Right Triangles

**Find x, y, and z.**

*x:*

Since $x$ is the measure of leg $\overline{AB}$, $x$ is the geometric mean of $AD$, the measure of the segment adjacent to $\overline{AB}$, and $AC$, the measure of the hypotenuse.

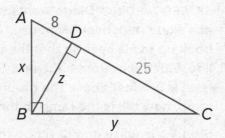

$x = \sqrt{AD \cdot AC}$   Geometric Mean (Leg) Theorem

$\phantom{x} = \sqrt{8 \cdot (\underline{\phantom{xxx}})}$   Substitution

$\phantom{x} = \sqrt{\underline{\phantom{xxxx}}}$   Simplify.

$\phantom{x} = 2\sqrt{\underline{\phantom{xx}}}$ or about 16.2   Use a calculator to simplify.

*y:*

Since $y$ is the measure of leg $\overline{BC}$, $y$ is the geometric mean of $DC$, the measure of the segment adjacent to $\overline{BC}$, and $AC$, the measure of the hypotenuse.

$y = \sqrt{DC \cdot AC}$   Geometric Mean (Leg) Theorem

$\phantom{y} = \sqrt{25 \cdot (\underline{\phantom{xxx}})}$   Substitution

$\phantom{y} = \sqrt{\underline{\phantom{xxx}}}$   Simplify.

$\phantom{y} = 5\sqrt{\underline{\phantom{xx}}}$ or about 28.7   Use a calculator to simplify.

*z:*

Since $z$ is the measure of the altitude drawn to the hypotenuse of right $\triangle ABC$, $z$ is the geometric mean of the lengths of the two segments that make up the hypotenuse, $AD$ and $DC$.

$z = \sqrt{AD \cdot DC}$   Geometric Mean (Altitude) Theorem

$\phantom{z} = \sqrt{8 \cdot \underline{\phantom{xx}}}$   Substitution

$\phantom{z} = \sqrt{\underline{\phantom{xx}}}$   Simplify.

$\phantom{z} = 10\sqrt{\phantom{x}}$ or about 14.1   Use a calculator to simplify.

## Check

**Find the values of x, y, and z.**

$x =$ _____, $y =$ _____, and $z =$ _____

**Go Online** You can complete an Extra Example online.

### 🌐 Example 4 Indirect Measurement

**SKATEBOARDING** Diego wants to find the height of a half pipe ramp at a skate park near his house. To find this height, Diego holds a book up to his eyes so that the top and bottom of the ramp are in line with the bottom edge and binding of the book. If Diego's eye level is 5.5 feet above the ground and he stands 6 feet from the ramp, how tall is the ramp to the nearest foot?

**Step 1 Visualize and describe the situation.**

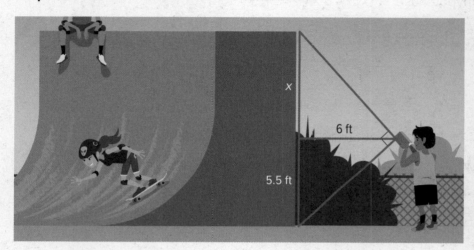

Draw a diagram to model the situation. The distance from Diego to the ramp is the altitude to the hypotenuse of a right triangle. The length of this altitude is the geometric mean between the lengths of the two segments that make up the hypotenuse. The shorter segment is equal to 5.5 feet, the height of Diego's eye level. Let the unknown measure be $x$ feet.

**Step 2 Find the height.**

You can use the Geometric Mean (Altitude) Theorem to find $x$.

$$6 = \sqrt{5.5 \cdot x}$$   Geometric Mean (Altitude) Theorem

$$\underline{\hspace{2cm}} = \underline{\hspace{2cm}}$$   Square each side.

$$\underline{\hspace{2cm}} \approx x$$   Divide each side by 5.5.

The height of the ramp is the total length of the hypotenuse, 6.55 + _____, or about 12 feet.

### Check

**EVENTS** Katelyn wants to make a banner for homecoming that will cover a wall in the cafeteria of her high school. To find the height of the wall, Katelyn holds a folder up to her eyes so that the top and bottom of the wall are in line with the edges of the folder. If Katelyn's eye level is 5.4 feet above the ground and she stands 11 feet from the wall, how tall is the wall to the nearest foot? _____ ft.

🐎 **Go Online** You can complete an Extra Example online.

# Practice

Go Online You can complete your homework online.

## Example 1

PRECISION **Find the geometric mean between each pair of numbers.**

**1.** 4 and 6

**2.** $\frac{1}{2}$ and 2

**3.** 4 and 25

**4.** 12 and 20

**5.** 17 and 3

**6.** 3 and 24

## Example 2

REGULARITY **Write a similarity statement identifying the three similar triangles in the figure.**

**7.**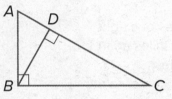

**8.**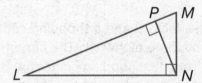

**9.**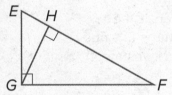

**10.**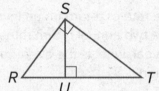

## Example 3

**Find x, y, and z.**

**11.**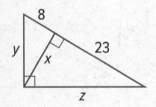

**12.**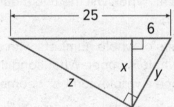

**13.**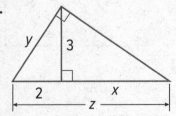

**14.**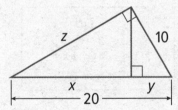

## Example 4

**15.** MODELING A museum has a famous statue on display. The curator places the statue in the corner of a rectangular room and builds a 15-foot-long railing in front of the statue. Use the information below to find how close visitors will be able to get to the statue.

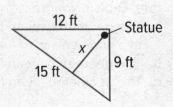

**16. VIEWING ANGLE** Noah wants to take a picture of a beach front. He wants to make sure two palm trees located at points *A* and *B* in the figure are just inside the edges of the photograph. He walks out on a walkway that goes over the ocean to get the shot. If his camera has a viewing angle of 90°, at what distance down the walkway should Noah stop to take his photograph?

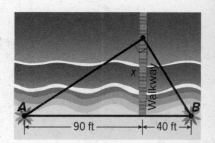

**Mixed Exercises**

**17. REASONING** The geometric mean of a number and four times the number is 22. What is the number?

**ARGUMENTS** Write a proof for each theorem.

**18.** Theorem 9.1

**19.** Theorem 9.2

**20.** Theorem 9.3

**21. CIVIL** An airport, a factory, and a shopping center are at the vertices of a right triangle formed by three highways. The airport and factory are 6.0 miles apart. Their distances from the shopping center are 3.6 miles and 4.8 miles, respectively. A service road will be constructed from the shopping center to the highway that connects the airport and factory. What is the shortest possible length for the service road? Round to the nearest hundredth.

**22. FINANCE** The average rate of return on an investment over two years is the geometric mean of the two annual return rates. If an investment returns 12% one year and 7% the next year, what is the average rate of return on this investment over the two-year period?

**23. PERSEVERE** Refer to the figure at the right. Find *x*, *y*, and *z*.

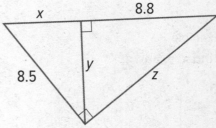

**24. WRITE** Compare and contrast the arithmetic and geometric means of two numbers. When will the two means be equal? Justify your reasoning.

**25. CREATE** Find two pairs of whole numbers with a geometric mean that is also a whole number. What condition must be met in order for a pair of numbers to produce a whole-number geometric mean?

**26. FIND THE ERROR** Aiden and Tia are finding the value of *x* in the triangle shown. Is either of them correct? Explain your reasoning.

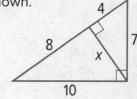

| Aiden | Tia |
|---|---|
| $\frac{4}{x} = \frac{x}{7}$ | $\frac{4}{x} = \frac{x}{10}$ |
| $x \approx 5.3$ | $x \approx 6.3$ |

**27. ANALYZE** Determine whether each statement is *always*, *sometimes*, or *never* true. Explain your reasoning.

**a.** The geometric mean for consecutive positive integers is the mean of the two numbers.

**b.** The geometric mean for two perfect squares is a positive integer.

**c.** The geometric mean for two positive integers is another integer.

# Pythagorean Theorem and Its Converse

**Today's Standards**
G.SRT.4, G.SRT.8
MP3, MP4, MP5, MP6

## Explore Proving the Pythagorean Theorem

**Today's Vocabulary**
Pythagorean triple

▶ **Online Activity** Use graphing technology to complete the Explore.

> ⓠ **INQUIRY** How can you use triangle similarity to prove the Pythagorean Theorem? ✕

## Learn The Pythagorean Theorem

The Pythagorean Theorem is perhaps one of the most famous theorems in mathematics. It relates the lengths of the hypotenuse and legs of a right triangle.

> **Theorem 9.4: Pythagorean Theorem**
>
> In a right triangle, the sum of the squares of the lengths of the legs is equal to the square of the length of the hypotenuse.

A **Pythagorean triple** is a set of three nonzero whole numbers $a$, $b$, and $c$, such that $a^2 + b^2 = c^2$. One common Pythagorean triple is 3, 4, 5; that is, the sides of a right triangle are in the ratio $3:4:5$. The most common Pythagorean triples are shown below in the first row. The triples below them are found by multiplying each number in the triple by the same factor.

💬 **Talk About It!**

Draw a right triangle with side lengths that form a Pythagorean triple. If you double the length of each side, is the resulting triangle *acute*, *right*, or *obtuse*? if you halve the length of each side? Justify your argument.

## Example 1 Find Missing Measures by Using the Pythagorean Theorem

**Find the value of $x$.**

The side opposite the right angle is the hypotenuse, so $c = x$.

$$a^2 + b^2 = c^2 \qquad \text{Pythagorean Theorem}$$
$$17^2 + 7^2 = x^2 \qquad a = \underline{\quad}, b = \underline{\quad}, c = \underline{\quad}$$
$$\underline{\quad} + \underline{\quad} = x^2 \qquad \text{Simplify.}$$
$$338 = x^2 \qquad \text{Add.}$$
$$\sqrt{338} = x \qquad \text{Take the positive square root of each side.}$$
$$13\sqrt{2} = x \qquad \text{Simplify.}$$

## Check

Find the value of $x$. Write your answer in simplest radical form, if necessary. _____

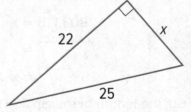

**Study Tip**

**Positive Square Root**
When finding the length of a side using the Pythagorean Theorem, use only the positive square root, because length cannot be negative.

▶ **Go Online** You can complete an Extra Example online.

**Go Online** An alternate solution method is available for this example.

## Example 2 Use a Pythagorean Triple

**Use a Pythagorean triple to find the value of $x$. Explain your reasoning.**

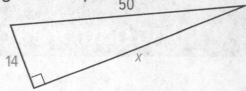

Notice that 14 and 50 are both multiples of 2, because $14 = 2 \cdot 7$ and $50 = 2 \cdot 25$. Since 7, 24, 25 is a Pythagorean triple, the missing leg length $x$ is $2 \cdot 24$ or _____.

## Check

Use a Pythagorean triple to find the value of $x$.

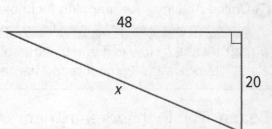

$x = $ _____

## Example 3 Use the Pythagorean Theorem

**ZIP LINING** **A summer camp is building a new zip lining course. The designer of the course wants the last zip line to start at a platform 450 meters above the ground and end 775 meters away from the base of the platform. How long must the zip line be to meet the designer's specifications?**

**Step 1 Visualize and describe the situation.**

The base of the platform and the ground should be approximately perpendicular. So, the height of the platform and the distance the end of the zip line is from the base of the platform make up the legs of a right triangle. We need to find the length of the zip line, which is the hypotenuse of the right triangle.

Draw a diagram that models the situation.

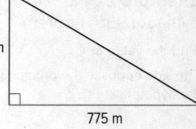

**Think About It!**

Is your answer reasonable? Use estimation to justify your reasoning.

**Step 2 Find the length of the zip line.**

Use the Pythagorean Theorem to find the length $x$ of the zip line.

$450^2 + \underline{\hspace{1cm}}^2 = x^2$     Pythagorean Theorem

$\underline{\hspace{1.5cm}} + \underline{\hspace{1.5cm}} = x^2$     Simplify.

$803{,}125 = x^2$     Add.

$\sqrt{\underline{\hspace{1cm}}} = x$     Take the positive square root of each side.

$25\sqrt{\underline{\hspace{1cm}}} = x$     Simplify.

So, the length of the zip line is $25\sqrt{1285}$ or about _____ meters.

**Go Online** You can complete an Extra Example online.

# Check

**RAMPS** Lincoln High School is installing more wheelchair ramps around the school. One of the new ramps needs to have a base that is 12 feet long and reaches a height of 1 foot. If the side of the ramp forms a right triangle, how long should the inclined surface of the ramp be?

## Part A

Find the exact length of the inclined surface of the ramp. Write your answer in simplest radical form, if necessary.

_____ ft

## Part B

Estimate the length of the inclined surface of the ramp. _____

**A.** 11 ft **B.** 11.96 ft **C.** 12 ft **D.** 12.04 ft

---

## **Learn** Converse of the Pythagorean Theorem

The converse of the Pythagorean Theorem also holds. You can use this theorem to determine whether a triangle is a right triangle given the measures of all three sides.

> **Theorem 9.5: Converse of the Pythagorean Theorem**
>
> In the sum of the squares of the lengths of the shortest sides of a triangle is equal to the square of the length of the longest side, then the triangle is a right triangle.

You can also use side lengths to classify a triangle as acute or obtuse.

> **Theorem 9.6: Pythagorean Inequality Theorem**
>
> In the square of the length of the longest side of a triangle is less than the sum of the squares of the lengths of the other two sides, then the triangle is an acute triangle.
>
>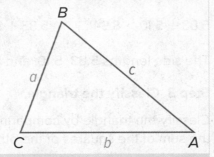
>
> **Example**
>
> If $c^2 < a^2 + b^2$, then $\triangle ABC$ is acute.

> **Theorem 9.7: Pythagorean Inequality Theorem**
>
> In the square of the length of the longest side of a triangle is greater than the sum of the squares of the lengths of the other two sides, then the triangle is an obtuse triangle.
>
>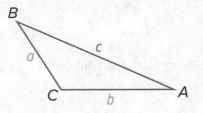
>
> **Example**
>
> If $c^2 > a^2 + b^2$, then $\triangle ABC$ is obtuse.

## Math History Minute

The Pythagorean Theorem is named for the Greek mathematician **Pythagoras (c. 570-c. 495 B.C.)**, although it is believed that Babylonian, Mesopotamian, Indian, and Chinese mathematicians all discovered the theorem independently. To date, over different 370 proofs of this theorem have been verified, including one written in 2006 by Iranian 14-year-old Sina Shiehyan.

## Study Tip

**Determining the Longest Side** If the measures of any of the sides of a triangle are expressed as radicals, you may wish to use a calculator to determine which side is the longest.

## Example 4  Classify Triangles

**Determine whether the points A(2, 2), B(5, 7), and C(10, 6) can be the vertices of a triangle. If so, classify the triangle as *acute*, *right*, or *obtuse*. Justify your answer.**

**Step 1  Calculate the measures of the sides.**

Use the Distance Formula to calculate the measures of $\overline{AB}$, $\overline{BC}$, and $\overline{AC}$.

$$AB = \sqrt{(5 - \underline{\phantom{xx}})^2 + (7 - \underline{\phantom{xx}})^2} \qquad BC = \sqrt{(10 - \underline{\phantom{xx}})^2 + (6 - \underline{\phantom{xx}})^2}$$

$$= \sqrt{3^2 + \underline{\phantom{xx}}^2} \qquad\qquad\qquad = \sqrt{5^2 + (\underline{\phantom{xx}})^2}$$

$$= \sqrt{\underline{\phantom{xx}}} \qquad\qquad\qquad\qquad = \sqrt{\underline{\phantom{xx}}}$$

$$AC = \sqrt{(10 - \underline{\phantom{xx}})^2 + (6 - \underline{\phantom{xx}})^2}$$

$$= \sqrt{8^2 + \underline{\phantom{xx}}^2}$$

$$= \sqrt{\underline{\phantom{xx}}}$$

$$= \underline{\phantom{xx}}\sqrt{5}$$

Use a calculator to approximate the measure of each side. So, $AB \approx 5.83$, $BC \approx \underline{\phantom{xx}}$, and $AC \approx \underline{\phantom{xx}}$.

**Step 2  Determine whether the measures can form a triangle.**

Use the Triangle Inequality Theorem to determine whether the measures 5.83, 5.10, and 8.94 can form a triangle.

$$5.83 + 5.10 > 8.94 \qquad 5.83 + 8.94 > 5.10 \qquad 5.10 + 8.94 > 5.83$$

The side lengths 5.83, 5.10, and 8.94 _____ form a triangle.

**Step 3  Classify the triangle.**

Classify the triangle by comparing the square of the longest side to the sum of the squares of the other two sides.

| | |
|---|---|
| $c^2 \overset{?}{=} a^2 + b^2$ | Compare $c^2$ and $a^2 + b^2$. |
| $8.94^2 \overset{?}{=} \underline{\phantom{xx}}^2 + 5.10^2$ | $c = 8.94$, $a = 5.83$, and $b = 5.10$ |
| $79.9 \underline{\phantom{xxx}}$ | Simplify and compare. |

Since $c^2 > a^2 + b^2$, the triangle is _____.

## Check

Determine whether the points J(1, 6), K(3, 2), and L(5, 3) can be the vertices of a triangle. If so, classify the triangle as *acute*, *right*, or *obtuse*. _____

🐾 **Go Online** You can complete an Extra Example online.

Study Tip:

**Approximations**
When finding the side lengths of a triangle using the Distance Formula, it may be easier to work with the side lengths after using a calculator to approximate their measures. However, when classifying a triangle, your final calculations will be more accurate if you keep the side lengths in radical form.

👁 **Think About It!**

Does your conclusion seem reasonable? Explain.

# Practice

◉ **Go Online** You can complete your homework online.

### Example 1

**Find the value of x.**

**1.**

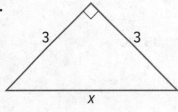

**2.**

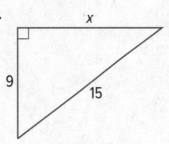

**3.**

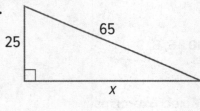

**4.**

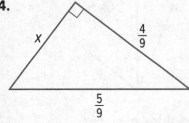

**5.**

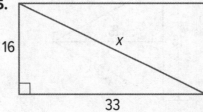

**6.**

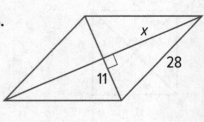

### Example 2

**Use a Pythagorean Triple to find the value of x.**

**7.**

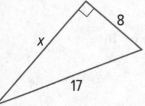

**8.**

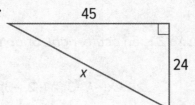

**9.**

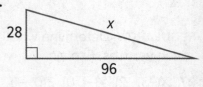

**10.**

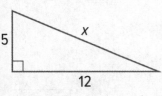

**11.**

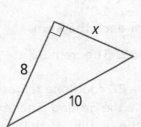

**12.**

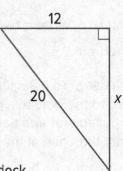

### Example 3

**13.** **CONSTRUCTION** The bottom end of a ramp at a warehouse is 10 feet from the base of the main dock and is 11 feet long. How high is the dock?

**14.** **FLIGHT** An airplane lands at an airport 60 miles east and 25 miles north of where it took off. How far apart are the two airports?

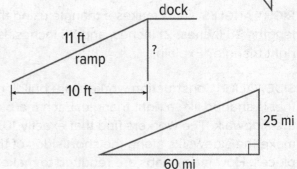

### Example 4

**Determine whether each set of measures can be the measures of the sides of a triangle. If so, classify the triangle as *acute*, *obtuse*, or *right*. Justify your answer.**

**15.** 30, 40, 50

**16.** 20, 30, 40

**17.** 18, 24, 30

**18.** 6, 8, 9

**19.** 6, 12, 18

**20.** 10, 15, 20

### Mixed Exercises

**Find the value of *x*.**

**21.**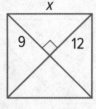

**22.**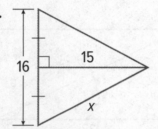

**23.**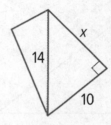

**Determine whether each set of measures can be the measures of the sides of a triangle. If so, classify the triangle as acute, obtuse, or right. Justify your answer.**

**24.** $\sqrt{5}, \sqrt{12}, \sqrt{13}$

**25.** $2, \sqrt{8}, \sqrt{12}$

**26.** 9, 40, 41

**REGULARITY** Determine whether △*XYZ* is an *acute*, *right*, or *obtuse* triangle for the given vertices. Explain.

**27.** $X(-3, -2), Y(-1, 0), Z(0, -1)$

**28.** $X(-7, -3), Y(-2, -5), Z(-4, -1)$

**29.** $X(1, 2), Y(4, 6), Z(6, 6)$

**30.** $X(3, 1), Y(3, 7), Z(11, 1)$

**ARGUMENTS Write a two-column proof for each theorem.**

**31.** Theorem 9.6

**32.** Theorem 9.7

**33. TETHERS** To help support a flag pole, a 50-foot-long tether is tied to the pole at a point 40 feet above the ground. The tether is pulled taut and tied to an anchor in the ground. How far away from the base of the pole is the anchor?

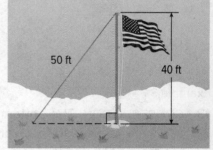

**34. RIGHT ANGLES** Cecil makes a triangle using three sticks of lengths 20 inches, 21 inches, and 28 inches. Is the triangle a right triangle? Explain.

**35. SIDEWALKS** Construction workers are building a marble sidewalk around a park that is shaped like a right triangle. Each marble slab adds 2 feet to the length of the sidewalk. The workers find that exactly 1071 and 1840 slabs are required to make the sidewalks along the short sides of the park, not counting corner pieces. How many slabs are required to make the sidewalk that runs along the long side of the park?

**Find the perimeter and area of each figure.**

36.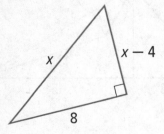

12

16

37.

13   13

|— 10 —|

38.

10

8

4

39. ALGEBRA The sides of a triangle have measures of $x$, $x + 5$, and 25. If the measure of the longest side is 25, what value of $x$ makes the triangle a right triangle?

40. PRECISION The sides of a triangle have measures of $2x$, 8, and 12. If the measure of the longest side is $2x$, what values of $x$ make the triangle acute?

41. REASONING A redwood tree in a national park is 20 meters tall. After it is struck by lightning, the tree breaks and falls over, as shown in the figure. The top of the tree lands at a point 16 feet from the centerline of the tree. A park ranger wants to know the height of the remaining stump of the tree.

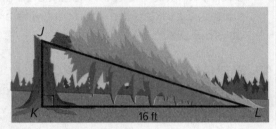

J

K          16 ft          L

a. The ranger lets $x$ represents the height of the stump, $\overline{JK}$. Explain how the rangers can write an expression for the length of $\overline{JL}$. Then write an equation that can be used to solve the problem.

b. Show how to solve the equation from **part a** to find the height of the stump.

42. Valeria and Sanjia are staking out a garden that has one pair of opposite sides measuring 30 feet and the other pair of sides measuring 40 feet. Using only a 60-foot-long tape measure, how can they be sure that their garden is a rectangle?

a. MODELING Draw a model of the garden with diagonal $t$. Let $p = 30$ and $q = 40$.

b. STRUCTURE If the garden is a rectangle, what must be true about $p$, $q$, and $t$? Why?

c. ARGUMENTS Walker measures the diagonal and finds that it is 50 feet long. Is there enough information to determine whether their garden is a rectangle? Explain.

**Find the value of $x$.**

43.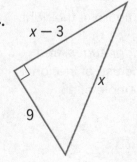

$x$   $x - 4$

8

44.

$x - 3$

$x$

9

45.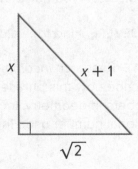

$x$   $x + 1$

$\sqrt{2}$

**46. HDTV** The screen aspect ratio, or the ratio of the width to the height, of a high-definition television is 16:9. The size of the television is given by the diagonal distance across the screen. If the height of Raj's HDTV screen is 32 inches, what is the screen size?

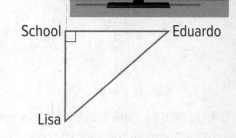

**47. MODELING** Eduardo and Lisa both leave school on their bikes at the same time. Eduardo rides due east at 18 miles per hour for 30 minutes and Lisa rides due south at 16 miles per hour for 30 minutes. Complete the diagram to represent the problem. To the nearest hundredth of a mile, how far apart are they when they stop riding their bikes?

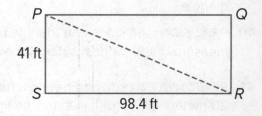

**48. OFFICE PARK** An office park has a rectangular lawn with the dimensions shown. Employees often take a shortcut by walking from *P* to *R*, rather than from *P* to *S* to *R*. What is the total distance an employee saves in a week by taking the shortcut twice a day for five days? Explain.

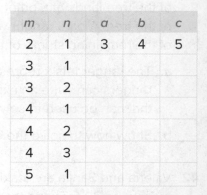

**49. PYTHAGOREAN TRIPLES** Ms. Jones assigned her fifth-period geometry class the following problem. Let *m* and *n* be two positive integers with $m > n$.

Let $a = m^2 - n^2$, $b = 2mn$, and $c = m^2 + n^2$.

  **a.** Show that there is a right triangle with side lengths *a*, *b*, and *c*.

  **b.** Complete the table shown at the right.

  **c.** Find a Pythagorean triple that corresponds to a right triangle with a hypotenuse $25^2 = 625$ units long. (*Hint:* Use the table you completed for part **b** to find two positive integers *m* and *n* with $m > n$ and $m^2 + n^2 = 625$.

| m | n | a | b | c |
|---|---|---|---|---|
| 2 | 1 | 3 | 4 | 5 |
| 3 | 1 |   |   |   |
| 3 | 2 |   |   |   |
| 4 | 1 |   |   |   |
| 4 | 2 |   |   |   |
| 4 | 3 |   |   |   |
| 5 | 1 |   |   |   |

**50. ANALYZE** *True* or *false*? Any two right triangles with the same hypotenuse have the same area. Explain your reasoning.

**51. CREATE** Draw a right triangle with side lengths that form a Pythagorean triple. If you double the length of each side, is the resulting triangle *acute*, *right*, or *obtuse*? If you halve the length of each side? Explain.

**52. PERSEVERE** Find the value of *x* in the figure at the right.

**53. WRITE** Research *incommensurable magnitudes*, and describe how this phrase relates to the use of irrational numbers in geometry. Include one example of an irrational number used in geometry.

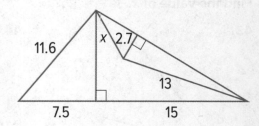

# Coordinates in Space

## Explore Proving the Distance Formula in Space

 **Online Activity** Use guiding exercises to complete the Explore.

×

@ **INQUIRY** How can you prove the Distance Formula in Space?

## Learn Coordinates in Space

You have used an ordered pair of two coordinates to describe the location of a point on the coordinate plane. Because space has three dimensions, an **ordered triple** is required to locate a point in space. An ordered triple is three numbers given in a specific order. A point in space is represented by $(x, y, z)$ where $x$, $y$, and $z$ are real numbers. A three-dimensional coordinate system has three axes: the $x$-, $y$-, and $z$-axes. The $x$- and $y$-axes lie on a horizontal plane, and the $z$-axis is vertical. The three axes divide space into eight **octants**. In octant 1, all of the coordinates of a point are positive.

### Key Concept · Distance and Midpoint Formulas in Space

$A$ has coordinates $A(x_1, y_1, z_1)$, and $B$ has coordinates $B(x_2, y_2, z_2)$.

**Distance Formula:**

$$AB = \sqrt{(x_2 - x_1)^2 + (y_2 - y_1)^2 + (z_2 - z_1)^2}$$

**Midpoint Formula:** $M\left(\dfrac{x_1 + x_2}{2}, \dfrac{y_1 + y_2}{2}, \dfrac{z_1 + z_2}{2}\right)$

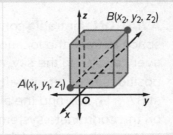

## Example 1 Graph a Rectangular Solid

**Graph a rectangular solid that contains $G(-3, 4, 2)$ and the origin as vertices. Label the coordinates of each vertex.**

**Step 1 Plot the $x$-coordinate.**
Plot the $x$-coordinate. Then draw a segment from the origin three units in the negative direction.

**Step 2 Plot the $y$-coordinate.**
To plot the $y$-coordinate, draw a segment four units in the positive direction from $(-3, 0, 0)$.

**Step 3 Plot the $z$-coordinate.**
To plot the $z$-coordinate, draw a segment two units in the positive direction from $(-3, 4, 0)$. Label the final point $G$.

*(continued on the next page)*

---

**Today's Standards**
MP3, MP4, MP6

**Today's Vocabulary**
ordered triple

octant

 **Go Online**
You may want to complete the Concept Check to check your understanding.

💭 **Think About It!**
How can you remember which part of the $x$-axis is positive and which is negative in three-dimensional coordinate space?

💬 Talk About It!

What is the relationship between the origin and the coordinates of a point?

🤔 Think About It!

What assumption did you make while solving this problem? Explain why this assumption caused your answer to be an approximation.

**Step 4 Draw the rectangular prism.**

Draw the rectangular prism and label each vertex: $G(-3, 4, 2)$, $F(-3, 0, 2)$, $E(0, 0, 2)$, $D(0, 4, 2)$, $H(0, 4, 0)$, $I(0, 0, 0)$, $J(-3, 0, 0)$, and $K(-3, 4, 0)$.

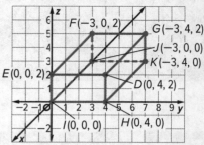

## 🌐 Example 2 Distance Formula in Space

**MEDICINE Doctors use three-dimensional coordinate systems for medical imaging. Medical imaging and positioning systems allow doctors to analyze a person's anatomy from a three-dimensional perspective. On an anatomical coordinate system the top of a man's spine is located at (5, 0, 65), and the bottom of his spine is located at (3, 0, −6). If each unit on the coordinate system represents a centimeter, what is the length of the man's spine?**

Write the coordinates to find the length of the man's spine.

$D = \sqrt{(x_2 - x_1)^2 + (y_2 - y_1)^2 + (z_2 - z_1)^2}$     Distance formula in Space

$= \sqrt{(5-3)^2 + (\underline{\quad} - \underline{\quad})^2 + (\underline{\quad} - (-\underline{\quad}))^2}$     Substitution

$= \sqrt{(2)^2 + (0)^2 + (71)^2}$     Subtract.

$= \sqrt{5045}$ or about 71.0     Use a calculator.

So, the length of the man's spine is about _____ centimeters.

### Check

**AVIATION Air traffic controllers use three-dimensional coordinate space to track the locations of aircraft. By assigning coordinates to every aircraft in the sky, air traffic controllers can describe the positions of other aircraft to pilots to prevent accidents. An airplane is at (17, −14, 23), and the air traffic control tower is at (0, 0, 0). If each unit on the coordinate system represents a kilometer, what is the distance between the airplane and the tower? Round your answer to the nearest tenth, if necessary.** _____ km

## Example 3 Midpoint Formula in Space

**Determine the coordinates of the midpoint $M$ of $\overline{DE}$ with endpoints $D(-4, -3, 5)$ and $E(6, 1, -9)$.**

$M = \left(\dfrac{x_1 + x_2}{2}, \dfrac{y_1 + y_2}{2}, \dfrac{z_1 + z_2}{2}\right)$     Midpoint Formula in Space

$= \left(\dfrac{-4 + 6}{2}, \dfrac{-3 + 1}{2}, \dfrac{5 + (-9)}{2}\right)$     Substitution

$= (\underline{\quad}, \underline{\quad}, \underline{\quad})$     Simplify.

### Check

**Determine the coordinates of the midpoint $M$ of $\overline{AB}$ with endpoints $A(-7, 9, 4)$ and $B(5, -3, -4)$.**

$M(\underline{\quad}, \underline{\quad}, \underline{\quad})$

🔗 **Go Online**

You can complete an Extra Example online.

Name _____ Period _____ Date _____

# Practice

**Go Online** You can complete your homework online.

### Example 1

**Graph a rectangular solid that contains the given point and the origin as vertices.
Label the coordinates of each vertex.**

**1.** $A(2, 1, 5)$

**2.** $P(-1, 4, 2)$

**3.** $C(-2, 2, 2)$

**4.** $R(3, -4, 1)$

**5.** $H(4, 5, -3)$

**6.** $G(4, 1, -3)$

### Example 2

**Determine the distance between each pair of points.**

**7.** $F(0, 0, 0)$ and $G(2, 4, 3)$

**8.** $X(-2, 5, -1)$ and $Y(9, 0, 4)$

**9.** $A(4, -6, 0)$ and $B(1, 0, 1)$

**10.** $C(8, 7, -2)$ and $D(0, 0, 0)$

**11. AIR TRAFFIC CONTROLLERS** An air traffic controller
knows the most recent position of an aircraft was at the
coordinates shown on the three-dimensional coordinate
system to the right. If the control tower is at $(0, 0, 0)$, then
what is the distance between the aircraft and the tower if
each unit on the coordinate system represents one mile?
Round your answer to the nearest tenth, if necessary.

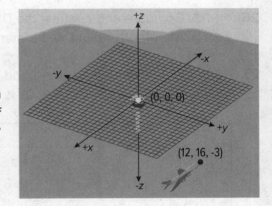

**12. ANIMATORS** An animator is using three-dimensional
software to create the character shown. She labels
the coordinates of the point, $R$, that represents the
end of the gavel and the point, $T$, that represents
the tip of the elbow as shown. What is the distance
between these two points of the animation? Round
your answer to the nearest tenth, if necessary.

### Example 3

**Determine the coordinates of the midpoint $M$ of the segment joining each pair
of points.**

**13.** $K(-2, -4, -4)$ and $L(4, 2, 0)$

**14.** $W(-1, -3, -6)$ and $Z(-1, 5, 10)$

**15.** $R(3, 3, 4)$ and $V(5, 4, 13)$

**16.** $A(4, 6, -8)$ and $B(0, 0, 0)$

**17.** $C(8, 7, 11)$ and $D(2, 1, 8)$

**18.** $T(-1, -7, 9)$ and $U(5, -1, -6)$

**Mixed Exercises**

REGULARITY **Determine the distance between each pair of points. Then determine the coordinates of the midpoint _M_ of the segment joining the pair of points.**

**19.** $P(-5, -2, -1)$ and $Q(-1, 0, 3)$

**20.** $J(1, 1, 1)$ and $K(-1, -1, -1)$

**21.** $F\left(\frac{3}{5}, 0, \frac{4}{5}\right)$ and $G(0, 3, 0)$

**22.** $G(1, -1, 6)$ and $H\left(\frac{1}{5}, -\frac{2}{5}, 2\right)$

**23.** $B(\sqrt{3}, 2, 2\sqrt{2})$ and $C(-2\sqrt{3}, 4, 4\sqrt{2})$

**24.** $S(6\sqrt{3}, 4, 4\sqrt{2})$ and $T(4\sqrt{3}, 5, \sqrt{2})$

**25.** ARGUMENTS  Write a coordinate proof of the Distance Formula in Space.

**Given:** _A_ has coordinates $A(x_1, y_1, z_1)$, and _B_ has coordinates $B(x_2, y_2, z_2)$.

**Prove:** $AB = \sqrt{(x_2 - x_1)^2 + (y_2 - y_1)^2 + (z_2 - z_1)^2}$

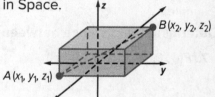

**26.** WRITE  Compare and contrast the Distance and Midpoint Formulas on the coordinate plane and in three-dimensional coordinate space.

**27.** FIND THE ERROR  Camilla and Teion were asked to find the distance between the points $A(2, 5, -8)$ and $B(3, -1, 0)$. Is either of them correct? Explain.

| Camilla | Teion |
|---|---|
| $AB = \sqrt{(x_2 - x_1)^2 + (y_2 - y_1)^2 + (z_2 - z_1)^2}$ | $AB = \sqrt{(x_2 - x_1)^2 + (y_2 - y_1)^2 + (z_2 - z_1)^2}$ |
| $= \sqrt{(2-3)^2 + (5-1)^2 + (-8-0)^2}$ | $= \sqrt{(2-3)^2 + (5-(-1))^2 + (-8-0)^2}$ |
| $= \sqrt{(-1)^2 + (4)^2 + (-8)^2}$ | $= \sqrt{(2-3)^2 + (5+1)^2 + (-8-0)^2}$ |
| $= \sqrt{1 + 16 + 64}$ | $= \sqrt{(-1)^2 + (6)^2 + (-8)^2}$ |
| $= \sqrt{81}$ | $= \sqrt{1 + 36 + 64}$ |
| $= 9$ | $= \sqrt{101}$ |
| | $\approx 10.05$ |

**28.** CREATE  Graph the cube that has all of the following characteristics:

- the origin is one of the vertices,
- one of the edges lies on the negative _y_-axis,
- one of the faces lies in the negative _xz_-plane.
  In which octants do your cubes lie?

**29.** PERSEVERE  Suppose the sphere has a radius of 9 units and passes through point _P_ which lies in Octant IV. Find the missing _z_-coordinate of point _P_. Justify your answer.

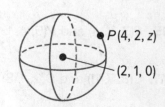

# Special Right Triangles

## Explore Properties of Special Right Triangles

**Today's Standards**
G.SRT.6
MP1, MP3, MP8

**Online Activity** Use graphing technology to complete the Explore.

> ⓠ **INQUIRY** What is the relationship between side lengths in 45°-45°-90° and 30°-60°-90° triangles? ✕

**Today's Vocabulary**
45°-45°-90° triangle

30°-60°-90° triangle

## Learn 45°-45°-90° Triangles

The diagonal of a square forms two congruent isosceles right triangles. Because the base angles of an isosceles triangle are congruent, the measure of each acute angle is 90° ÷ 2 or 45°. Such a special right triangle is known as a **45°-45°-90° triangle**.

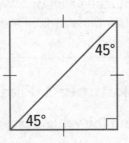

> 💬 **Talk About It!**
> Why do all 45°-45°-90° triangles have the same side length ratios? Use similarity to justify your reasoning.

### Theorem 9.8: 45°-45°-90° Triangle Theorem

In a 45°-45°-90° triangle, the legs $\ell$ are congruent and the length of the hypotenuse $h$ is $\sqrt{2}$ times the length of a leg.

**Go Online** A proof of Theorem 9.8 is available.

## Example 1 Find the Hypotenuse Length Given an Angle Measure

**Find the value of x.**

The acute angles of a right triangle are complementary, so the measure of the third angle is 90 − 45 or 45°. Because this is a 45°-45°-90° triangle, use the 45°-45° -90° Triangle Theorem.

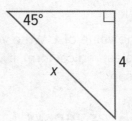

$h = \ell\sqrt{2}$      45°-45°-90° Triangle Theorem

$x = \underline{\hspace{2em}}\sqrt{2}$      Substitution

> 🤔 **Think About It!**
> How can you remember the ratios of the side lengths of a 45°-45°-90° triangle?

## Check

Find the value of x. Write your answer in simplest radical form, if necessary.

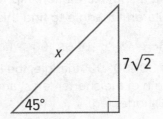

\_\_\_\_\_

**Go Online** You can complete an Extra Example online.

## Example 2 Find the Hypotenuse Length Given a Side Measure

**Find the value of x.**

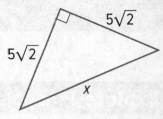

The legs of this right triangle have the same measure, so it is isosceles. Because this is a 45°-45°-90° triangle, use the 45°-45°-90° Triangle Theorem.

| $h = \ell\sqrt{2}$ | 45°-45°-90° Triangle Theorem |
|---|---|
| $x = 5\sqrt{2} \cdot \sqrt{2}$ | Substitution |
| $x = 5 \cdot$ ___ or ___ | $\sqrt{2} \cdot \sqrt{2} =$ ___ |

### Check

Find the value of x. Write your answer in simplest radical form, if necessary. _____

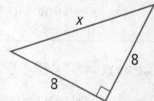

## Example 3 Find Leg Lengths in a 45°-45°-90° Triangle

**Find the value of x.**

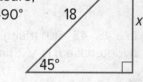

The legs of this right triangle have the same measure, x, so it is a 45°-45°-90° triangle. Use the 45°-45°-90° Triangle Theorem to find the value of x.

| $h = \ell\sqrt{2}$ | 45°-45°-90° Triangle Theorem |
|---|---|
| ___ $= x\sqrt{2}$ | Substitution |
| $\dfrac{18}{\phantom{x}} = x$ | Divide each side by $\sqrt{2}$. |

The value of x is $\dfrac{18}{\sqrt{2}}$ or ___ $\sqrt{2}$.

### Check

Find the value of x. Write your answer in simplest radical form, if necessary. _____

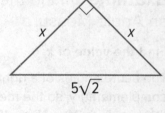

## Learn 30°-60°-90° Triangles

A **30°-60°-90° triangle** is a special right triangle or right triangle with side lengths that share a special relationship. You can use an equilateral triangle to find this relationship.

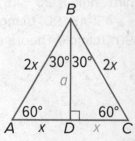

**Theorem 9.9: 30°-60°-90° Triangle Theorem**

In a 30°-60°-90° triangle, the length of the hypotenuse h is 2 times the length of the shorter leg s, and the longer leg ℓ is $\sqrt{3}$ times the length of the shorter leg.

🔵 **Go Online** You can complete an Extra Example online.

---

### Study Tip

**Rationalizing the Denominator** You can use the properties of square roots to rationalize the denominator of a fraction with a radical. This involves multiplying the numerator and denominator by a factor that eliminates radicals in the denominator. In the example above, $\dfrac{18}{\sqrt{2}}$ can be simplified by multiplying the numerator and denominator by $\sqrt{2}$.

🔵 **Go Online** A proof of Theorem 9.9 is available.

## Example 4 Find Leg Lengths in a 30°-60°-90° Triangle

**Find the values of *x* and *y*.**

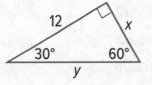

Use the 30°-60°-90° Triangle Theorem to find the value of *x*, the length of the shorter side.

$s\sqrt{3} = \ell$          30°-60°-90° Triangle Theorem

$x\sqrt{3} =$ _____         Substitution

$x =$ _____ or _____    Divide each side by $\sqrt{3}$.

Now use the 30°-60°-90° Triangle Theorem to find *y*, the length of the hypotenuse.

$h = 2s$           30°-60°-90° Triangle Theorem

$y = 2$_____       Substitution

$y =$ _____ or _____    Simplify.

## Check

Find the value of each variable.

**a. *x*** _____

**A.** 5    **B.** $\frac{15}{2}$    **C.** $\frac{15}{\sqrt{3}}$    **D.** $\frac{15\sqrt{3}}{2}$    **E.** $10\sqrt{3}$

**b. *y*** _____

**A.** 5    **B.** $\frac{15}{2}$    **C.** $5\sqrt{3}$    **D.** $\frac{15\sqrt{3}}{2}$    **E.** $\frac{30}{\sqrt{3}}$

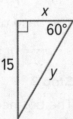

## Example 5 Use Properties of 30°-60°-90° Triangles

JEWELRY **Destiny makes and sells upcycled earrings. The earrings shown are made from congruent equilateral triangles. Each triangle has a height of 2 centimeters. The hooks attached to the top of the earrings are 1 centimeter tall. Destiny needs to mail this pair of earrings to a customer. If she mails the earrings in a rectangular box, what width and length must the base of the box have so the earrings will fit?**

Understand

What do you know?

_____

_____

What do you need to find?

_____

🄑 **Go Online** You can complete an Extra Example online.

*(continued on the next page)*

*(continued on the next page)*

---

🧠 Think About It!

Ian states that in a 30°-60°- 90° triangle, sometimes the 30° angle is opposite the longer leg and the 60° angle is opposite the shorter leg. Do you *agree* or *disagree* with Ian? Justify your answer.

Study Tip

**Use Ratios** The lengths of the sides of a 30°-60°-90° triangle are in a ratio of 1 to $\sqrt{3}$ to 2 or $1 : \sqrt{3} : 2$.

**Step 1** _____

**Step 2** _____

**Step 3** _____

SOLVE

**Step 1**

Draw one of the four equilateral triangles. Then draw an altitude of the triangle. The altitude represents the height of the triangle. So, the altitude is _____ centimeters.

**Step 2**

The altitude of the equilateral triangle forms the longer leg of two 30°-60°-90° triangles. Use the 30°-60°-90° Triangle Theorem to find the length of the shorter leg of one of the right triangles.

$\ell = s\sqrt{3}$          30°-60°-90° Triangle Theorem

$\underline{\phantom{xx}} = s\sqrt{3}$          Substitution

$\underline{\phantom{xx}} = s$          Divide each side by $\sqrt{3}$.

The base of the triangle is equal to $2s$. So, the length of the base is

$\underline{\phantom{xx}} \dfrac{2}{\sqrt{3}}$ or $\dfrac{}{\sqrt{3}}$ centimeters.

**Step 3**

The box needs to contain two earrings that are positioned next to each other. So, the width of the box must be at least 2 times the length of the base of one equilateral triangle.

$\underline{\phantom{xx}} \dfrac{4}{\sqrt{3}}$ or $\dfrac{}{\sqrt{3}}$

The width of the box must be at least $\dfrac{8}{\sqrt{3}}$ or about 4.62 centimeters wide.

The box must also be long enough to hold the earrings. One earring is made from two stacked equilateral triangles and a hook. The height of each hook is _____ centimeter. So, the length of the box must be at least _____ times the height of one equilateral triangle plus the height of the hook. The length of the box must be 2 · 2 + 1 or _____ centimeters.

Destiny must use a box that is at least _____ centimeters by _____ centimeters to mail the earrings.

Check

Is your solution is reasonable? Justify your reasoning.

_____

_____

_____

**Think About It!**

What assumption did you make while solving this problem?

Go Online You can complete an Extra Example online.

Name _____ Period _____ Date _____

# Practice

**Go Online** You can complete your homework online.

## Example 1

**REGULARITY** **Find the value of x.**

**1.**
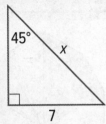
45°
x
7

**2.**
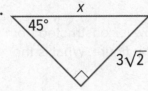
x
45°
$3\sqrt{2}$

**3.**
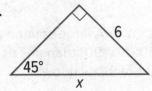
6
45°
x

**4.**
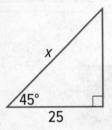
x
45°
25

**5.**
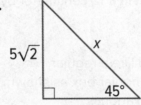
$5\sqrt{2}$
x
45°

**6.**
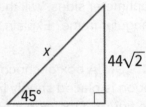
x
$44\sqrt{2}$
45°

## Example 2

**Find the value of x.**

**7.**
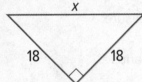
x
18   18

**8.**
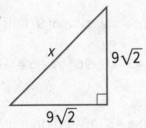
x
$9\sqrt{2}$
$9\sqrt{2}$

**9.**
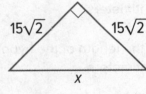
x
25
25

**10.**
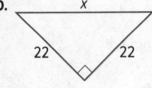
x
22   22

**11.**
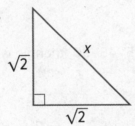
x
$\sqrt{2}$
$\sqrt{2}$

**12.**
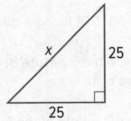
$15\sqrt{2}$   $15\sqrt{2}$
x

## Example 3

**Find the value of x.**

**13.**
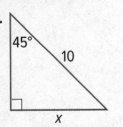
45°
10
x

**14.**
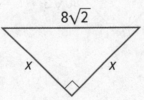
$8\sqrt{2}$
x   x

**15.**
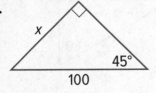
4
x
x

**16.**

x
45°
100

**17.**
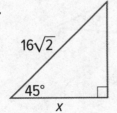
$16\sqrt{2}$
45°
x

**18.**
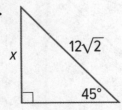
$12\sqrt{2}$
x
45°

**Example 4**

**19. ESCALATORS** A 40-foot-long escalator rises from the first floor to the second floor of a shopping mall. The escalator makes a 30° angle with the horizontal. How high above the first floor is the second floor?

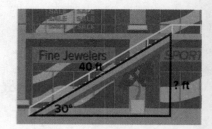
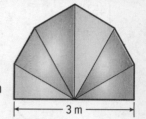

**20. WINDOWS** A large stained glass window is constructed from six 30°-60°-90° triangles as shown in the figure. What is the height of the window?

**21. MODELING** An award certificate is in the shape of an equilateral triangle with 12-centimeter sides. Will the certificate fit in a 12-centimeter by 10-centimeter rectangular frame? Explain.

**22. PRECISION** A box of chocolates shaped like a regular hexagon is placed snugly inside a rectangular box as shown in the figure. If the side length of the hexagon is 3 inches, what are the dimensions of the rectangular box?

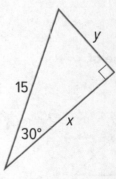

**Mixed Exercises**

**23.** If a 45°-45°-90° triangle has a hypotenuse measure of 9, find the leg measure.

**24.** Determine the measure of the leg of a 45°-45°-90° triangle with a hypotenuse measure of 11.

**25.** What is the length of the hypotenuse of a 45°-45°-90° triangle if the leg length is 6 centimeters?

**26.** Find the length of the hypotenuse of a 45°-45°-90° triangle with a leg length of 8 centimeters.

**Find the values of $x$ and $y$.**

**27.**

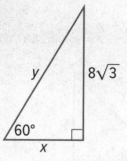

**28.**

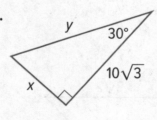

**29.**

**30.**

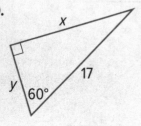

**31.**

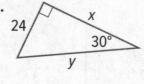

**32.**

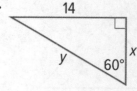

**33.** An equilateral triangle has an altitude length of 18 feet. Determine the length of a side of the triangle.

**34.** Find the length of the side of an equilateral triangle that has an altitude length of 24 feet.

**35. BOTANICAL GARDENS** One of the displays at a botanical garden is an herb garden planted in the shape of a square. The square measures 6 yards on each side. Visitors can view the herbs from a diagonal pathway through the garden. How long is the pathway?

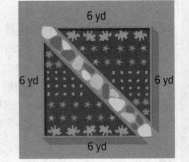

**36. ORIGAMI** A square piece of paper 150 millimeters on a side is folded in half along a diagonal. The result is a 45°-45°-90° triangle. What is the length of the hypotenuse of this triangle?

**37. REASONING** Kim and Yolanda are watching a movie in a movie theater. Yolanda is sitting $x$ feet from the screen and Kim is 15 feet behind Yolanda. The angle that Kim's line of sight to the top of the screen makes with the horizontal is 30°. The angle that Yolanda's line of sight to the top of the screen makes with the horizontal is 45°.

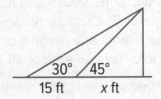

**a.** How high is the top of the screen in terms of $x$?

**b.** What is $\frac{x + 15}{x}$?

**c.** How far is Yolanda from the screen? Round your answer to the nearest tenth.

**38. STRUCTURE** Each triangle in the figure is a 45°-45°-90° triangle. Find the value of $x$.

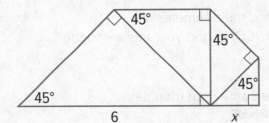

**Find the values of $x$ and $y$.**

**39.**

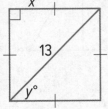

**40.**

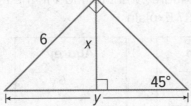

**41.**

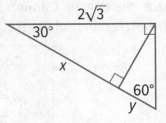

**42.**

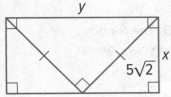

**43.**

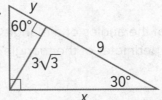

**44.**

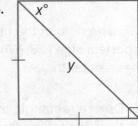

**45. COORDINATE GEOMETRY** △XYZ is a 45°-45°-90° triangle with right angle Z. Find the coordinates of X in Quadrant I for Y(−1, 2) and Z(6, 2).

**46. COORDINATE GEOMETRY** △EFG is a 30°-60°-90° triangle with m∠F = 90°. Find the coordinates of E in Quadrant III for F(−3, −4) and G(−3, 2). $\overline{FG}$ is the longer leg.

**47. USE TOOLS** Tenisha is in charge of building a ramp for a loading dock. According to the plan, the ramp makes a 30° angle with the ground, as shown. Also, the plan states that $\overline{ST}$ is 4 feet longer than $\overline{RS}$. Use a calculator to find the lengths of the three sides of the ramp to the nearest thousandth.

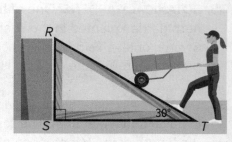

**48. STATE YOUR ASSUMPTION** Liling is making a quilt. She starts with two small squares of material and cuts them along the diagonal. Then she arranges the four resulting triangles to make a large square quilt block. She wants the large quilt block to have an area of 36 square inches.

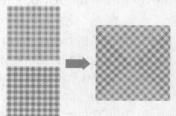

   a. What side lengths should she use for the two small squares of material? Explain.

   b. What assumption are you making so that you know your answer to part **a** is accurate?

**49. PERSEVERE** Find the perimeter of quadrilateral *ABCD*. Round your answer to the nearest tenth.

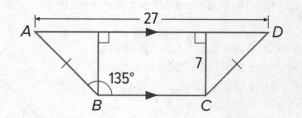

**50. WRITE** Why are some right triangles considered *special*?

**51. FIND THE ERROR** Carmen and Audrey want to find *x* in the triangle shown. Is either of them correct? Explain.

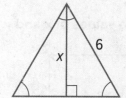

| Carmen | Audrey |
|---|---|
| $x = \dfrac{6\sqrt{3}}{2}$ | $x = \dfrac{6\sqrt{2}}{2}$ |
| $x = 3\sqrt{3}$ | $x = 3\sqrt{2}$ |

**52. ANALYZE** The ratio of the measure of the angles of a triangle is 1:2:3. The length of the shortest side is 8. What is the perimeter of the triangle? Round your answer to the nearest tenth.

**53. CREATE** Draw a rectangle that has a diagonal twice as long as its width. Then write an equation to find the length of the rectangle.

# Trigonometry

## Explore Sine and Cosine Complementary Angles

▶ **Online Activity** Use graphing technology to complete the Explore.

> ⓠ **INQUIRY** What can trigonometric ratios tell
> you about the relationship between the
> complementary angles in a right triangle?

## Explore Trigonometry and Similarity

▶ **Online Activity** Use graphing technology to complete the Explore.

> ⓠ **INQUIRY** If two right triangles have the same
> angle measure, what do you know about the
> trigonometric ratios of the angle?

## Learn Trigonometry

The word **trigonometry** comes from the Greek terms *Trigon*, meaning
triangle, and *metron*, meaning measure. So the study of trigonometry
involves triangle measurement. A **trigonometric ratio** is a ratio of the
lengths of two sides of a right triangle.

The names of the three most common trigonometric ratios are given
below.

| Key Concept • Trigonometric Ratios |
| --- |

**sine:**

$\sin A = \frac{\text{opp}}{\text{hyp}}$ or $\frac{a}{c}$; $\sin B = \frac{\text{opp}}{\text{hyp}}$ or $\frac{b}{c}$

**cosine:**

$\cos A = \frac{\text{adj}}{\text{hyp}}$ or $\frac{b}{c}$; $\cos B = \frac{\text{adj}}{\text{hyp}}$ or $\frac{a}{c}$

**tangent:**

$\tan A = \frac{\text{opp}}{\text{adj}}$ or $\frac{a}{b}$; $\tan B = \frac{\text{opp}}{\text{adj}}$ or $\frac{b}{a}$

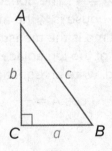

## Example 1 Find Trigonometric Ratios

Find sin *J*, cos *J*, tan *J*, sin *K*, cos *K*, and tan *K*.
Express each ratio as a fraction and as a
decimal to the nearest hundredth.

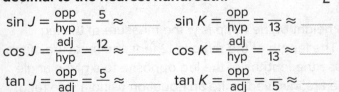

$\sin J = \frac{\text{opp}}{\text{hyp}} = \frac{5}{} \approx$ _____ 　　 $\sin K = \frac{\text{opp}}{\text{hyp}} = \frac{}{13} \approx$ _____

$\cos J = \frac{\text{adj}}{\text{hyp}} = \frac{12}{} \approx$ _____ 　　 $\cos K = \frac{\text{adj}}{\text{hyp}} = \frac{}{13} \approx$ _____

$\tan J = \frac{\text{opp}}{\text{adj}} = \frac{5}{} \approx$ _____ 　　 $\tan K = \frac{\text{opp}}{\text{adj}} = \frac{}{5} \approx$ _____

▶ **Go Online** You can complete an Extra Example online.

### Today's Standards
G.SRT.6; G.SRT.7
MP5, MP7

### Today's Vocabulary
trigonometry
trigonometric ratio
sine
cosine
tangent
inverse sine
inverse cosine
inverse tangent

### Study Tip:

**Memorizing
Trigonometric Ratios**
SOH-CAH-TOA is a
mnemonic device for
learning the ratios for
sine, cosine, and
tangent using the first
letter of each word in
the ratios.

$\sin A = \frac{\text{opp}}{\text{hyp}}$ 　$\cos A = \frac{\text{adj}}{\text{hyp}}$

$\tan A = \frac{\text{opp}}{\text{adj}}$

### 💭 Think About It!

How are sin *J* and
cos *K* related?

Special right triangles can be used to find the sine, cosine, and tangent of 30°, 45°, and 60° angles.

## Example 2 Use Special Right Triangles to Find Trigonometric Ratios

**Use a special right triangle to express the sine of 60° as a fraction and as a decimal to the nearest hundredth.**

Using the 30°-60°-90° Triangle Theorem, write the correct side lengths for each leg of the right triangle with $x$ as the length of the shorter leg.

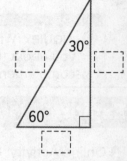

$$\sin 60° = \frac{opp}{hyp} \qquad \text{Definition of sine ratio}$$

$$= \frac{x\sqrt{3}}{\quad} \qquad \text{Substitution}$$

$$= \underline{\quad} \qquad \text{Divide the numerator and denominator by } x.$$

$$\approx \underline{\quad} \qquad \text{Use a calculator.}$$

## 🌐 Example 3 Estimate Measures by Using Trigonometry

**ACCESSIBILITY Mathias builds a ramp so his sister can access the back door of their house. The 7-foot ramp to the house slopes upward from the ground at a 4° angle. What is the horizontal distance between the foot of the ramp and the house? What is the height of the ramp? Find the horizontal distance.**

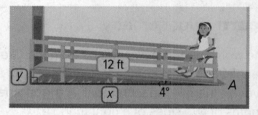

Let $m\angle A = 4°$. The horizontal distance between the foot of the ramp and the house is $x$, the measure of the leg adjacent to $\angle A$. The length of the ramp is the measure of the hypotenuse, 12 feet. Because the lengths of the leg adjacent to a given angle and the hypotenuse are involved, write an equation using the cosine ratio.

$$\cos A = \frac{adj}{hyp} \qquad \text{Definition of cosine ratio}$$

$$\cos 4° = \frac{x}{\quad} \qquad \text{Substitution}$$

$$\underline{\quad} \cos 4° = x \qquad \text{Multiply each side by 12.}$$

$$x \approx \underline{\quad} \qquad \text{Use a calculator.}$$

The horizontal distance between the foot of the ramp and the house is about 11.97 feet.

**Find the height.**

Let $m\angle A = 11°$. The height of the ramp is $y$, the measure of the leg opposite from $\angle A$. The length of the ramp is 12 feet, the measure of the hypotenuse. Because the lengths of the leg opposite to a given angle and the hypotenuse are involved, write an equation using a sine ratio.

🧭 **Go Online** You can complete an Extra Example online.

**Study Tip**

**Graphing Calculator**
Be sure that your graphing calculator is in degree mode rather than radian mode. Then, use the trigonometric functions [cos] and [sin] to find $x$ and $y$, respectively.

12 [cos] 11 [enter] 11.795262

12 [sin] 11 [enter]
2.289707945

$$\sin A = \frac{\text{opp}}{\text{hyp}} \qquad \text{Definition of sine ratio}$$

$$\sin 4° = \frac{y}{\underline{\quad}} \qquad \text{Substitution}$$

$$\underline{\quad} \sin 4° = y \qquad \text{Multiply each side by 12.}$$

$$y \approx \underline{\quad} \qquad \text{Use a calculator.}$$

The height $y$ of the ramp is about 0.84 feet or about ____ inches.

## Learn Inverse Trigonometric Ratios

If you know the value of a trigonometric ratio for an acute angle, you can use a calculator to find the measure of the angle, which is the inverse of the trigonometric ratio.

| Key Concept • Inverse Trigonometric Ratios | | |
|---|---|---|
| Inverse Sine | Inverse Cosine | Inverse Tangent |
| **Words** | | |
| If $\angle A$ is an acute angle and the sine of $A$ is $x$, then the **inverse sine** of $x$ is the measure of $\angle A$. | If $\angle A$ is an acute angle and the cosine of $A$ is $x$, then the **inverse cosine** of $x$ is the measure of $\angle A$. | If $\angle A$ is an acute angle and the cosine of $A$ is $x$, then the **inverse cosine** of $x$ is the measure of $\angle A$. |
| **Symbols** | | |
| If $\sin A = x$, then $\sin^{-1}x = m\angle A$. | If $\cos A = x$, then $\cos^{-1}x = m\angle A$. | If $\tan A = x$, then $\tan^{-1}x = m\angle A$. |

## Example 4 Find Angle Measures Using Inverse Trigonometric Ratios

**Use a calculator to find $m\angle A$ to the nearest tenth.**

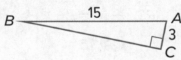

**Step 1:**

The measures given are those of the leg adjacent to $\angle A$ and the hypotenuse, so write an equation using the _____ ratio.

**Step 2:**

$$\cos A = \frac{3}{15} \text{ or } \underline{\quad} \qquad \cos A = \frac{\text{adj}}{\text{hyp}}$$

If $\cos A = \frac{1}{5}$, then $\cos^{-1}\frac{1}{5} = m\angle\underline{\quad}$.

**Step 3:**

**Use a calculator.**

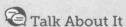 78.46304097

So, $m\angle A \approx \underline{\quad}$.

## Check

Use a calculator to find $m\angle Z$ to the nearest tenth.
$m\angle Z = \underline{\quad}°$

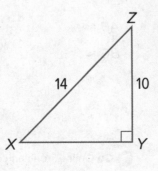

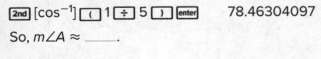 **Go Online** You can complete an Extra Example online.

Study Tip

**Inverse Trigonometric Ratios** The expression $\sin^{-1}x$ is read *the inverse sine of x* and is interpreted as the angle with sine $x$. Be careful not to confuse this notation with the notation for negative exponents. That is, $\sin^{-1}x \neq \frac{1}{\sin x}$. Instead, this notation is similar to the notation for an inverse function, $f^{-1}(x)$.

Study Tip

**Graphing Calculators** The second functions of the $\boxed{\sin}$ $\boxed{\cos}$ and $\boxed{\tan}$ keys are usually their inverses.

💬 Talk About It

What other method could you use to find $m\angle A$? Explain.

Lesson 9-5 • Trigonometry **531**

## Example 5 Solve a Right Triangle

When you are given measurements to find the unknown angle and side measures of a right triangle, this is known as **solving a right triangle**. To solve a right triangle, you need to know:

- two side lengths or
- one side length and the measure of one acute angle

**Solve the right triangle. Round side and angle measures to the nearest tenth.**

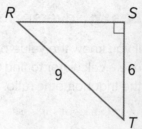

**Find $m\angle R$. Use the sine ratio.**

$$\sin R = \frac{6}{\underline{\phantom{0}}} \qquad \sin X = \frac{opp}{hyp}$$

$$\sin^{-1}\frac{6}{9} = m\angle\underline{\phantom{000}} \qquad \text{Use a calculator.}$$

$$\underline{\phantom{0000000}} \approx m\angle R$$

So, $m\angle R \approx \underline{\phantom{00}}°.$

**Find $m\angle T$. Use the known angles.**

$$m\angle R + m\angle T = 90 \qquad \text{Acute } \angle\text{s of rt } \triangle\text{s are comp.}$$

$$\underline{\phantom{00}} + m\angle T \approx 90 \qquad m\angle R \approx 41.8$$

$$m\angle T \approx \underline{\phantom{00}} \qquad \text{Subtract 41.8 from each side.}$$

So, $m\angle T \approx \underline{\phantom{00}}°.$

**Find $RS$. Use the Pythagorean Theorem.**

$$(RS)^2 + (ST)^2 = (RT)^2 \qquad \text{Pythagorean Theorem}$$

$$(RS)^2 + 6^2 = 9^2 \qquad \text{Substitution}$$

$$(RS)^2 = \underline{\phantom{00}} \qquad \text{Simplify.}$$

$$RS = \underline{\phantom{00}} \qquad \text{Take the positive square root of each side.}$$

$$RS \approx \underline{\phantom{00}} \qquad \text{Use a calculator.}$$

So, $RS \approx \underline{\phantom{000}}.$

## Check

Solve the right triangle. Round side and angle measures to the nearest tenth.

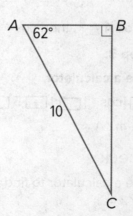

$$m\angle C \approx \underline{\phantom{000}}°$$

$$AB \approx \underline{\phantom{000}}$$

$$BC \approx \underline{\phantom{000}}$$

 **Go Online** You can complete an Extra Example online.

---

Go Online
to watch a video to see
how to solve a right
triangle using
trigonometry.

**Study Tip:**

**Alternative Methods**
Right triangles can often
be solved by using
different methods. In
the example, $m\angle T$ could
have been found first
using a cosine ratio,
and $m\angle T$ and a tangent
ratio could have been
used to find $RS$.

**Study Tip:**

**Approximation** Is using
calculated measures to
find other measures in a
right triangle, be careful
not to round until the
last step.

# Practice

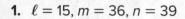

Go Online You can complete your homework online.

### Example 1

**Find sin L, cos L, tan L, sin M, cos M, and tan M. Express each ratio as a fraction and as a decimal to the nearest hundredth.**

1. $\ell = 15$, $m = 36$, $n = 39$

2. $\ell = 12$, $m = 12\sqrt{3}$, $n = 24$

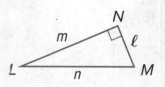

3. Find sin R, cos R, tan R, sin S, cos S, and tan S. Express each ratio as a fraction and as a decimal to the nearest hundredth.

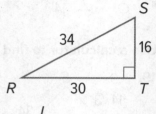

4. Find sin J, cos J, tan J, sin L, cos L, and tan L. Express each ratio as a fraction and as a decimal to the nearest hundredth if necessary.

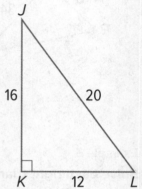

### Example 2

**Use a special right triangle to express each trigonometric ratio as a fraction and as a decimal to the nearest hundredth if necessary.**

5. sin 30°

6. tan 45°

7. cos 60°

8. sin 60°

9. tan 30°

10. cos 45°

### Example 3

**Find the value of x. Round to the nearest hundredth.**

11.

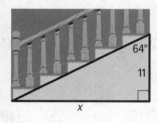

12.

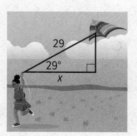

13.

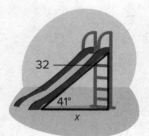

14. **GEOGRAPHY** Diego used a theodolite to map a region of land for his class in geomorphology. To determine the elevation of a vertical rock formation, he measured the distance from the base of the formation to his position and the angle between the ground and the line of sight to the top of the formation. The distance was 43 meters and the angle was 36°. What is the height of the formation to the nearest meter?

15. **RAMPS** A 60-foot ramp rises from the first floor to the second floor of a parking garage. The ramp makes a 15° angle with the ground. How high above the first floor is the second floor? Express your answer to the nearest tenth of a foot.

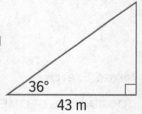

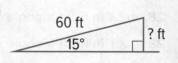

**Example 4**

**Use a calculator to find the measure of ∠B to the nearest tenth.**

**16.**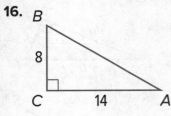

**17.**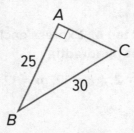

**18.**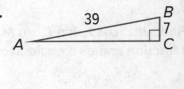

**Use a calculator to find the measure of ∠T to the nearest tenth.**

**19.**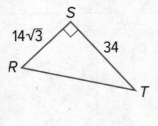

**20.**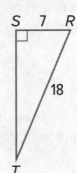

**21.**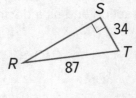

**Example 5**

**Solve each right triangle. Round side measures to the nearest tenth and angle measures to the nearest degree.**

**22.**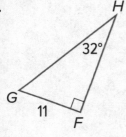

**23.**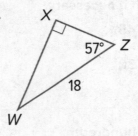

**24.**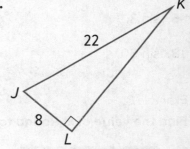

**25.**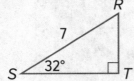

**26.**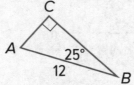

**27.**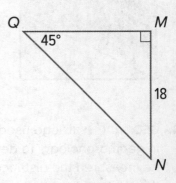

**Mixed Exercises**

COORDINATE GEOMETRY **Find each angle measure to the nearest tenth of a degree using the Distance Formula and an inverse trigonometric ratio.**

**28.** m∠K in right triangle JKL with vertices J(−2, −3), K(−7, −3), and L(−2, 4)

**29.** m∠Y in right triangle XYZ with vertices X(4, 1), Y(−6, 3), and Z(−2, 7)

Name _____ Period _____ Date _____

**30.** $m\angle A$ in right triangle $ABC$ with vertices $A(3, 1)$, $B(3, -3)$, and $C(8, -3)$

**31. LINES** Melah draws line $m$ on a coordinate plane. What angle does $m$ make with the $x$-axis? Round your answer to the nearest degree.

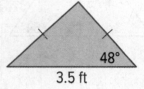

**REASONING Find the perimeter and area of each triangle. Round to the nearest hundredth.**

**32.**

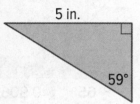

5 in.

59°

**33.**

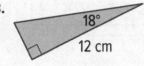

18°

12 cm

**34.**

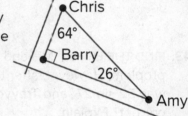

48°

3.5 ft

**35. NEIGHBORS** Amy, Barry, and Chris live in the same neighborhood. Chris lives up the street and around the corner from Amy, and Barry lives at the corner between Amy and Chris. The three homes are the vertices of a right triangle.

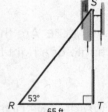

Chris

64°

Barry

26°

Amy

a. Give two trigonometric expressions for the ratio of Barry's distance from Amy to Chris' distance from Amy.

b. Give two trigonometric expressions for the ratio of Barry's distance from Chris to Amy's distance from Chris.

c. Give a trigonometric expression for the ratio of Amy's distance from Barry to Chris' distance from Barry.

**36. TOWERS** A cell phone tower is supported by a guy wire as shown. Chilam wants to determine the height of the tower. She finds that the guy wire makes an angle of 53° with the ground and it is attached to the ground 65 feet from the base of the tower.

a. **MODELING** Let the height of the tower be $x$. Which trigonometric ratio should you use to write an equation that you can solve for $x$? Justify your choice.

b. **REASONING** Write and solve an equation for the height of the tower. Round to the nearest tenth of a foot.

c. **ARGUMENTS** Suppose Chilam had wanted to find the length of the guy wire. What would you have done differently to solve the problem? Explain.

**37. COMPLEMENTARY ANGLES** In the right triangle shown, $\sin \alpha = 0.6428$ and $\cos \alpha = 0.7660$. Find $\sin \beta$ and $\cos \beta$ and explain your reasoning.

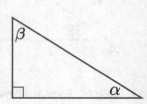

β

α

**38. REASONING** A right triangle with legs of length $a$ and $2a$ and an angle $\theta$ is shown. If $a$ is a positive real number, does $\theta$ depend on the value of $a$? If not, find the measure of $\theta$. Explain.

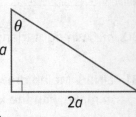

**Find the values of x and y. Round to the nearest tenth.**

**39.**

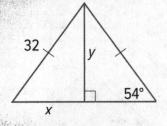

**40.**

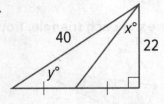

**41.**

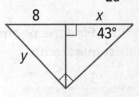

**42. STRUCTURE** Explain how you can use only the table at the right to find the value of cos 20°.

| m∠A | sin A |
|-----|-------|
| 65° | 0.9063 |
| 70° | 0.9397 |
| 75° | 0.9659 |
| 80° | 0.9848 |
| 85° | 0.9962 |

**43. FIND THE ERROR** Mia and Treyvon were both solving the same trigonometry problem. However, after they finished their computations, Mia said the answer was 52 sin 27° and Treyvon said the answer was 52 cos 63°. Could they both be correct? Explain.

**44. PERSEVERE** Solve △ABC. Round to the nearest whole number.

**45. ANALYZE** Are the values of sine and cosine for an acute angle of a right triangle always less than 1? Explain.

**46. WHICH ONE DOESN'T BELONG** If the directions say to *Solve the right triangle*, then which of the triangles shown does not belong? Justify your reasoning.

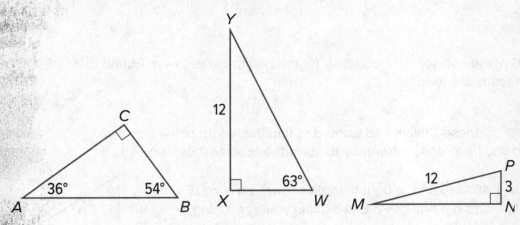

**47. WRITE** Explain how you can use ratios of the side lengths to find the angle measures of the acute angles in a right triangle.

# Applying Trigonometry

## Explore Measuring Angles of Elevation

🔗 **Online Activity** Use a concrete tool to complete the Explore.

> ⓠ **INQUIRY** When sighting an object from a given distance, how does the tangent of the angle of elevation relate to the height of the object?

## Explore Angles of Elevation and Depression

🔗 **Online Activity** Use a spreadsheet to complete the Explore.

> ⓠ **INQUIRY** How can angles of elevation and depression be used to find measurements?

## Learn Angles of Elevation and Depression

Often the only way to measure an object is through the use of **indirect measurement**, which involves using similar figures and proportions to measure an object. In these cases, you measure the object by measuring something else. Indirect measurements use special types of angles. An **angle of elevation** is the angle formed by a horizontal line and an observer's line of sight to an object above the horizontal line. An **angle of depression** is the angle formed by a horizontal line and an observer's line of sight to an object below the horizontal line.

## 🌐 Example 1 Angle of Elevation

**DRONES Rakeem is flying his drone at the park. He spots the drone at an angle of elevation that he estimates to be 30°. The remote control tells Rakeem that his drone is 102 feet above the ground. If Rakeem is 6 feet tall, how far is he from the drone to the nearest foot?**

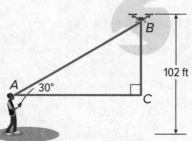

Since Rakeem is 6 feet tall, $BC = 102 - 6$ or 96 feet. Let $x$ represent the distance from Rakeem to the drone $AB$.

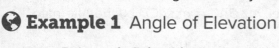

$$\sin A = \frac{BC}{\quad}$$  $\qquad \sin = \frac{\text{opposite}}{\text{hypotenuse}}$

$$\sin 30° = \frac{\quad}{x}$$  $\qquad m\angle A = 30°, BC = 96, AB = x$

$$x = \frac{96}{\quad}$$  $\qquad$ Solve for $x$.

$$x \approx \underline{\qquad}$$  $\qquad$ Use a calculator.

Rakeem is about _____ feet from his drone.

🔗 **Go Online** You can complete an Extra Example online.

---

**Today's Standards**
G.SRT.8; G.SRT.9
MP4, MP5

**Today's Vocabulary**
indirect measurement
angle of elevation
angle of depression

💬 **Talk About It!**
What determines whether an angle is an angle of depression or an angle of elevation?

💭 **Think About It!**
If Rakeem set the drone's camera to view himself, what would be the angle of depression? Explain.

## Check

**SEARCH AND RESCUE** A flare is shot vertically into the air approximately 200 meters from the base camp. The angle of elevation to the maximum height of the flare is 35°. The group at base camp need to know the altitude of the flare.

**Part A** Write the equation that represents the situation if *a* represents the height of the flare.

**Part B** What is the maximum height of the flare to the nearest meter?

_____

**Study Tip:**

**Angles of Elevation and Depression** To avoid mislabeling, remember that angles of elevation and depression are always formed with a horizontal line, never with a vertical line.

🌐 **Example 2** Angle of Depression

**SIGHTSEEING** **Cottonwood, Idaho's Dog Bark Park Inn in is a popular tourist attraction featuring a hotel in the shape of a 30-foot wood-carved beagle. Pedro looks out the window 30 feet from the ground and spots a fire hydrant on the ground at an estimated angle of depression of 40°.**

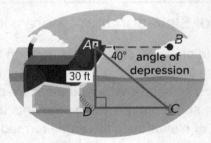

**What is the horizontal distance from Pedro to the hydrant to the nearest foot?**

Since $\overrightarrow{AB}$ and $\overline{DC}$ are parallel, $m\angle BAC = m\angle ACD$ by the Alternate Interior Angles Theorem.

Let *x* represent the _____ distance from the Pedro to the hydrant.

$$\tan C = \frac{AD}{DC}$$          $\tan = \frac{\text{opposite}}{\text{adjacent}}$

$$\tan 40° = \frac{\quad}{x}$$          $C = 40°$, $AD = 30$, and $DC = x$

$$\underline{\quad\quad} = 30$$          Multiply each side by *x*.

$$x = \frac{30}{\underline{\quad}}$$          Divide each side by tan 24°.

$$x \approx \underline{\quad}$$          Use a calculator.

The horizontal distance from Pedro to the hydrant is _____ feet.

🧠 **Think About It!**

Chaz hikes to the top of Mount Elbert in Colorado. He uses binoculars to sight his car in the parking lot at the base of the mountain at an angle of depression of 52°. Use available resources to find the total height Mount Elbert. Then calculate the distance Chaz is from his car to the nearest foot.

Chaz is about _____ feet from his car.

## Check

**LIFEGUARDING** Braylen stands on an 8-foot platform and sights a swimmer at an angle of depression of 5°. If Braylen is 6 feet tall, how far away from the swimmer is Braylen to the nearest foot? _____ ft

🖱 **Go Online** You can complete an Extra Example online.

## Example 3 Use Two Angles of Elevation and Depression

**MALL** Wei is estimating the height of the second floor in the mall. She sights the second floor at a 10° angle of elevation. She then steps forward 50 feet, until she is 5.5 feet from the wall and sights the second floor again. If Wei's line of sight is 66 inches above the ground, at what angle of elevation does she sight the second floor?

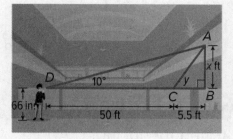

$\triangle ABC$ and $\triangle ABD$ are right triangles. In order to find the angle of elevation, we need to first find the height of the second floor of the mall. This height is the sum of Wei's height and $AB$, the length of the vertical line extending from Wei's sight on the second floor.

Since this length is not given, write and solve a system of equations using both triangles. Let $AB = x$ and let $\angle ACB = y$. $BD = 5.5 +$ _____ or _____, and the height of the second floor is $x +$ _____, since 66 in. = 5.5 ft.

Use $\triangle ABD$ to find $x$.

$$\tan 10° = \frac{x}{\rule{2cm}{0.4pt}} \qquad \tan = \frac{\text{opposite}}{\text{adjacent}}; m\angle ADB = 10; BD = 50 + 5.5$$

$$\rule{3cm}{0.4pt} = x \qquad \text{Multiply each side by 55.5}$$

Use $\triangle ABC$.

$$\tan y° = \frac{x}{\rule{2cm}{0.4pt}} \qquad \tan = \frac{\text{opposite}}{\text{adjacent}}; BC = 5.5$$

$$\rule{3cm}{0.4pt} = x \qquad \text{Multiply each side by 5.5}$$

Use the value for $x$ from $\triangle ABD$ in the equation for $\triangle ABC$ and solve for $y$.

$$5.5 \tan y° = x$$

$$5.5 \tan y° = \rule{3cm}{0.4pt}$$

$$\tan y° = \frac{55.5 \tan 10°}{\rule{2cm}{0.4pt}}$$

Use a calculator and the inverse tangent ratio to find that $y \approx$ _____°.

Using the equation from $\triangle ABC$, $x = 5.5 \tan 10°$ or about _____. The height of the second floor of the mall is about $9.79 + 5.5$ or $15.29$, which is about _____ feet.

## Check

**SIGHTSEEING** Two skyscrapers are sighted from the viewing deck of the Empire State Building at 1250 feet up. One skyscraper is sighted at a 20° angle of depression and a second skyscraper is sighted at a 30° angle of depression. How far apart are the two skyscrapers to the nearest foot?

_____ ft

🌐 **Go Online** You can complete an Extra Example online.

**Study Tip:**

**Units of Measure**
When solving real-world problems using trigonometric ratios and inverses, be sure to convert all measurements to the same units to avoid unnecessary errors.

💭 **Think About It!**
How can you use the value you found for $y$ to check your solution?

### Think About It!

Describe the relationship between $a$, $b$, and $C$ in a triangle when finding the area using Area $= \frac{1}{2}ab \sin C$.

## Learn Trigonometry and Areas of Triangles

**Key Concept • Area of a Triangle**

To find the area of a triangle when the height is not known, you can use Area $= \frac{1}{2}ab \sin C$, where $a$ and $b$ area side lengths and $C$ is the included angle.

### Example 4 Find the Area of a Triangle Given the Included Angle

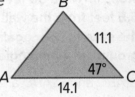

**Use trigonometry to find the area of △ABC to the nearest tenth.**

Area $= \frac{1}{2}ab$ _____    Area of a triangle

Area $= \frac{1}{2}(11.1)($____$)$ sin ____°    $a = 11.1, b = 14.1, m\angle C = 47°$

Area $\approx$ _____

The area of △ABC is about _____ units$^2$.

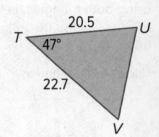

### Check

**Use trigonometry to find the area of △TUV to the nearest tenth.**

_____ units$^2$

### Example 5 Find the Area of Any Triangle

**Use trigonometry to find the area of △DEF to the nearest tenth.**

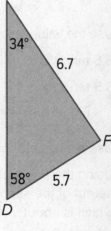

Since you do not know the measure of the included angle $F$, add the measures of angles $D$ and $E$ and subtract the total from _____.

$m\angle F =$ _____°

Area $= \frac{1}{2}de$ _____

Area $= \frac{1}{2}(6.7)($_____$)$ sin 88°

Area $\approx$ _____

The area of △DEF is about _____ units$^2$.

### Check

**Use trigonometry to find the area of △JKL to the nearest tenth.**

_____ units$^2$

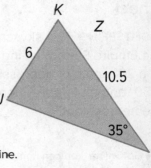

 **Go Online** You can complete an Extra Example online.

# Practice

🔿 **Go Online** You can complete your homework online.

**Example 1**

1. **LEDGE** The angle of elevation from point A to the top of a ledge is 34°. If point A is 1000 feet from the base of the cliff, how high is the ledge?

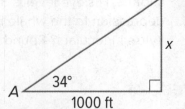

2. **WATER TOWERS** A student can see a water tower from the closest point of the soccer field at San Lobos High School. The edge of the soccer field is about 110 feet from the water tower and the water tower stands at a height of 32.5 feet. What is the angle of elevation if the eye level of the student viewing the tower from the edge of the soccer field is 6 feet above the ground? Round to the nearest tenth.

3. **CONSTRUCTION** A roofer props a ladder against a wall so that the top of the ladder just reaches a 30-foot roof that needs repair. If the angle of elevation from the bottom of the ladder to the roof is 55°, how far is the ladder from the base of the wall? Round your answer to the nearest foot.

4. **SUN** Find the angle of elevation of the Sun when a 12.5-meter-tall telephone pole casts an 18-meter-long shadow. Round to the nearest degree.

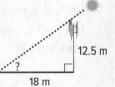

5. **MOUNTAIN BIKING** On a mountain bike trip along the Gemini Bridges Trail in Moab, Utah, Nabuko stopped on the canyon floor to get a good view of the twin sandstone bridges. Nabuko is standing about 60 meters from the base of the canyon cliff, and the natural arch bridges are about 100 meters up the canyon wall. If her line of sight is 5 meters above the ground, what is the angle of elevation to the top of the bridges? Round to the nearest tenth of a degree.

6. **SHADOWS** If the angle of elevation to the Sun is 60°, how long is the shadow cast by a 35-foot building to the nearest tenth of a foot?

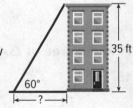

**Example 2**

7. **SKIING** A ski run is 1000 yards long with a vertical drop of 208 yards. Find the angle of depression from the top of the ski run to the bottom to the nearest degree.

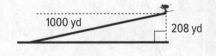

8. **AIR TRAFFIC** From the top of the 120-foot-high tower, an air traffic controller observes an airplane on the runway at an angle of depression of 19°. How far from the base of the tower is the airplane? Round to the nearest tenth of a foot.

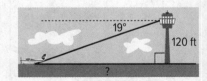

9. **AVIATION** Due to a storm, a pilot flying at an altitude of 528 feet has to land. If he has a horizontal distance of 2000 feet to the landing strip, at what angle of depression should he land? Round to the nearest tenth of a degree.

10. **INDIRECT MEASUREMENT** Kyle is sitting at the end of a pier 30 feet above the ocean using binoculars to watch a whale surface. His eye level is 3 feet above the pier. If the angle of depression to the whale is 20°, how far is the whale from Kyle's binoculars? Round to the nearest tenth of a foot.

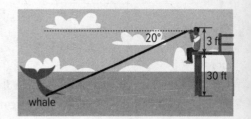

## Example 3

11. **GARAGE** To estimate the height of a garage, Jason sights the top of the garage at a 42° angle of elevation. He then steps back 20 feet and sights the top at a 10° angle. If Jason is 6 feet tall, how tall is the garage to the nearest foot?

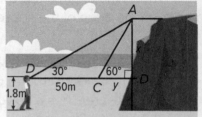

12. **CLIFF** Sarah stands on the ground and sights the top of a steep cliff at a 60°angle of elevation. She then steps back 50 meters and sights the top of the cliff at a 30°angle. If Sarah is 1.8 meters tall, how tall is the cliff to the nearest meter?

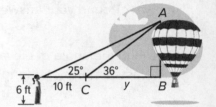

13. **BALLOON** The angle of depression from a hot air balloon to a person on the ground is 36°. When the person steps back 10 feet, the new angle of depression is 25°. If the person is 6 feet tall to the nearest foot, how far above the ground is the hot air balloon?

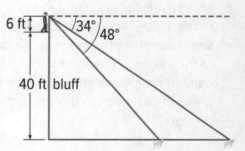

14. **INDIRECT MEASUREMENT** Mr. Dominguez is standing on a 40-foot ocean bluff near his home. He can see his two dogs on the beach below. If his line of sight is 6 feet above the ground and the angles of depression to his dogs are 34° and 48°, how far apart are the dogs to the nearest foot?

## Example 4

**Find the area of △ABC to the nearest tenth.**

15.
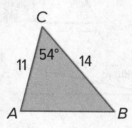

16.
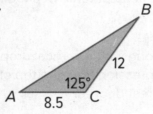

17.
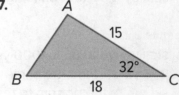

18.
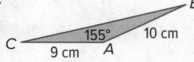

19.
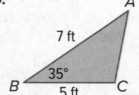

20.
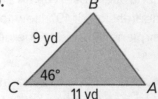

## Example 5

**Find the area of △ABC to the nearest tenth.**

21.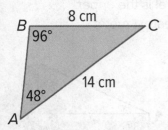
8 cm
B 96°
C
14 cm
48°
A

22.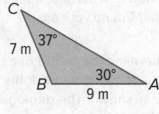
C
37°
7 m
30°
B 9 m A

23.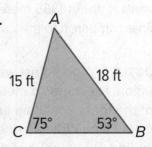
A
15 ft
18 ft
C 75°
53° B

## Mixed Exercises

24. **LIGHTHOUSES** Sailors on a ship at sea spot the light from a lighthouse. at an angle of elevation of 25°. The light of the lighthouse is 30 meters above sea level. How far from the shore is the ship? Round your answer to the nearest meter.

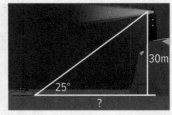
30m
25°
?

25. **AIRPLANES** The angle of elevation to an airplane viewed from the control tower at an airport is 7°. The tower is 200 feet tall and the pilot reports that the altitude of the airplane is 5200 feet. How far away from the control tower is the airplane? Round your answer to the nearest foot.

26. **RESCUE** A hiker dropped his backpack over one side of a canyon onto a ledge below. Because of the shape of the cliff, he could not see exactly where it landed. From the other side, the park ranger reports that the angle of depression to the backpack is 32°. If the width of the canyon is 115 feet, how far down did the backpack fall? Round your answer to the nearest foot.

27. **ROOFTOP** Lucia is 5.5 feet tall. She is standing on the roof of a building that is 80 feet tall. She spots a fountain at ground level that she knows to be 122 feet away from the base of the building. What is the measure of the angle of depression formed by Lucia's horizontal line of sight and her line of sight to the fountain? Round to the nearest degree.

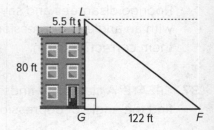
L
5.5 ft
80 ft
G
122 ft
F

28. **REASONING** Jermaine and John are standing 10 meters apart watching a helicopter hover above the ground.

   a. Find two different expressions that can be used to find $h$, the height of the helicopter.

   b. Equate the two expressions you found for part a to solve for $x$. Round your answer to the nearest hundredth.

   c. How high above the ground is the helicopter? Round your answer to the nearest hundredth.

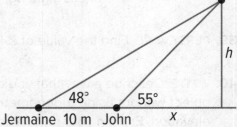
h
48°
55°
Jermaine 10 m John
x

29. **PEAK TRAM** The Peak Tram in Hong Kong connects two terminals, one at the base of a mountain, and the other at the summit. The angle of elevation of the upper terminal from the lower terminal is about 15.5°. The distance between the two terminals is about 1365 meters. About how much higher above sea level is the upper terminal compared to the lower terminal? Round your answer to the nearest meter.

30. **STRUCTURE** A geologist wants to determine the height of a rock formation. He stands *d* meters from the formation and sights the top of the formation at an angle of *x*°, as shown. The geologist's height is 1.8 m. Write a general formula that the geologist can use to find the height *h* of the rock formation if he knows the values of *d* and *x*.

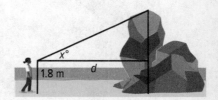

**REGULARITY** Find the area of △*ABC* to the nearest tenth.

31. *A* = 20°, *c* = 4 cm, *b* = 7 cm

32. *C* = 55°, *a* = 10 m, *b* = 15 *m*

33. *B* = 42°, *c* = 9 ft, *a* = 3 ft

34. *c* = 15 in., *b* = 13 in., *A* = 53°

35. **MODELING** Alex is helping to build the set for a play. One piece of scenery is a large triangle that will be constructed out of wood and be painted to represent a mountain. Alex would like to know the area of the piece of scenery so that he can buy the right amount of paint. What is the area of this triangle? Round your answer to the nearest tenth of a foot.

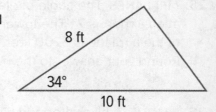

36. **FIND THE ERROR** Terrence and Rodrigo are trying to determine the relationship between angles of elevation and depression. Terrence says that if you are looking up at someone with an angle of elevation of 35°, then they are looking down at you with an angle of depression of 55°, which is the complement of 35°. Rodrigo disagrees and says that the other person would be looking down at you with an angle of depression equal to your angle of elevation or 35°. Is either of them correct? Explain.

37. **CREATE** A classmate finds the angle of elevation of an object, but she is trying to find the angle of depression. Write a question to help her solve the problem.

38. **ANALYZE** Classify the statement below as *true* or *false*. Explain your reasoning.

> As a person moves closer to an object he or she is sighting, the angle of elevation increases.

39. **PERSEVERE** Find the value of *x*. Round to the nearest tenth.

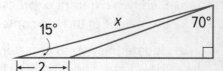

40. **WRITE** Describe a way that you can estimate the height of an object without using trigonometry by choosing your angle of elevation. Explain your reasoning.

# The Law of Sines

**Today's Standards**
G.SRT.10, G.SRT.11
MP3 MP5

**Today's Vocabulary**
ambiguous case

## Explore Trigonometric Ratios in Nonright Triangles

 **Online Activity** Use graphing technology to complete the Explore.

> ☓
>
> @ **INQUIRY** How can you use trigonometric ratios to solve for missing side lengths in nonright triangles?

## Learn Law of Sines

The **Law of Sines** can be used to find side lengths and angle measurements for any triangle.

> **Theorem 9.10: Law of Sines**
>
> If △*ABC* has lengths *a*, *b*, and *c*, representing the lengths of the sides opposite the angles with measures *A*, *B*, and *C*, then
>
> $\frac{\sin A}{a} = \frac{\sin B}{b} = \frac{\sin C}{c}$.

 **Go Online** A proof of Theorem 9.10 is available.

**Go Online**
You may want to complete the Concept Check to check your understanding.

## Example 1 Law of Sines (AAS)

You can use the Law of Sines to solve a triangle if you know the measures of two angles and any side (AAS or ASA).

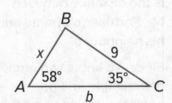

**Find the value of *x* to the nearest tenth.**

Because we are given the measures of two angles and a nonincluded side, use the Law of Sines to write a proportion.

$\frac{\sin A}{a} = \frac{\sin C}{c}$      Law of Sines

$\frac{\sin 58°}{\rule{1cm}{0.4pt}} = \frac{\sin \rule{0.5cm}{0.4pt}}{x}$      *m∠A* = 58°, *a* = 9, *m∠C* = 35°, *c* = *x*

$x \sin 58° = \rule{1cm}{0.4pt} \sin \rule{0.5cm}{0.4pt}°$      Cross Products Property

$x = \frac{9 \sin 35°}{\rule{1cm}{0.4pt}}$      Divide each side by sin 58°.

$x \approx \rule{1cm}{0.4pt}$      Use a calculator.

🫧 **Think About It!**

Think About It How could you find the value of *b*?

## Check

Consider triangle *CDF*.

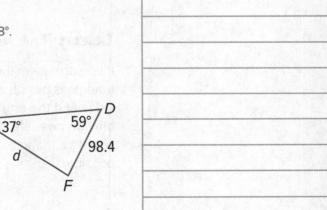

**Part A** What value completes the proportion?

$\frac{\sin 37°}{?} = \frac{\sin 59°}{d}$    ? = $\rule{1cm}{0.4pt}$

**Part B** Find the value of *d* to the nearest tenth.

*d* = $\rule{1cm}{0.4pt}$

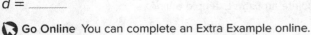 **Go Online** You can complete an Extra Example online.

🐸 Think About It!

Could you use
$\frac{\sin D}{d} = \frac{\sin F}{f}$ to find the
value of $x$? Justify your
reasoning.

## Example 2 The Law of Sines (ASA)

**Find the value of $x$ to the nearest tenth.**

By the Triangle Angle Theorem,

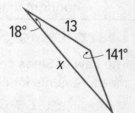

$$m\angle G = \underline{\quad} - (60 + 55) \text{ or } \underline{\quad}.$$

| | |
|---|---|
| $\dfrac{\sin D}{d} = \dfrac{\sin G}{g}$ | Law of Sines |
| $\dfrac{\sin 60°}{x} = \dfrac{\sin \underline{\quad}}{\underline{\quad}}$ | $m\angle D = 60,\ g = 73,\ m\angle G = 65,\ d = x$ |
| $73 \sin \underline{\quad}° = \underline{\quad} \sin 65°$ | Cross Products Property |
| $\dfrac{73 \sin 60°}{\underline{\quad}} = x$ | Divide each side by sin 65°. |
| $\underline{\quad} \approx x$ | Use a calculator. |

### Check

Find the value of $x$ to the nearest tenth. _____

**A.** 6.4   **B.** 7.4   **C.** 8.2   **D.** 22.8

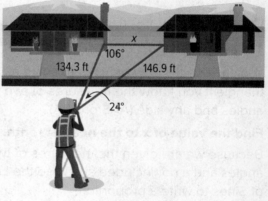

## 🌐 Example 3 Indirect Measurement and Law of Sines

**SURVEYING Mr. Fortunado
is having a boundary survey
done on his property. What
is the distance between
Mr. Fortunado's home and
his neighbor's?**

Since we know two angles
of a triangle and one
nonincluded side, use the
Law of Sines.

$$\frac{\sin 106°}{146.9} = \frac{\sin 24°}{\underline{\quad}}$$   Law of Sines

$x \sin 106° = \underline{\quad} \sin \underline{\quad}°$   Cross Products Property

$\qquad x \approx 62.2$   Use a calculator.

## Learn The Ambiguous Case

If you are given the measures of two angles and a side, exactly one
triangle is possible. However, if you are given the measures of two
sides and the angle opposite one of them, zero, one, or two triangles
may be possible. This is known as the **ambiguous case**. So, when
solving a triangle using the SSA case, zero, one or two solutions are
possible.

 **Go Online** You can complete an Extra Example online.

## Key Concept • Possible Triangles in SSA Case

Consider a triangle in which $a$, $b$, and $m\angle A$ are given. Shown below are the triangles that are possible when $\angle A$ is acute and when $\angle A$ is right or obtuse.

**Angle A is acute:**

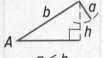
$a < h$
no solution

$a = h$
one solution

$h < a < b$
two solutions

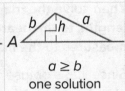
$a \geq b$
one solution

**Angle A is right or obtuse:**

$a \leq b$
no solution

$a \leq b$
no solution

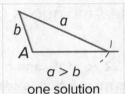
$a > b$
one solution

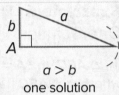
$a > b$
one solution

Solving a triangle with an obtuse angle sometimes requires finding sine ratios for measures greater than 90°. The sine ratios for obtuse angles are defined based on their supplementary angles.

### Postulate 9.1

The sine of an obtuse angle is defined to be the sine of its supplement.

## Example 4 The Ambiguous Case with One Solution

Because $\sin A = \frac{h}{b}$, you can use $h = b \sin A$ to find $h$ in acute triangles.

**In $\triangle MNP$, $N = 32°$, $n = 7$, and $p = 4$. Determine whether $\triangle MNP$ has no solution, one solution, or two solutions. Then solve the triangle. Round side lengths to the nearest tenth and angle measures to the nearest degree.**

Because $\angle N$ is acute, and $n > p$, you know that one solution exists.

**Step 1** Use the Law of Sines to find $m\angle P$.

$$\frac{\sin \underline{\quad}°}{7} = \frac{\sin P}{\quad} \qquad \text{Law of Sines}$$

$$\frac{\sin \underline{\quad}°}{7} = \sin P \qquad \text{Multiply each side by 4.}$$

$$\underline{\qquad} \approx \sin P \qquad \text{Use a calculator.}$$

$$\underline{\qquad}° \approx P \qquad \text{Use the } \sin^{-1} \text{ function.}$$

**Step 2** Use the Triangle Angle-Sum Theorem to find $m\angle M$.

$$m\angle M = 180 - (\underline{\quad} + \underline{\quad}) \text{ or } \underline{\quad}°$$

**Step 3** Use the Law of Sines to find $m$.

$$\frac{\sin \underline{\quad}°}{m} \approx \frac{\sin 32°}{\quad} \qquad \text{Law of Sines}$$

$$m \approx \frac{\sin 130°}{\sin \underline{\quad}} \qquad \text{Solve for } m.$$

$$m \approx \underline{\quad} \qquad \text{Use a calculator.}$$

So, $P \approx \underline{\quad}°$, $M \approx \underline{\quad}°$, and $m \approx \underline{\quad}$.

**Study Tip**

**A is Acute** In the figures, the altitude $h$ is compared to $a$ because $h$ is the minimum distance from $C$ to $\overline{AB}$ when $A$ is acute.

$$\sin A = \frac{opposite}{hypotenuse}$$

$$\sin A = \frac{h}{b}$$

**Talk About It!**

If the given angle is a right angle and there is one solution to the triangle, how can you find the third side?

## Think About It!

If *R* were acute and *r* < *s*, how could you find the number of possible solutions?

## Example 5 The Ambiguous Case with No Solution

In, △RST, **R** = 95°, **r** = 10, and **s** = 12. Determine whether △RST has no solution, one solution, or two solutions. Then solve the triangle. Round side lengths to the nearest tenth and angle measures to the nearest degree.

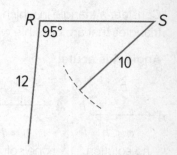

Since ∠*R* is obtuse, and 10 < 12, there is _____ .

## Example 6 The Ambiguous Case with More than One Solution

In, △ABC, **m∠A** = 32°, **a** = 15, and **b** = 18. Determine whether △ABC has no solution, one solution, or two solutions. Then solve the triangle. Round side lengths to the nearest tenth and angle measures to the nearest degree.

Because ∠*A* is acute, and 15 < 18, find *h* and compare it to *a*.

$b \sin A = 18 \sin 32°$        *b* = 18 and *A* = 32°

     ≈ _____        Use a calculator.

Since 9.5 < 15 < 18, or *h* < *a* < *b*, there are _____

| ∠*B* is acute. | ∠*B* is obtuse. |
|---|---|
| **Find m∠B.** | **Find m∠B.** |
| $\frac{\sin}{18} = \frac{\sin 32°}{}$   Law of Sines | Find an obtuse angle *B* for which sin *B* ≈ 0.6359. |
| $\sin B = \frac{\sin}{15}°$   Solve for *B*. | $m\angle B \approx 180° - 39°$ Postulate 9.1 |
| sin *B* = _____   Use a calculator. |     or _____° |
| *B* ≈ _____°   Use the sin⁻¹ function. | |
| **Find m∠C.** | **Find m∠C.** |
| $m\angle C \approx 180 - ($ ___ + ___ $)$ or ___°. | $m\angle C \approx 180 - ($ ___ + ___ $)$ or ___°. |
| **Find c.** | **Find c.** |
| $\frac{\sin}{c}$ $= \frac{\sin 32°}{}$   Law of Sines | $\frac{\sin}{c} = \frac{\sin 32°}{}$   Law of Sines |
| $C = \frac{\sin°}{\sin 32°}$   Solve for *c*. | $c = \frac{\sin°}{\sin 32°}$ Solve for *c*. |
| *c* ≈ _____   Use a calculator. | *c* ≈ _____   Use a calculator. |

So, one solution is *B* ≈ 39°, *C* ≈ _____°, and *c* ≈ _____, and another solution is *B* ≈ _____°, *C* ≈ _____°, and *c* = 3.4.

🌀 **Go Online** You can complete an Extra Example online.

Name _____ Period _____ Date _____

# Practice

⟲ **Go Online** You can complete your homework online.

### Examples 1 and 2

**Find the value of *x* to the nearest tenth.**

**1.**
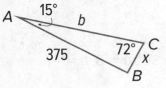

**2.**
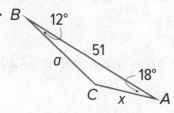

**3.**
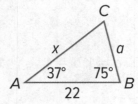

**4.**
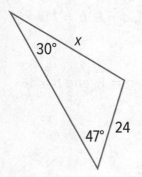

**5.**
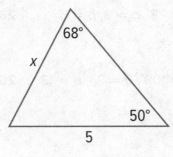

**6.**
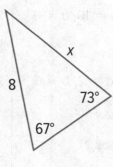

**7.**
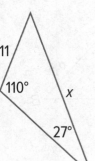

**8.**
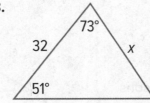

**9.**
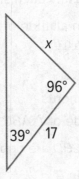

**10.**
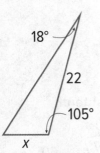

**11.**
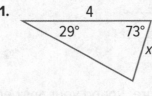

**12.**
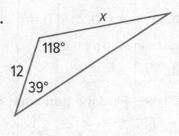

### Example 3

**13. WILDLIFE** Sarah Phillips, an officer for the Department of Fisheries and Wildlife, checks boaters on a lake to make sure they do not disturb two osprey nesting sites. She leaves a dock and heads due north in her boat to the first nesting site. From here, she turns 5° north of due west and travels an additional 2.14 miles to the second nesting site. She then travels 6.7 miles directly back to the dock. How far from the dock is the first osprey nesting site? Round to the nearest tenth.

**14. HEAD** Observers at two shoreline towers 100 feet apart measure the angle to an incoming ship. Find the distance *d* that the ship is from Tower A to the nearest tenth of a foot.

**Examples 4–6**

Determine whether each triangle has *no* solution, *one* solution, or *two* solutions. Then solve the triangle. Round side lengths to the nearest tenth and angle measures to the nearest degree.

**15.** $A = 50°, a = 34, b = 40$    **16.** $A = 24°, a = 3, b = 8$    **17.** $A = 125°, a = 22, b = 15$

**18.** $A = 30°, a = 1, b = 4$    **19.** $A = 30°, a = 2, b = 4$    **20.** $A = 30°, a = 3, b = 4$

**21.** $A = 38°, a = 10, b = 9$    **22.** $A = 78°, a = 8, b = 5$    **23.** $A = 133°, a = 9, b = 7$

**24.** $A = 127°, a = 2, b = 6$    **25.** $A = 109°, a = 24, b = 13$    **26.** $A = 48°, a = 11, and\ b = 16$

**Mixed Exercises**

**27. ARGUMENTS** Justify each statement for the derivation of the Law of Sines.

Given: $\overline{CD}$ is an altitude of $\triangle ABC$.

Prove: $\frac{\sin A}{a} = \frac{\sin B}{b}$

Proof:

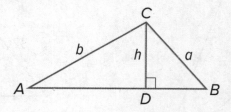

| Statements | Reasons |
|---|---|
| $\overline{CD}$ is an altitude of $\triangle ABC$ | Given |
| $\triangle ACD$ and $\triangle CBD$ are right | Def. of altitude |
| **a.** $\sin A = \frac{h}{b}, \sin B = \frac{h}{a}$ | **a.** ___?___ |
| **b.** $b \sin A = h, a \sin B = h$ | **b.** ___?___ |
| **c.** $b \sin A = a \sin B$ | **c.** ___?___ |
| **d.** $\frac{\sin A}{a} = \frac{\sin B}{b}$ | **d.** ___?___ |

**STRUCTURE** Find the perimeter of each figure. Round your answer to the nearest tenth.

**28.**

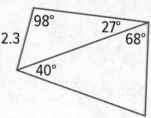

**29.**

**30.**

**31. WALKING** Alliya is taking a walk along a straight road. She leaves the road and walks on a path that makes an angle of 35° with the road. After walking for 450 meters, she turns 75° and heads back toward the road.

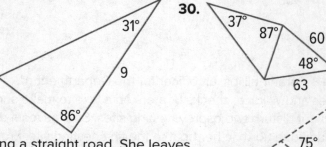

  **a.** How far does Alliya walk on her current path to get back to the road?

  **b.** When Alliya returns to the road, how far along the road is she from where she started?

**32. SAILING** A spinnaker is a large triangular sail that swings out opposite the mainsail, and is used when running with the wind. The *luff* is the leading edge of the sail, the *leach* is the edge away from the wind, and the *foot* is the bottom edge. Find the missing measure for each sail.

| Luff (ft) | Leach | Foot | Angle | Area |
|---|---|---|---|---|
| 22 | 20 | 14 | 38° | |
| 48 | 23.8 | 18 | | 214.1 |
| 45 | | 21 | 27° | 357.5 |

**33. CAMERAS** A security camera is located on top of a building at a certain distance from the sidewalk. The camera revolves counterclockwise at a steady rate of one revolution per minute. At one point in the revolution it directly faces a point on the sidewalk that is 20 meters from the camera. Four seconds later, it directly faces a point 10 meters down the sidewalk.

a. How many degrees does the camera rotate in 4 seconds?

b. To the nearest tenth of a meter, how far is the security camera from the sidewalk?

**34. FISHING** A fishing pole is resting against the railing of a boat making an angle of 22° with the deck. The fishing pole is 5 feet long, and the hook hangs 3 feet from the tip of the pole. The movement of the boat causes the hook to sway back and forth. Determine which angles the fishing line must make with the pole in order for the hook to be level with the boat's deck.

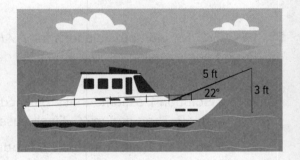

**35. USE A SOURCE** Go online to research the individual who is credited with the development of the Law of Sines. Provide a brief description.

**36. REASONING** How many triangles can be formed if $a = b$? if $a < b$? if $a > b$?

**REGULARITY Determine whether the given measures define 0, 1, or 2 triangles. Justify your answers.**

**37.** $a = 14, b = 16, m\angle A = 55$    **38.** $a = 7, b = 11, m\angle A = 68$    **39.** $a = 22, b = 25, m\angle A = 39$

**40.** $a = 13, b = 12, m\angle A = 81$    **41.** $a = 10, b = 10, m\angle A = 45$    **42.** $a = 17, b = 15, m\angle A = 128$

**43.** $a = 13, b = 17, m\angle A = 52$    **44.** $a = 5, b = 9, c = 6$    **45.** $a = 10, b = 15, m\angle A = 33$

**46. TOWERS** Cell towers $A$, $B$, and $C$ form a triangular region in one of the suburban districts of Fairfield County. Towers $A$ and $B$ are 8 miles apart. The angle formed at tower $A$ is 112°, and the angle formed at tower $B$ is 40°. How far apart are towers $B$ and $C$?

**47. REASONING** A guy wire attached to the top of a pole makes a 71° angle with the ground. At a point 25 meters farther from the pole than the guy wire, the angle of elevation of the top of the pole is 37°. What is the height of the pole? Round to the nearest tenth.

**48. FIND THE ERROR** In $\triangle RST$, $R = 56°$, $r = 24$, and $t = 12$. Cameron and Gabriela are using the Law of Sines to find $T$. Is either of them correct? Explain your reasoning.

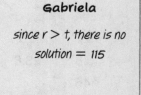

| Cameron | Gabriela |
|---|---|
| $\dfrac{\sin T}{12} = \dfrac{\sin 56°}{24}$ | since $r > t$, there is no solution = 115 |
| $\sin T = 0.4145$ | |
| $T = 24.5$ | |

**49. CREATE** Create an application problem involving right triangles and the Law of Sines. Then solve your problem, drawing diagrams if necessary.

**50. WRITE** Keshawn and Lacy were asked to find the value of $x$ in the figure at the right. Keshawn used a trigonometric ratio to write $\sin 35° = \dfrac{x}{10}$, and then he solved for $x$. Lacy used the Law of Sines to write $\dfrac{\sin 35°}{x} = \dfrac{\sin 90°}{10}$, and then she solves for $x$. Is either student correct? Write an explanation to justify your reasoning.

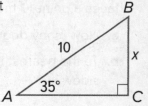

**51. PERSEVERE** Find both solutions for $\triangle ABC$ if $a = 15$, $b = 21$, $m\angle A = 42°$. Round angle measures to the nearest degree and side measures to the nearest tenth.

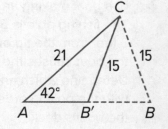

- For Solution 1, assume that $\angle B$ is acute, and use the Law of Sines to find $m\angle B$. Then find $m\angle C$. Finally, use the Law of Sines again to find $c$.

- For Solution 2, assume that $\angle B$ is obtuse. Let this obtuse angle be $\angle B'$. Use $m\angle B$ you found in Solution 1 and the diagram shown to find $m\angle B'$. Then find $m\angle C$. Finally, use the Law of Sines to find $c$.

**52. FIND THE ERROR** Colleen and Mike are planning a party. Colleen wants to sew triangular decorations and needs to know the perimeter of one of the triangles to buy enough trim. The triangles are isosceles with angle measurements of 64° at the base and side lengths of 5 inches. Colleen thinks the perimeter is 15.7 inches and Mike thinks it is 15 inches. Is either of them correct?

**53. CREATE** Give measures for $a$, $b$, and an acute $\angle A$ that define

    **a.** 0 triangles         **b.** exactly one triangle         **c.** two triangles

# The Law of Cosines

## Explore Trigonometric Relationships in Nonright Triangles

**Today's Standards**
G.SRT.10, G.SRT.11

MP3, MP6

 **Online Activity** Use guiding exercises to complete the Explore.

> ✕
>
> @ **INQUIRY** When can the Law of Cosines be used to solve triangles?

## Learn Law of Cosines

When the Law of Sines cannot be used to solve a triangle, the **Law of Cosines** may apply. You can use the Law of Cosines to find the length of the third side of a triangle when the measures of two sides and their enclosed angle are known, or to find the angle measures of a triangle if the lengths of all three sides are known.

**Theorem 9.11: Law of Cosines**

If $\triangle ABC$ has lengths $a$, $b$, and $c$, representing the lengths of the sides opposite the angles with measures $A$, $B$, and $C$, then

$a^2 = b^2 + c^2 - 2bc \cos A$,

$b^2 = a^2 + c^2 - 2ac \cos B$, and

$c^2 = a^2 + b^2 - 2ab \cos C$

 **Go Online** A proof of Theorem 9.11 is available.

## Example 1 Law of Sines (SAS)

You can use the **Law of Cosines** to solve a triangle if you know the measures of two sides and the included angle (SAS).

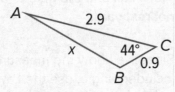

**Find the value of x to the nearest tenth.**

We are given the measures of two sides and their included angle. Use the Law of Cosines to write an equation.

| | |
|---|---|
| $c^2 = a^2 + b^2 - 2ab \cos C$ | Law of Cosines |
| $x^2 = \underline{\quad}^2 + 2.9^2 - 2(0.9)\underline{\quad\quad} \cos 44°$ | Substitution |
| $x^2 = \underline{\quad\quad} - 2.61 \cos \underline{\quad}°$ | Simplify. |
| $x = \sqrt{9.22 - \underline{\quad} \cos 44°}$ | Take the positive square root of each side. |
| $x \approx \underline{\quad}$ | Use a calculator. |

 **Go Online** You can complete an Extra Example online.

💬 **Talk About It!**

What happens when you apply the Law of Cosines to find the missing measures of a right triangle?

💭 **Think About It!**

Why can't you use the Law of Sines to find the value of $x$?

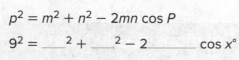

## Example 2 The Law of Cosines (SSS)

You can also use the Law of Cosines if you know the three side lengths.

**Find the value of x to the nearest tenth.**

| | |
|---|---|
| $p^2 = m^2 + n^2 - 2mn \cos P$ | Law of Cosines |
| $9^2 = \underline{\hspace{0.5em}}^2 + \underline{\hspace{0.5em}}^2 - 2\underline{\hspace{1.5em}} \cos x°$ | Substitution |
| $81 = \underline{\hspace{1em}} - 110 \cos x°$ | Simplify. |
| $\underline{\hspace{1em}} = -110 \cos x°$ | Subtract 146 from each side. |
| $\dfrac{-65}{\underline{\hspace{1em}}} = \cos x°$ | Divide each side by −110. |
| $x = \cos^{-1} \underline{\hspace{1em}}$ | Use the inverse cosine function. |
| $x \approx \underline{\hspace{1em}}°$ | Use a calculator. |

### Check

Find the value of x to the nearest degree.

$x = \underline{\hspace{2em}}$

## 🌐 Example 3 Indirect Measurement with the Law of Cosines

**GOLF Wei is golfing and uses a distance measuring tool to determine that the tee box where she is standing is 378 yards from the hole. To avoid a water hazard, Wei turns 32° and hits a shot 261.5 yards up the fairway. Complete the diagram with the correct values. Then find the distance between Wei's ball and the hole to the nearest yard.**

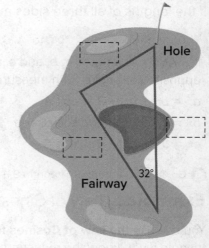

Since we know the measures of two sides of the triangle and the included angle, use the Law of Cosines to find the remaining distance.

| | |
|---|---|
| $x^2 = 261.5^2 + \underline{\hspace{1em}}^2 - 2(\underline{\hspace{1.5em}})(378) \cos \underline{\hspace{1em}}°$ | Law of Cosines |
| $x^2 = \underline{\hspace{3em}} - 197{,}316 \cos 32°$ | Simplify. |
| $x = \sqrt{211{,}266.25 - \underline{\hspace{2em}} \cos 32°}$ | Take the positive square root of each side. |
| $x \approx \underline{\hspace{1em}}°$ | Use a calculator. |

Wei's ball is about ___ yards from the hole.

🔵 **Go Online** You can complete an Extra Example online.

Your Notes ↘

## Check

**CELL PHONE TOWERS** A cell phone company builds two towers that are 2 miles apart. They choose a random location 1.1 miles from tower A and 1.5 miles from tower B to test the towers' signal strengths. Find the value of x to the nearest degree.

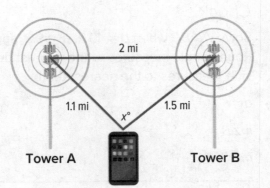

Tower A                    Tower B

$x = \underline{\quad}°$

---

## Example 4 Solve a Nonright Triangle with Law of Cosines

When solving right triangles, you can use sine, cosine, or tangent. When solving any triangle, you can use the Law of Sines or the Law of Cosines, depending on what information is given.

**Solve △ABC. Round to the nearest degree.**

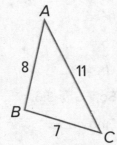

Because $7^2 + 8^2 \neq 11^2$, this is not a right triangle. The measures of all three sides are given (SSS), so decide which angle measure you want to find. Then use the Law of Cosines.

$a^2 = b^2 + c^2 - 2bc \cos A$     Law of Cosines

$7^2 = \underline{\quad} + \underline{\quad} - 2(11)(8) \cos A$     Substitute.

$\underline{\quad} = 185 - \underline{\quad} \cos \underline{\quad}$     Simplify.

$\underline{\quad\quad} = -176 \cos A$     Subtract 185 from each side.

$\dfrac{-136}{\underline{\quad}} = \cos A$     Divide each side by −176.

$m\angle A = \underline{\quad} \left( \dfrac{-136}{-176} \right)$     Use the inverse cosine function.

$m\angle A = \underline{\quad}$     Use a calculator.

Use the Law of Sines to find $m\angle B$.

$\dfrac{\sin A}{a} = \dfrac{\sin B}{b}$     Law of Sines

$\dfrac{\sin \underline{\quad}}{7} = \dfrac{\sin B}{\underline{\quad}}$     $m\angle A \approx 39°$, $a = 7$, and $b = 11$

$11 \sin \underline{\quad}° = \underline{\quad} \sin B$     Cross Products Property

$\left( \dfrac{11 \sin 39°}{7} \right) = \sin B$     Divide each side by 7.

$m\angle B = \underline{\quad} \left( \dfrac{11 \sin 39°}{7} \right)$     Use the inverse sine function.

$m\angle B \approx \underline{\quad}°$     Use a calculator.

By the Triangle Angle-Sum Theorem, $m\angle C \approx 180 - (39 + 81)$ or 60°.

**Go Online** You can complete an Extra Example online.

> **Study Tip**
>
> **Rounding** When you round a numerical solution and then use it in later calculations, your answers may be inaccurate. Wait until after you have completed all of your calculations to round.

## Check

Solve $\triangle ABC$ when $b = 10.2$, $c = 9.3$, and $m\angle A = 26°$.
Round angle measures to the nearest degree and
side measures to the nearest tenth.

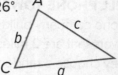

$a = $ _____

$m\angle B = $ _____ °

$m\angle C = $ _____ °

---

## Example 5 Solve a Right Triangle with Law of Cosines

**Solve $\triangle FGH$. Round angle measures to the nearest degree and side measures to the nearest tenth.**

Find $FG$.

| | |
|---|---|
| $h^2 = f^2 + g^2$ | Pythagorean Theorem |
| $h^2 = \underline{\quad}^2 + \underline{\quad}^2$ | Substitution |
| $h^2 = \underline{\quad}$ | Simplify. |
| $h = \underline{\quad}$ | Take the positive square root of each side. |
| $h \approx \underline{\quad}$ | Use a calculator. |

So, $FG \approx 3.6$.

Find $m\angle F$.

| | |
|---|---|
| $f^2 = g^2 + h^2 - 2gh \cos F$ | Law of Cosines |
| $2^2 = \underline{\quad}^2 + \underline{\quad\quad} - \underline{\quad} (3) \left(\sqrt{13}\right) \cos \underline{\quad}$ | Substitution |
| $\underline{\quad} = 22 - \underline{\quad\quad} \cos \underline{\quad}$ | Simplify. |
| $\underline{\quad} = -6\sqrt{13} \cos F$ | Subtract 22 from each side. |
| $\dfrac{-18}{\underline{\quad}} = \cos F$ | Divide each side by $-6\sqrt{13}$. |
| $\underline{\quad} \approx F$ | Use a calculator. |

So, $m\angle F \approx 34°$.

Find $m\angle G$.

Because we know that $m\angle H = 90°$ and $m\angle F \approx 34°$, find $m\angle G$.

$m\angle G = 90 - $ _____ or _____ °

## Check

Solve $\triangle EFG$. Round angle measures to the nearest
degree and side measures to the nearest tenth.

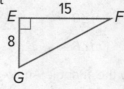

$e = $ _____

$m\angle F = $ _____ °

$m\angle G = $ _____ °

**Go Online** You can complete an Extra Example online.

**Go Online**

to practice what you've
learned in Lessons 9–7
through 9–8.

### Study Tip

**Obtuse Angles** There
are also values for sin $A$,
cos $A$, and tan $A$ when
$A \geq 90°$. Values of the
ratios for these angles
can be found by using
the trigonometric
functions on your
calculator.

# Practice

Go Online You can complete your homework online.

## Examples 1 and 2

**Find the value of _x_ to the nearest tenth for side lengths and nearest degree for angle measures.**

**1.**

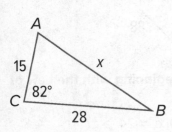

**2.**

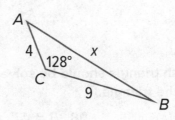

**3.**

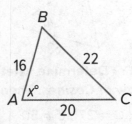

**4.**

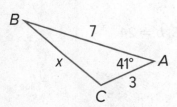

**5.**

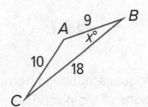

**6.**

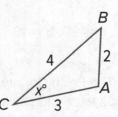

## Example 3

**7. RADAR** Two radar stations 2.4 miles apart are tracking an airplane. The straight-line distance between Station _A_ and the plane is 7.4 miles. The straight-line distance between Station _B_ and the plane is 6.9 miles. What is the angle of elevation from Station _A_ to the plane? Round to the nearest degree.

**8. DRAFTING** Marion is using a computer-aided drafting program to produce a drawing for a client. She begins a triangle by drawing a segment 4.2 inches long from point _A_ to point _B_. From _B_, she draws a second segment that forms a 42° angle with $\overline{AB}$ and is 6.4 inches long, ending at point _C_. To the nearest tenth, how long is the segment from _C_ to _A_?

## Examples 4 and 5

**REASONING Solve each triangle. Round side lengths to the nearest tenth and angle measures to the nearest degree.**

**9.**

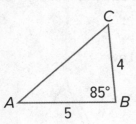

**10.**

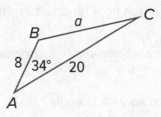

**11.**

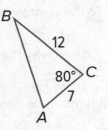

**12.**

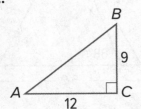

**13.**

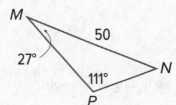

**14.**

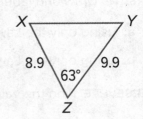

**Mixed Exercises**

**STRUCTURE** Find the perimeter of each figure.

**15.**

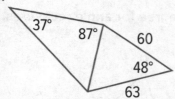

**16.**

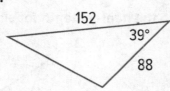

**REGULARITY** Determine whether each triangle should be solved by beginning with the Law of *Sines* or *Law of Cosines*. Then solve the triangle.

**17.** $A = 11°$, $C = 27°$, $c = 50$

**18.** $B = 47°$, $a = 20$, $c = 24$

**19.** $A = 37°$, $a = 20$, $b = 18$

**20.** $C = 35°$, $a = 18$, $b = 24$

**21. POOLS** The Perth County pool has a lifeguard station in both the deep water and shallow water sections of the pool. The distance between each station and the bottom of the slide is known, but the manager would like to calculate more information about the pool setup.

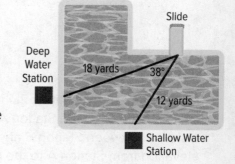

   **a.** When the lifeguards switch positions, the lifeguard at the deep-water station swims to the shallow water station. How far does the lifeguard swim?

   **b.** If the lifeguard at the deep-water station is directly facing the bottom of the slide, what angle does she need to turn in order to face the lifeguard at the shallow water station?

**22. CAMPING** At Shady Pines Campground, Campsites A and B are situated 80 meters apart. The camp office is 95 meters from Campsite A and 115 meters from Campsite B. When a rotating siren is placed on the roof of the office, how many degrees does the siren have to rotate to directly be projected on both Campsites A and B? Round your answer to the nearest tenth of a degree.

**23. SKATING** During a figure skating routine, Jackie and Peter begin at the same point and then skate apart with an angle of 15° between them. Jackie skates for 5 meters and Peter skates for 7 meters. To the nearest tenth of a meter, how far apart are the skaters?

**24. ANALYZE** Explain why the Pythagorean Theorem is a specific case of the Law of Cosines.

**25. WRITE** What methods can you use to solve a triangle?

**26. CREATE** Draw and label a triangle that can be solved:

   **a.** using only the Law of Sines.

   **b.** using only the Law of Cosines.

**27. PERSEVERE** Find the value of $x$ in the figure at the right.

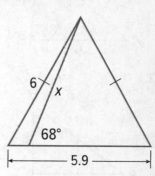

## Essential Question
How are right triangle relationships useful in solving real-world problems?

## Module Summary

### Lesson 9-1

**Geometric Mean**

- The geometric mean of two positive numbers $a$ and $b$ is $x$ such that $\frac{a}{x} = \frac{x}{b}$.

  So, $x^2 = ab$ and $x = \sqrt{ab}$.

### Lesson 9-2

**Pythagorean Theorem**

- If $\triangle ABC$ is a right triangle with right angle $C$, then $a^2 + b^2 = c^2$.

### Lesson 9-3

**Coordinates in Space**

- A point in space is represented by $(x, y, z)$ where $x, y,$ and $z$ are real numbers.
- A three-dimensional coordinate system has three axes: the $x$-, $y$-, and $z$-axes. The $x$- and $y$-axes lie on a horizontal plane, and the $z$-axis is vertical.

### Lesson 9-4

**Special Right Triangles**

- In a 45°-45°-90° triangle, the legs $\ell$ are congruent and the length of the hypotenuse $h$ is $\sqrt{2}$ times the length of a leg.
- In a 30°-60°-90° triangle, the length of the hypotenuse $h$ is 2 times the length of the shorter leg $s$, and the longer leg $\ell$ is $\sqrt{3}$ times the length of the shorter leg.

### Lesson 9-5

**Trigonometry**

- $\sin A = \frac{\text{opp}}{\text{hyp}}$, $\cos A = \frac{\text{adj}}{\text{hyp}}$, and $\tan A = \frac{\text{opp}}{\text{adj}}$.
- If $\sin A = x$, then $\sin^{-1} x = m\angle A$. If $\cos A = x$, then $\cos^{-1} x = m\angle A$. If $\tan A = x$, then $\tan^{-1} x = m\angle A$.

### Lessons 9-6 through 9-8

**Applications of Trigonometry**

- Area $= \frac{1}{2}ab \sin C$, where $a$ and $b$ are side lengths and $C$ is the included angle.
- Law of Sines: If $\triangle ABC$ has lengths $a$, $b$, and $c$, representing the lengths of the sides opposite the angles with measures $A$, $B$, and $C$, then $\frac{\sin A}{A} = \frac{\sin B}{B} = \frac{\sin C}{C}$.
- Law of Cosines: If $\triangle ABC$ has lengths $a$, $b$, and $c$, representing the lengths of the sides opposite the angles with measures $A$, $B$, and $C$, then $a^2 = b^2 + c^2 - 2bc \cos A$, $b^2 = a^2 + c^2 - 2ac \cos B$, and $c^2 = a^2 + b^2 - 2ab \cos C$.

### Study Organizer

📖 **Foldables**

Use your Foldable to review this module. Working with a partner can be helpful. Ask for clarification of concepts as needed.

# Test Practice

**1. MULTIPLE CHOICE** What is the geometric mean between 6 and 12? (Lesson 9-1)

Ⓐ $6\sqrt{2}$

Ⓑ $3\sqrt{2}$

Ⓒ 9

Ⓓ 36

**2. MULTI-SELECT** Which of the following triangles is similar to △JKL? Select all that apply. (Lesson 9-1)

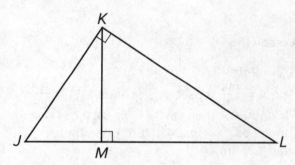

Ⓐ △JMK

Ⓑ △MKL

Ⓒ △JKM

Ⓓ △KML

**3. GRIDDED RESPONSE** A right triangle has legs with measures of 20 and 48. Use a Pythagorean triple to find the measure of the hypotenuse. (Lesson 9-2)

**4. TABLE ITEM** In △DEF, DE = 8 and EF = 15. Match the length of the third side to the type of triangle it would create. (Lesson 9-2)

| Third Side | Type of Triangle | | |
|---|---|---|---|
| | Acute | Right | Obtuse |
| 15 | | | |
| 16 | | | |
| 17 | | | |
| 18 | | | |

**5. OPEN RESPONSE** Point C is the midpoint of the line segment with endpoints at A(5, −8, −9) and B(13, −6, 9). What are the coordinates of point C? (Lesson 9-3)

**6. MULTIPLE CHOICE** This rectangular solid has one vertex at the origin.

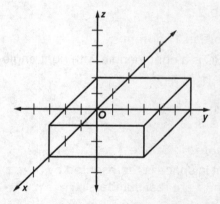

Which of the following points is also a vertex of the solid? (Lesson 9-3)

Ⓐ (2, 6, 3)

Ⓑ (3, 2, 6)

Ⓒ (3, 6, 2)

Ⓓ (6, 3, 2)

**Name** _____ **Period** _____ **Date** _____

**7. GRIDDED RESPONSE** A shipping container is in the shape of a rectangular prism. The coordinates of one corner are (0, 0, 0) and the opposite corner is (12, 8, 8), where each coordinate unit is 1 foot. What is the length, rounded to the nearest tenth of a foot, of the diagonal between the two opposite corners?
(Lesson 9-3)

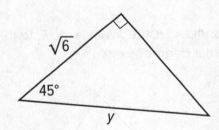

**8. MULTIPLE CHOICE** What is the value of $y$?
(Lesson 9-4)

(A) $\sqrt{2}$

(B) $\sqrt{3}$

(C) $2\sqrt{3}$

(D) $3\sqrt{2}$

**9. OPEN RESPONSE** A 10-foot-long ladder leans against a wall so that the ladder and the wall form a 30° angle. What is the distance to the nearest tenth of a foot from the ground to the point where the ladder touches the wall?
(Lesson 9-4)

**10. TABLE ITEM** Match the trigonometric ratio with its decimal value. (Lesson 9-5)

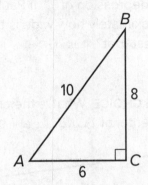

| Ratio | Decimal Value | | |
|---|---|---|---|
| | 0.6 | 0.75 | 0.8 |
| sin B | | | |
| cos B | | | |
| tan B | | | |

**11. OPEN RESPONSE** Use special right triangles to evaluate sin 45°. Give an exact answer.
(Lesson 9-5)

**12. MULTIPLE CHOICE** A support wire for an electric pole makes a 50° angle with the ground. If the bottom of the support wire is 8 feet from the base of the pole, which of the following equations can be used to find the height at which the support wire is attached to the pole? (Lesson 9-6)

(A) $\frac{x}{8} = \tan 50°$

(B) $\frac{8}{x} = \tan 50°$

(C) $\frac{x}{8} = \sin 50°$

(D) $\frac{8}{x} = \sin 50°$

**13. OPEN RESPONSE** Paula stands on the side of a river and sights the opposite bank at an angle of depression of 7°. If Paula is 5.5 feet tall, approximately how wide is the river, to the nearest foot? (Lesson 9-6)

**14. MULTIPLE CHOICE** What is the area, to the nearest tenth, of △JKL? (Lesson 9-6)

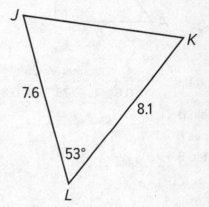

Ⓐ 18.5 ft²

Ⓑ 24.6 ft²

Ⓒ 30.8 ft²

Ⓓ 49.2 ft²

**15. MULTI-SELECT** In △DEF, e = 7, f = 5, and m∠F = 44°. What are the possible measures of ∠E, to the nearest degree? Select all that apply. (Lesson 9-7)

☐ 32°

☐ 44°

☐ 60°

☐ 77°

☐ 103°

☐ 120°

**16. OPEN RESPONSE** What is the value of a, to the nearest tenth? (Lesson 9-7)

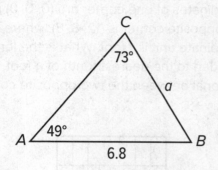

**17. MULTIPLE CHOICE** Two ships leave from the same location. The first ship travels 13 miles along a bearing of 0°. The second ship travels 11 miles along a bearing of 50°. What is the distance between the ships? (Lesson 9-8)

Ⓐ 6.9 miles

Ⓑ 9.2 miles

Ⓒ 10.3 miles

Ⓓ 10.9 miles

**18. MULTIPLE CHOICE** What is m∠E, to the nearest degree? (Lesson 9-8)

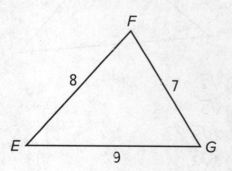

Ⓐ 40°

Ⓑ 42°

Ⓒ 45°

Ⓓ 48°

# Circles

## ⓔ Essential Question
How can circles and parts of circles be used to model situations in the real world?

G.CO.13, G.C.1, G.C.2, G.C.3, G.C.4, G.C.5, G.GMD.1, G.GPE.1, G.GPE.2, G.GPE.4

## What will you learn?
Place a checkmark (✓) in each row that corresponds with how much you already know about each topic **before** starting this module.

| KEY | Before | | | After | | |
|---|---|---|---|---|---|---|
| 👎 — I don't know.  👍 — I've heard of it.  👍 — I know it! | 👎 | 👍 | 👍 | 👎 | 👍 | 👍 |
| use the formula for circumference of a circle | | | | | | |
| prove all circles are similar | | | | | | |
| find measures of angles and arcs using the properties of circles | | | | | | |
| solve problems using the relationships between arcs, chords, and diameters | | | | | | |
| solve problems using inscribed angles | | | | | | |
| solve problems using inscribed polygons | | | | | | |
| solve problems using relationships between circles, tangents, and secants | | | | | | |
| construct inscribed and circumscribed circles | | | | | | |
| use equations of circles to solve problems | | | | | | |
| graph equations of parabolas | | | | | | |

📓 **Foldables** Make this Foldable to help you organize your notes about circles. Begin with nine sheets of notebook paper.

1. **Trace** an 8-inch circle on each paper using a compass.

2. **Cut** out each of the circles.

3. **Staple** an inch from the left side of the papers.

4. **Label** each tab with a lesson number and title.

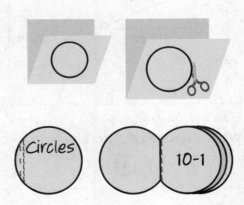

# What Vocabulary Will You Learn?

Check the box next to each vocabulary term that you may already know.

- ☐ adjacent arcs
- ☐ arc
- ☐ arc length
- ☐ center
- ☐ central angle of a circle
- ☐ chord of a circle

- ☐ circumscribed
- ☐ common tangent
- ☐ congruent arcs
- ☐ diameter
- ☐ directrix
- ☐ focus of a parabola
- ☐ hyperbola

- ☐ inscribed angle
- ☐ inscribed circle
- ☐ inscribed polygon
- ☐ intercepted arc
- ☐ major and minor arc
- ☐ parabola
- ☐ point of tangency

- ☐ radian
- ☐ radius
- ☐ secant
- ☐ semicircle
- ☐ tangent

## Are You Ready?

Complete the Quick Review to see if you are ready to start this module.
Then complete the Quick Check.

### Quick Review

**Example 1**

**Find the value of $x$. Round to the nearest tenth if necessary.**

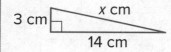

| $a^2 + b^2 = c^2$ | Pythagorean theorem |
| $3^2 + 14^2 = x^2$ | Substitute the side lengths. |
| $9 + 196 = x^2$ | Evaluate exponents. |
| $205 = x^2$ | Add. |
| $14.3 \approx x$ | Take the square root of each side. |

**Example 2**

**Find the distance between (5, 3) and (−2, 7). Round to the nearest tenth if necessary.**

$$d = \sqrt{(x_2 - x_1)^2 + (y_2 - y_1)^2}$$
$$= \sqrt{(-2 - 3)^2 + (7 - 5)^2}$$
$$= \sqrt{(-5)^2 + (2)^2}$$
$$= \sqrt{25 + 4}$$
$$= \sqrt{29}$$

So, the distance is about 5.4 units.

### Quick Check

**Solve for $x$.**

**1.**

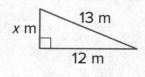

**2.**

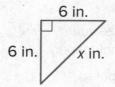

**Find the distance between each pair of points. Round to the nearest tenth if necessary.**

**3.** (2, 1) and (4, 10)

**4.** (−3, 7) and (5, 4)

**5.** (−4, −2) and (−1, −2)

**6.** (0, −6) and (3, 1)

### How Did You Do?

Which exercises did you answer correctly in the Quick Check? Shade those exercise numbers below.

# Circles and Circumference

## Explore Discovering the Formula for Circumference

Today's Standards
G.C.1, G.GMD.1
MP4, MP6, MP7

◥ **Online Activity** Use guiding exercises to complete the Explore.

> ⊘ **INQUIRY** Why is the circumference of a circle
> equal to $2\pi r$?

Today's Vocabulary
circle
center of a circle
radius of a circle
chord of a circle
diameter of a circle
pi
concentric circles

## Learn Parts of Circles

A **circle** is the set of all points in a plane that are the same distance from a given point called the **center** of the circle. The center of the circle below is $C$.

A **radius** (plural radii) is a line segment from the center to a point on a circle. $\overline{CD}$, $\overline{CE}$, and $\overline{CF}$ are radii of circle $C$.

A **chord of a circle** is a segment with endpoints on the circle. $\overline{AB}$ and $\overline{DE}$ are chords of circle $C$.

A **diameter** is a chord that passes through the center of a circle. $\overline{DE}$ Is a diameter of circle $C$.

All radii $r$ of a circle are congruent. Because a diameter $d$ is composed of two radii, all diameters of a circle are also congruent.

The words *radius* and *diameter* are used to describe lengths as well as segments.

### Key Concept • Radius and Diameter Relationships

If a circle has radius $r$ and diameter $d$, the following are true.

| | |
|---|---|
| **Radius Formula** $r = \frac{d}{2}$ or $r = \frac{1}{2}d$ | **Diameter Formula** $d = 2r$ |

The **circumference** of a circle is the distance around the circle. By definition, the ratio $\frac{C}{d}$ is an irrational number called **pi ($\pi$)**. Two formulas for circumference can be derived by using this definition.

| | |
|---|---|
| $\frac{C}{d} = \pi$ | Definition of pi |
| $C = \pi d$ | Multiply each side by $d$. |
| $C = \pi(2r)$ | $d = 2r$ |
| $C = 2\pi r$ | Simplify. |

### Key Concept • Circumference Formula

| | |
|---|---|
| **Words** | If a circle has diameter $d$ or radius $r$, the circumference $C$ equals the diameter times pi or twice the radius time pi. |
| **Symbols** | $C = \pi d$ or $C = 2\pi r$ |

**Math History Minute**

Chinese mathematician Tsu **Ch'ung-chih (429 AD-500 AD)** was particularly interested in finding the value of p. He approximated its value to be about the same as the ratio 355:113, which is actually correct to about 6 decimal places.

## Your Notes

🍥 **Think About It!**

Identify two congruent concentric circles in the figure above. If the radius of one circle is 9 centimeters, what do you know about the radius of the other circle?

_____
_____
_____
_____
_____
_____
_____
_____
_____

🍥 **Think About It!**

What assumption did you make while solving this problem?

_____
_____
_____
_____
_____
_____

**Use a Source**

Find the diameter of a famous traffic circle. Then calculate the circumference of the traffic circle.

---

## Example 1 Identify Segments in a Circle

**Name the circle and identify a radius, a chord, and a diameter of the circle.**

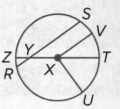

The circle has a center at _____, so it is named _____ or ⊙_____.

Four radii are shown: $\overline{XV}$, $\overline{XT}$ _____, and _____.

Two chords are shown: $\overline{RS}$ and _____.

$\overline{TZ}$ contains the center, so $\overline{TZ}$ is the _____.

## Example 2 Use Radius and Diameter Relationships

**If *TU* = 14 feet, what is the radius of ⊙Q?**

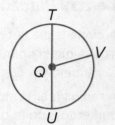

$r = \dfrac{d}{2}$      Radius Formula

$r = \dfrac{}{2}$ or _____      Substitute and simplify.

The radius of ⊙Q is _____ feet.

### Check

If *LM* = 11 inches, what is the diameter ⊙L?

_____ in.

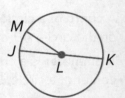

---

## 🌐 Example 3 Find Circumference

**TRAFFIC CIRCLES** Traffic circles, also known as roundabouts, are circular roadways that reflow traffic in one direction around an island. A car enters a traffic circle and is 18 meters away from the center of the island. If the car drives around the traffic circle until it is back to its original position, what is the circumference of the car's path?

Since the car is 18 meters away from the center of the island, the radius of the car's path is _____ meters.

$C = 2\pi r$      Circumference Formula

$= 2\pi(\_\_\_)$      Substitution

$= \_\_\_ \pi$      Simplify.

$\approx \_\_\_$      Use a calculator.

The circumference of the car's path is _____ $\pi$ or about 113.1 meters.

🔵 **Go Online** You can complete an Extra Example online.

## Check

**AUTOMOBILES** Many automobiles have customized rims that are attached to wheels and align with the inner edges of tires. The rim of a tire has a radius of 7.5 inches, and the width of the tire is 6 inches.

**Part A** Find the circumference of the rim. Select all that apply. _____

**A.** $\frac{15\pi}{2}$ in.

**B.** $15\pi$ in.

**C.** $27\pi$ in.

**D.** 23.56 in.

**E.** 47.12 in.

**F.** 84.82 in.

**Part B** What assumptions did you make while solving this problem? Select all that apply. _____

**A.** I assumed the radius of the tire rim.

**B.** I assumed that the tire rim was a perfect circle.

**C.** I assumed the width of the tire.

**D.** I assumed there was no space between the tire and the rim.

---

## Example 4 Find Diameter and Radius

**Find the diameter and radius of a circle to the nearest hundredth if the circumference of the circle is 77.8 centimeters.**

| | |
|---|---|
| $C = \pi d$ | Circumference Formula |
| _____ $= \pi d$ | Substitution |
| $\frac{77.8}{\quad} = d$ | Divide each side by $\pi$. |
| _____ $\approx d$ | Use a calculator. |

The diameter of the circle is about 24.76 centimeters. Use the diameter of the circle to find the radius.

| | |
|---|---|
| $r = \frac{1}{2}d$ | Radius Formula |
| $\approx \frac{1}{2}($_____$)$ | $d \approx 24.76$ |
| $\approx$ _____ | Use a calculator. |

So, the radius of the circle is about _____ centimeters.

## Check

Find the diameter and radius of a circle if the circumference of the circle is 94.2 yards.

**Part A** Select the most appropriate estimates of the diameter and radius. _____

**A.** 15.7 yd; 7.85 yd.

**B.** 26.91 yd; 13.46 yd

**C.** 18.84 yd; 9.42 yd

**D.** 31.4 yd; 15.7 yd

**Part B** Find the exact diameter and radius of the circle. Round your answers to the nearest hundredth.

diameter = _____ yd

radius = _____ yd

🅑 **Go Online** You can complete an Extra Example online.

**Study Tip:**

**Levels of Accuracy**
Since $\pi$ is irrational, its value cannot be given as a terminating decimal. Using a value of 3 for $\pi$ provides a quick estimate in calculations. Using a value of 3.14 or $\frac{22}{7}$ provides a closer approximation. For the most accurate approximation, use the $\pi$ key on a calculator. Unless stated otherwise, assume that in this course, a calculator with a $\pi$ key was used to generate answers.

## Learn Pairs of Circles

As with other figures, pairs of circles can be congruent, similar, or share other special relationships.

| Postulate 10.1 | |
| --- | --- |
| Words | Two circles are congruent if and only if they have congruent radii. |
| Example | 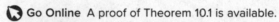 $\overline{GH} \cong \overline{JK}$ so $\odot G \cong \odot J$. |

| Theorem 10.1 | |
| --- | --- |
| Words | All circles are similar. |
| Example | $\odot X \sim \odot Y$. |

**Go Online** A proof of Theorem 10.1 is available.

**Concentric circles** are coplanar circles that have the same center.

$\odot A$ with radius $\overline{AB}$ and $\odot A$ with radius $\overline{AC}$ are concentric.

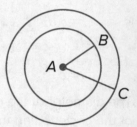

## Example 5 Find Measures in Intersecting Circles

**The diameter of $\odot K$ is 12 units, the diameter of $\odot J$ is 20 units, and $JD = 8$ units. Find $EK$.**

Because the diameter of $\odot J$ is 20, $JE = 10$.

$\overline{DE}$ is a part of radius $\overline{JE}$.

| | |
| --- | --- |
| $JD + DE = JE$ | Segment Addition Postulate |
| _____ $+ DE =$ _____ | Substitution |
| $DE =$ _____ | Subtract _____ from each side. |

Because the diameter of $\odot K$ is 12, $DK = 6$. $\overline{DE}$ and $\overline{EK}$ for radius $\overline{DK}$.

| | |
| --- | --- |
| $DE + EK = DK$ | Segment Addition Postulate |
| $2 + EK =$ _____ | Substitution |
| $EK =$ _____ | Subtract _____ from each side. |

So, $EK$ is _____ units.

**Go Online** You can complete an Extra Example online.

 **Talk About It!**

What is the relationship between two concentric circles with congruent radii? Justify your argument.

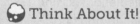 **Think About It!**

Is the distance from the center of a circle to a point in the interior of a circle *sometimes, always,* or *never* less than the radius of the circle? Justify your reasoning.

# Practice

**Go Online** You can complete your homework online.

**Example 1**

**For Exercises 1–3, refer to the circle at the right.**

1. Name the circle.

2. Name the radii of the circle.

3. Name the chords of the circle.

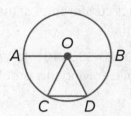

**For Exercises 4–8, refer to the circle at the right.**

4. Name the circle.

5. Name the radii of the circle.

6. Name the chords of the circle.

7. Name a diameter of the circle.

8. Name a radius not drawn as part of a diameter.

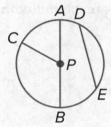

**Example 2**

**For Exercises 9–11, refer to $\odot R$.**

9. If $AB = 18$ millimeters, find $AR$.

10. If $RY = 10$ inches, find $AR$ and $AB$.

11. Is $\overline{AB} \cong \overline{XY}$? Explain.

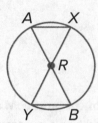

**For Exercises 12–14, refer to $\odot L$.**

12. Suppose the radius of the circle is 3.5 yards. Find the diameter.

13. If $RT = 19$ meters, find $LW$.

14. If $LT = 4.2$ inches, what is the diameter of $\odot L$?

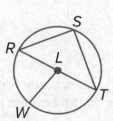

**Example 3**

15. **TIRES** A bicycle has tires with a diameter of 26 inches. Find the radius and circumference of a tire. Round to the nearest hundredth, if necessary.

**16. SUNDIALS** Herman purchased a sundial to use as the centerpiece for a garden. The diameter of the sundial is 9.5 inches.

a. Find the radius of the sundial.

b. Find the circumference of the sundial to the nearest hundredth.

c. **STATE YOUR ASSUMPTION** What assumption did you make while solving this problem?

### Example 4

**Find the diameter and radius of a circle with the given circumference. Round to the nearest hundredth.**

**17.** $C = 40$ in.

**18.** $C = 256$ ft

**19.** $C = 15.62$ m

**20.** $C = 9$ cm

**21.** $C = 79.5$ yd

**22.** $C = 204.16$ m

### Example 5

**The diameters of $\odot F$ and $\odot G$ are 5 and 6 units, respectively. Find each measure.**

**23.** $BF$

**24.** $AB$

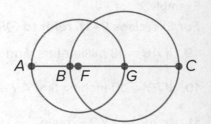

**The diameters of $\odot L$ and $\odot M$ are 20 and 13 units, respectively, and $QR = 4$. Find each measure.**

**25.** $LQ$

**26.** $RM$

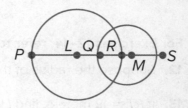

### Mixed Exercises

**REASONING The radius, diameter, or circumference of a circle is given. Find each missing measure to the nearest hundredth.**

**27.** $d = 8\frac{1}{2}$in., $r = \underline{\ ?\ }$, $C = \underline{\ ?\ }$

**28.** $r = 11\frac{2}{5}$ft, $d = \underline{\ ?\ }$, $C = \underline{\ ?\ }$

**29.** $C = 628$ m, $d = \underline{\ ?\ }$, $r = \underline{\ ?\ }$

**30.** $d = \frac{3}{4}$yd, $r = \underline{\ ?\ }$, $C = \underline{\ ?\ }$

**31.** $C = 35x$ cm, $d = \underline{\ ?\ }$, $r = \underline{\ ?\ }$

**32.** $r = \frac{x}{8}$, $d = \underline{\ ?\ }$, $C = \underline{\ ?\ }$

**Determine whether the circles in the figures below appear to be *congruent*, *concentric*, or *neither*.**

33.

34.

35.

**Find the exact circumference of each circle.**

36.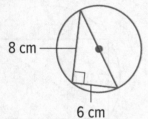
8 cm

6 cm

37.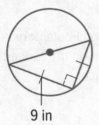
9 in

38.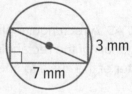
3 mm

7 mm

39.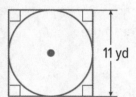
11 yd

40.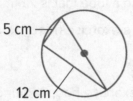
5 cm

12 cm

41.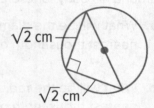
√2 cm

√2 cm

42. **ARGUMENTS** Determine whether each statement is *always*, *sometimes*, or *never* true. Explain.

   a. If points *G* and *H* lie on ⊙*C*, then *CG* = *CH*.

   b. If points *P*, *Q*, and *R* lie on ⊙*C*, then points *P*, *Q* and *R* are coplanar.

   c. If points *A* and *B* lie on ⊙*C*, then the line segment with the endpoints *A* and *B* is a diameter.

   d. If points *X* and *Y* lie on ⊙*C*, then △*XYC* is a scalene triangle.

43. **USE A SOURCE** Research and write about the history of pi and its importance to the study of geometry.

**44. WHEELS** Zack is designing wheels for a concept car. The diameter of the wheel is 18 inches. Zack wants to make spokes in the wheel that run from the center of the wheel to the rim. In other words, each spoke is a radius of the wheel. How long are these spokes?

**45. MODELING** Kathy slices through a circular cake. The cake has a diameter of 14 inches. The slice that Kathy made is straight and has a length of 11 inches. Did Kathy cut along a *radius*, a *diameter*, or a *chord* of the circle?

**46. COINS** Three identical circular coins are lined up in a row as shown. The distance between the centers of the first and third coins is 3.2 centimeters. What is the radius of one of these coins?

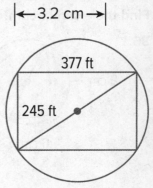

**47. PLAZAS** A rectangular plaza has a surrounding circular fence. The diagonals of the rectangle pass from one point on the fence through the center of the circle to another point on the fence.

**48. EXERCISE HOOPS** Taiga wants to make a circular loop that he can twirl around his body for exercise. He will use a tube that is 2.5 meters long.

a. What will be the diameter of Taiga's exercise hoop? Round your answer to the nearest thousandth of a meter.

b. What will be the radius of Taiga's exercise hoop? Round your answer to the nearest thousandth of a meter.

**49. PERSEVERE** The sum of the circumferences of circles $H$, $J$, and $K$ shown at the right is $56\pi$ units. Find $KJ$.

**50. ANALYZE** Is the distance from the center of a circle to a point in the interior of a circle *sometimes*, *always*, or *never* less than the radius of the circle? Explain.

**51. CREATE** Design a sequence of transformations that can be used to prove that $\odot D$ is similar to $\odot E$.

**52. WRITE** How can we describe the relationships that exist between circles and lines?

# Measuring Angles and Arcs

## Explore Relationships Between Arc Lengths and Radii

**Today's Standards**
G.C.2, G.C.5
MP1, MP3, MP7

🔾 **Online Activity** Use graphing technology to complete the Explore.

> @ **INQUIRY** How is the radian measure of a central angle related to the length of its arc and the radius of the circle?

## Learn Measuring Angles and Arcs

A **central angle of a circle** is an angle with a vertex at the center of a circle and sides that are radii.

A **degree** is $\frac{1}{360}$ of the circular rotation about a point. This leads to the following relationship.

**Today's Vocabulary**
central angle of a circle
degree
arc
minor arc
major arc
semicircle
congruent arcs
adjacent arcs
arc length
radian

| Key Concept • Sum of Central Angles | |
|---|---|
| Words | The sum of the measures of the central angles of a circle with no interior points in common is 360°. |
| Example | $m\angle 1 + m\angle 2 + m\angle 3 = 360°$  |

An **arc** is part of a circle that is defined by two endpoints. A central angle separates the circle into two arcs with measures related to the measure of the central angle.

A **minor arc** has a measure less than 180°.
A **major arc** has a measure greater than 180°.
A **semicircle** is an arc that measures exactly 180°.
**Congruent arcs** are arcs in the same or congruent circles that have the same measure.

| Theorem 10.2 | |
|---|---|
| Words | In the same circle or in congruent circles, two minor arcs are congruent if and only if their central angles are congruent. |
| Example | If $\angle 1 \cong \angle 2$, then $\widehat{FG} \cong \widehat{HJ}$. <br> If $\widehat{FG} \cong \widehat{HJ}$, then $\angle 1 \cong \angle 2$.  |

Arcs in a circle that have exactly one point in common are called **adjacent arcs**.

**Study Tip**

**Naming Arcs** Minor arcs can be named by just their endpoints. Major arcs and semicircles are named by their endpoints and another point on the arc that lies between these endpoints.

**Postulate 10.2: Arc Addition Postulate**

The measure of an arc formed by two adjacent arcs is the sum of the measures of the two arcs.

$m\angle \widehat{XYZ} = m\widehat{XY} + m\widehat{YZ}$.

## Example 1 Find Measures of Central Angles

**Find the value of *x*.**

$m\angle EAB + m\angle BAC + m\angle CAD + m\angle DAE = 360$   Sum of Central Angles

$90 + 40 + 85 + x = 360$   Substitution

$\underline{\hspace{1cm}} + x = 360$   Simplify.

$x = \underline{\hspace{1cm}}$   Subtract \_\_\_\_\_ from each side.

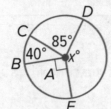

## Example 2 Classify Arcs and Find Arc Measures

$\overline{PM}$ **is a diameter of ⊙*R*. Identify each arc as a *major arc*, *minor arc*, or *semicircle*. Then find its measure.**

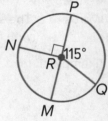

a. $\widehat{MQ}$

$\widehat{MQ}$ is a minor arc, so $m\widehat{MQ} = m\angle MRQ$. Because $\angle MRQ$ and $\angle PRQ$ are linear pairs, $m\angle MRQ = 180 - 115$ or \_\_\_\_\_°. So, $m\widehat{MQ} = $ \_\_\_\_\_°.

b. $\widehat{MNP}$

$\widehat{MNP}$ is a semicircle, so $m\widehat{MNP} = $ \_\_\_\_\_°.

c. $\widehat{MNQ}$

$\widehat{MNQ}$ is a major arc that shares the same endpoints as minor arc $\widehat{MQ}$.

$m\widehat{MNQ} = 360 - m\widehat{MQ}$

$= 360 - 65$ or \_\_\_\_\_°

## 🌐 Example 3 Find Arc Measures in Circle Graphs

**ONLINE The circle graph shows how teenagers spend their time online. Find $m\widehat{DE}$.**

$\widehat{DE}$ is a _____ arc.
So, $m\widehat{DE} = m\angle$_____.

**Go Online** You can complete an Extra Example online.

∠DPE represents 12% of the whole, or 12% of the circle.

$m∠DPE = \underline{\quad}(360)$   Find 12% of 360.

$= \underline{\quad}°$   Simplify.

So, $m\widehat{DE}$ is $\underline{\quad}°$.

## Example 4 Use Arc Addition to Find Measures of Arcs

Find $m\widehat{WY}$ in ⊙V.

$m\widehat{WY} = m\widehat{WX} + m\underline{\quad}$   Arc Addition Postulate

$= m∠WVX + \underline{\quad}$   $m\widehat{WX} = m∠WVX, m\widehat{XY} = m∠XVY$

$= 90 + \underline{\quad}$ or $\underline{\quad}$   Substitution

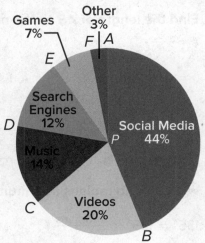

**Teenagers' Online Activities**

Games 7%
Other 3%
Search Engines 12%
Social Media 44%
Music 14%
Videos 20%

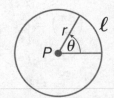

## Learn Arc Length and Radian Measure

**Arc length** is the distance between the endpoints of an arc measured along the arc in linear units. Because an arc is a portion of a circle, its length is a fraction of the circumference.

### Key Concept • Arc Length in Degrees

| Words | The ratio of the length of an arc $\ell$ to the circumference of the circle is equal to the ratio of the degree measure of the arc to 360°. |
|---|---|
| Proportion | $\frac{\ell}{2\pi r} = \frac{x}{360}$ |
| Equation | $\ell = \frac{x}{360} \cdot 2\pi r$ |

### Key Concept • Radian Measure

Angles can be measured in radians. A **radian** is a unit of angular measurement equal to $\frac{180°}{\pi}$ or about 57.296°.

The radian measure $\theta$ of a central angle is the ratio of the arc length to the radius of the circle: $\theta = \frac{\ell}{r}$ radians.

### Key Concept • Degree and Radian Conversion Rules

1. To convert a degree measure to radians, multiply by $\frac{\pi \text{ radians}}{180°}$.

2. To convert a radian measure to radians, multiply by $\frac{180°}{\pi \text{ radians}}$.

🧭 **Go Online** You can complete an Extra Example online.

**Watch Out!**

**Arc Length** The *length* of an arc is given in linear units, such as centimeters. The *measure* of an arc is given in degrees.

🖱 **Go Online** You can watch a video to see how to find the length of an arc of a circle.

💭 **Think About It!**

If the measure of a central angle $\theta = \frac{\ell}{r}$, then what is the length of the arc with the related central angle?

## Problem Solving Tip

**Use Reasoning Skills**
If you have difficulty remembering an equation, use your reasoning skills to create the equation. An arc is a portion of a circle, so its length is a fraction of the circle's circumference. You can find the length of an arc by multiplying the ratio of the arc's measure to 360° by the circumference of the circle. So, the length of an arc is equal to $\frac{x}{360} \cdot 2\pi r$, where $x$ is the measure of the arc.

 **Think About It!**

**Check your answer to the problem.**

What method did you use to check your answer?

## Example 5 Find Arc Length Using Degrees

**Find the length of $\overset{\frown}{AB}$ to the nearest hundredth.**

$\ell = \frac{x}{\underline{\quad}} \cdot \underline{\quad} \pi r$      Arc Length Equation

$\phantom{\ell} = \frac{\underline{\quad}}{360} \cdot 2\pi(\underline{\quad})$      Substitution

$\phantom{\ell} \approx \underline{\quad}$      Use a calculator.

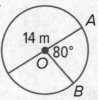

## Example 6 Convert From Degrees to Radian Measure

**Write 135° in radians as a multiple of $\pi$.**

$135° = 135° \times \frac{\pi \text{ radians}}{\underset{\circ}{\phantom{x}}}$      Multiply by $\frac{\pi \text{ radians}}{\underset{\circ}{\phantom{x}}}$

$\phantom{135°} = \frac{135\pi}{180}$ or $\frac{\phantom{x}}{4}$ radians      Simplify.

## Example 7 Convert From Radian Measure to Degrees

**Write $\frac{11\pi}{6}$ radians in degrees.**

$\frac{11\pi}{6}$ radians $= \frac{\pi}{\underline{\quad}}$ radians $\times \frac{\circ}{\pi \text{ radians}}$      Multiply by $\frac{\circ}{\pi \text{ radians}}$.

$\phantom{\frac{11\pi}{6} \text{ radians}} = \frac{\underline{\quad}}{6}$ or $\underline{\quad}°$      Simplify.

## Check

Write $\frac{2\pi}{3}$ radians in degrees.

$\underline{\quad}°$

## Example 8 Find Arc Length Using Radian Measure

**Find the length of $\overset{\frown}{ZY}$ to the nearest hundredth.**

$\theta = \frac{\underline{\quad}}{\underline{\quad}}$      Arc length Equation

$\frac{3\pi}{4} = \frac{\ell}{9}$      $\theta = \frac{3\pi}{4}, r = \underline{\quad}$

$\underline{\quad}\left(\frac{3\pi}{4}\right) = \ell$      Multiply each side by 9.

$\frac{\underline{\quad}\pi}{\phantom{x}} = \ell$      Use a calculator.

$\underline{\quad} \approx \ell$

So, the length of $\overset{\frown}{ZY}$ is about $\underline{\quad}$ centimeters.

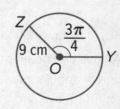

## Check

Find the length of $\overset{\frown}{MN}$ to the nearest hundredth.

$\underline{\quad}$ in.

🧭 **Go Online** You can complete an Extra Example online.

# Practice

Go Online You can complete your homework online.

### Example 1

**Find the value of x.**

1.
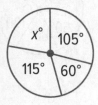

2.
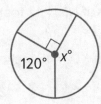

3.
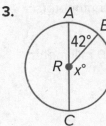

### Example 2

$\overline{AC}$ and $\overline{EB}$ are diameters of $\odot R$. Identify each arc as a *major arc, minor arc,* or *semicircle* of the circle. Then find its measure.

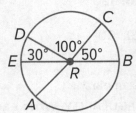

4. $m\widehat{EA}$

5. $m\widehat{CB}$

6. $m\widehat{DC}$

7. $m\widehat{DEB}$

8. $m\widehat{AB}$

9. $m\widehat{CDA}$

### Example 3

10. **SURVEYS** A survey asked students at Westwood High School their preference for the new school mascot. The results are shown in the circle graph. Find $m\widehat{AB}$.

11. **SPORTS** The circle graph shows the favorite spectator sport among a group of teens at an area high school. Find $m\widehat{AD}$.

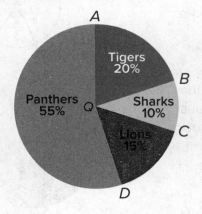

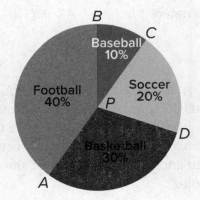

### Example 4

$\overline{PR}$ and $\overline{QT}$ are diameters of $\odot A$. Find each measure.

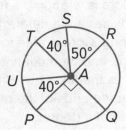

12. $m\widehat{UPQ}$

13. $m\widehat{PQR}$

14. $m\widehat{UTS}$

15. $m\widehat{RS}$

16. $m\widehat{RSU}$

17. $m\widehat{STP}$

18. $m\widehat{PQS}$

19. $m\widehat{PRU}$

### Example 5

**MODELING** Use ⊙D to find the length of each arc. Round to the nearest hundredth.

20. $\overarc{LM}$ if the radius is 5 inches

21. $\overarc{MN}$ if the diameter is 3 yards

22. $\overarc{KL}$ if $JD = 7$ centimeters

23. $\overarc{NJK}$ if $NL = 12$ feet

24. $\overarc{KLM}$ if $DM = 9$ millimeters

25. $\overarc{JK}$ if $KD = 15$ inches

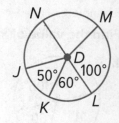

### Example 6

**PRECISION** Write each degree measure in radians as a multiple of π.

26. 120°

27. 45°

28. 30°

29. 90°

30. 180°

31. 225°

### Example 7

**REGULARITY** Write each radian measure in degrees.

32. $\frac{3\pi}{4}$ radians

33. $\frac{3\pi}{2}$ radians

34. $\frac{\pi}{3}$ radians

35. $\frac{5\pi}{6}$ radians

36. $2\pi$ radians

37. $\frac{\pi}{12}$ radians

### Example 8

Use ⊙Z to find the length of each arc. Round to the nearest hundredth.

38. $\overarc{QR}$, if $PZ = 12$ feet

39. $\overarc{ST}$, if $SZ = 8$ inches

40. $\overarc{PQ}$, if $TZ = 14$ centimeters

41. $\overarc{PT}$, if $TR = 20$ inches

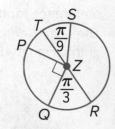

### Mixed Exercises

42. **CLOCKS** Shiatsu is a Japanese massage technique. One of the beliefs is that various body functions are most active at various times during the day. To illustrate this, they use a Chinese clock that is based on a circle divided into 12 equal sections by radii. What are the degree and radian measures of any one of the 12 equal central angles?

43. **RIBBONS** Cora is wrapping a ribbon around a cylinder-shaped gift box. The box has a diameter of 15 inches and the ribbon is 60 inches long. Cora is able to wrap the ribbon all the way around the box once, and then continue so that the second end of the ribbon passes the first end. What is the central angle formed by the arc between the ends of the ribbon? Round your answer to the nearest tenth of a degree.

**44. PIES** Yolanda has divided a circular apple pie into 4 slices by cutting the pie along 4 radii. The central angles of the 4 slices are $3x°$, $(6x - 10)°$, $(4x + 10)°$, and $5x°$. What are the measures of the central angles?

**45. BIKE WHEELS** Lucy has to buy a new wheel for her bike. The bike wheel has a diameter of 20 inches.

a. If Lucy rolls the wheel one complete rotation along the ground, how far will the wheel travel? Round your answer to the nearest hundredth of an inch.

b. If the bike wheel is rolled along the ground so that it rotates 45°, how far will the wheel travel? Round your answer to the nearest hundredth of an inch.

c. If the bike wheel is rolled along the ground for 10 inches, through what angle does the wheel rotate? Round your answer to the nearest tenth of a degree.

**46. USE TOOLS** Refer to the table, which shows the number of hours students at Leland High School say they spend on homework each night.

a. If you were to construct a circle graph of the data, how many degrees would be allotted to each category?

b. Describe the types of arcs associated with each category.

| Homework | |
|---|---|
| **Less than 1 hour** | 8% |
| 1–2 hours | 29% |
| 2–3 hours | 58% |
| 3–4 hours | 3% |
| Over 4 hours | 2% |

**47. ARGUMENTS** Write a two-column proof of Theorem 10.2.

**Given:** $\angle BAC \cong \angle DAE$

**Prove:** $\overarc{BC} \cong \overarc{DE}$

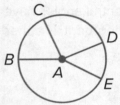

**REASONING** Find each measure. Round each linear measure to the nearest hundredth and each arc measure to the nearest degree.

**48.** circumference of $\odot S$

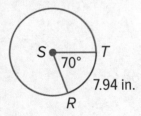

**49.** $m\overarc{CD}$

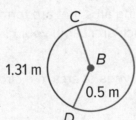

**50.** radius of $\odot S$

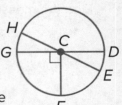

**ALGEBRA** In $\odot C$, $m\angle HCG = 2x°$ and $m\angle HCD = (6x + 28)°$. Find each measure.

**51.** $m\overarc{EF}$   **52.** $m\overarc{HD}$   **53.** $m\overarc{HGF}$

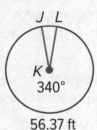

**54. STRUCTURE** An arc is intercepted by a central angle of 60°. Complete the table by finding the length of the arc in terms of $\pi$ for each given radius.

| Radius of circle, $r$ | 3 | 5 | 11 | 15 | $r$ |
|---|---|---|---|---|---|
| Length of arc, $\ell$ | | | | | |

**55. ARCHITECT** An architect is designing the seating area for a theater. The area is formed by a region that lies between two circles, as shown in the figure. The architect is planning to place a brass rail in front of the first row of seats. She wants to know the length of the rail.

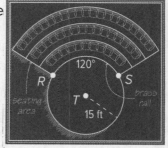

**a.** The architect wants to know the length of the rail. Express the length in terms of $\pi$ and to the nearest tenth of a foot.

**b.** The architect is considering changing the radius of $\odot T$ or changing the measure of $\widehat{RS}$. Describe a general method she can use to find the length of $\widehat{RS}$.

### 🧁 Higher-Order Thinking Skills

**ANALYZE Determine whether each statement is *sometimes*, *always*, or *never* true. Explain your reasoning.**

**56.** The measure of a minor arc is less than 180°.

**57.** If a central angle is obtuse, its corresponding arc is a major arc.

**58.** The sum of the measures of adjacent arcs of a circle depends on the measure of the radius.

**59. FIND THE ERROR** Brody says that $\widehat{WX}$ and $\widehat{YZ}$ are congruent because their central angles have the same measure. Selena says they are not congruent. Is either of them correct? Explain your reasoning.

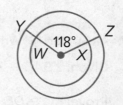

**60. PERSEVERE** The time shown on an analog clock is 8:10. What is the measure of the angle formed by the hands of the clock?

**61. CREATE** Draw a circle and locate three points on the circle. Estimate the measures of the three nonoverlapping arcs that are formed. Then use a protractor to find the measure of each arc. Label your circle with the arc measures.

**62. WRITE** Describe the three different types of arcs in a circle and the method for finding the measure of each one.

**63. PERSEVERE** The measures of $\widehat{LM}$, $\widehat{MN}$, and $\widehat{NL}$ are in the ratio of 5:3:4. Find the measure of each arc.

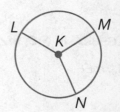

# Arcs and Chords

## Explore Chords in Circle

Today's Standards
G.C.2
MP1, MP3, MP8

⬤ **Online Activity** Use graphing technology to complete the Explore.

×

@ **INQUIRY** What relationships exist between chords and arcs in circles?

## Learn Arcs and Chords

| Theorems | |
| --- | --- |
| **10.3: Congruent Chords** | In the same circle or in congruent circles, two minor arcs are congruent if and only if their corresponding chords are congruent. |
| **10.4: Bisecting Arcs and Chords** | If a diameter (or radius) of a circle is perpendicular to a chord, then it bisects the chord and its arc. |
| **10.5: Bisecting Arcs and Chords** | The perpendicular bisector of a chord is a diameter (or radius) of the circle. |
| **10.6: Congruent Chords** | In the same circle or in congruent circles, chords are congruent if and only if they are equidistant from the center. |

⬤ **Go Online**
Proofs are available for Theorems 10.3 through 10.6

💭 **Think About It!**
What is the relationship between a chord and a radius perpendicular to the chord?

## Example 1 Use Congruent Arcs to Find Arc Measures

$\overline{JK} \cong \overline{LM}$ and $m\angle LM = 75°$. Find $m\angle JK$.

$\overline{JK}$ and $\overline{LM}$ _____ congruent chords, so the corresponding arcs $\widehat{JK}$ and $\widehat{LM}$ _____ congruent.

$m\widehat{JK} = m\widehat{LM} =$ _____ °

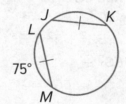

## Example 2 Use Congruent Arcs to Find Chord Length

In ⊙C, $\widehat{DE} \cong \widehat{FG}$. Find FG.

$\widehat{DE}$ and $\widehat{FG}$ are congruent arcs in the same circle, the corresponding chords $\overline{DE}$ and $\overline{FG}$ are congruent.

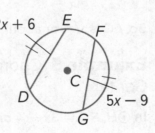

| | |
| --- | --- |
| $DE = FG$ | Definition of congruent segments |
| $2x + 6 = 5x - 9$ | Substitution |
| _____ = _____ − 9 | Subtract 2x from each side. |
| _____ = 3x | Add 9 to each side. |
| _____ = x | Divide each side by 3. |

So, $FG = 5(5) - 9$ or _____.

⬤ **Go Online** You can complete an Extra Example online.

💬 **Talk About It!**
In ⊙P, chords $\overline{DE}$ and $\overline{FG}$ are congruent. What do you know about $\angle DPE$ and $\angle FPG$? Explain.

**Think About It!**

In a circle, $\overline{AB}$ is a diameter and $\overline{QR}$ is a chord that intersects $\overline{AB}$ at point $X$. Is it *sometimes*, *always*, or *never* true that $QX = XR$? Explain.

**Study Tip**

**Assumptions**
Assuming that objects can be modeled by perfect circles allows you to find reasonable measures in objects.

**Go Online**
You may want to complete the construction activities for this lesson.

## Example 3 Use a Radius Perpendicular to a Chord

In ⊙D, $m\widehat{EFG} = 120°$. Find $m\widehat{FG}$ and $EG$.

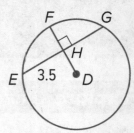

Radius $\overline{DF}$ is perpendicular to chord $\overline{EG}$. So by Theorem 10.4, $\overline{DF}$ bisects $\widehat{EFG}$ and $\overline{EG}$.

Therefore, $m\widehat{EF} = m\widehat{FG}$ and $EH = $ _____.

By Substitution, $m\widehat{EF} = \frac{120}{2}$ or _____°.

By the **Segment Addition Postulate** and substitution, $EG = 2 \cdot EH$. So, $EG = 2 \cdot$ _____ or _____ units.

## 🌐 Example 4 Use a Diameter Perpendicular to a Chord

RECORDS **The record shown can be modeled by a circle. Diameter $\overline{CD}$ is 12 inches long, and chord $\overline{FH}$ is 10 inches long. Find $EG$.**

**Step 1 Draw radius $\overline{EF}$.**
Radius $\overline{EF}$ forms right $\triangle EFG$.

**Step 2 Find $EF$ and $FG$.**
Because $CD = $ _____ inches, $ED = $ _____ inches. All radii of a circle are congruent, so $EF = $ _____ inches. Because diameter $\overline{CD}$ is perpendicular to $\overline{FH}$, $\overline{CD}$ bisects chord $\overline{FH}$ by Theorem 10.4.
So $FG = \frac{1}{2}(10)$ or _____ inches.

**Step 3 Use the Pythagorean Theorem to find $EG$.**

| | |
|---|---|
| $EG^2 + FG^2 = EF^2$ | Pythagorean Theorem |
| $EG^2 + 5^2 = 6^2$ | $FG = $ _____ and $EF = $ _____ |
| $EG^2 + $ _____ $= $ _____ | Simplify. |
| $EG^2 = $ _____ | Subtract 25 from each side. |
| $EG = $ _____ | Take the positive square root of each side. |

So, $EG$ is $\sqrt{11}$ or about _____ inches long.

## Example 5 Chords Equidistant from Center

In ⊙H, $PQ = 3x - 4$ and $RS = 14$. Find $x$.

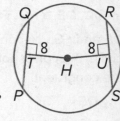

Because chords $\overline{PQ}$ and $\overline{RS}$ are equidistant from $H$, they are congruent. So, $PQ = RS$.

| | |
|---|---|
| $PQ = RS$ | Definition of congruent segments |
| _____ $-$ _____ $= $ _____ | Substitution |
| _____ $= $ _____ | Add 4 to each side. |
| $x = $ _____ | Divide each side by 3. |

**Go Online** You can complete an Extra Example online.

# Practice

⬆ **Go Online** You can complete your homework online.

**Examples 1 and 2**

REGULARITY **Find the value of *x*.**

**1.**
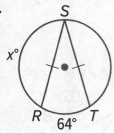

**2.**
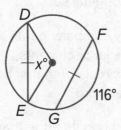

**3.**
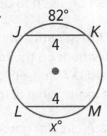

**4.**
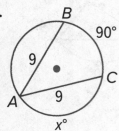

**5.**
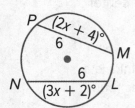

**6.**
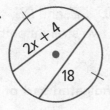

**7.**
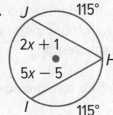

**8.** ⊙*M* ≅ ⊙*P*
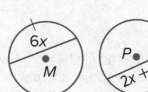

**9.** ⊙*V* ≅ ⊙*W*
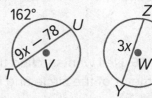

**Examples 3 and 4**

In ⊙*P*, the radius is 13 and *RS* = 24. Find each measure. Round to the nearest hundredth.

**10.** *RT*

**11.** *PT*

**12.** *TQ*

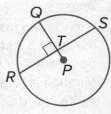

PRECISION  In ⊙*A*, the diameter is 12, *CD* = 8, and m$\widehat{CD}$ = 90. Find each measure. Round to the nearest hundredth.

**13.** m$\widehat{DE}$

**14.** *FD*

**15.** *AF*

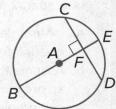

**Example 5**

**16.** In ⊙*R*, *TS* = 21 and
  *UV* = 3*x*. What is the value of *x*?

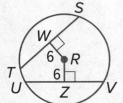

**17.** In ⊙*Q*, *CD* ≅ *CB*, *GQ* = *x* + 5 *and*
  *EQ* = 3*x* − 6. What is the value of *x*?

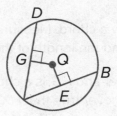

## Mixed Exercises

**18. USE TOOLS** A one piece of a broken plate is found during an archaeological dig. Use the sketch of the pottery piece shown to demonstrate how constructions with chords and perpendicular bisectors can be used to draw the plate's original size.

**19. MODELING** For security purposes a jewelry company prints a hidden watermark on the logo of all its official documents. The watermark is a chord located 0.7 cm from the center of a circular ring that has a 2.5 cm radius. To the nearest tenth, what is the length of the chord?

**20. REASONING** A circular garden has paths around its edge that are identified by the given arc measures. It also has four straight paths, identified by segments $\overline{AC}$, $\overline{AD}$, $\overline{BE}$, and $\overline{DE}$, that cut through the garden's interior. Which two straight paths have the same length?

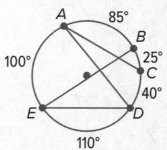

**ARGUMENTS** **Write a two-column proof of the indicated part of Theorem 10.6.**

**21.** In a circle, if two chords are equidistant from the center, then they are congruent.

**22.** In a circle, if two chords are congruent, then they are equidistant from the center.

**23. WRITE** Neil wants to find the center of a large circle. He draws what he thinks is a diameter of the circle and then marks its midpoint and declares that he has found the center. His teacher asks Neil how he knows that the line he drew is the diameter of the circle and not a smaller chord. Neil realizes that he does not know for sure. What can Neil do to determine if it is an actual diameter?

**24. PERSEVERE** Miranda is following directions for a quilt pattern. The directions are intended to make a rectangle. They are "In a 10-inch diameter circle, measure 3 inches from the center of the circle and mark a chord $\overline{AB}$ perpendicular to the radius of the circle. Then cut along the chord." Miranda is to repeat this for another chord, $\overline{CD}$. Finally, she is to cut along chord $\overline{DB}$ and $\overline{AC}$. The result should be four curved pieces and one quadrilateral.

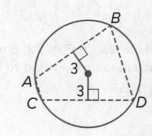

   **a.** If Miranda follows the directions, why might the resulting quadrilateral not be a rectangle? Explain you to adjust the directions.

   **b.** Assume the resulting quadrilateral is a rectangle. One of the curved pieces has an arc measure of 74°. What are the measures of the arcs on the other three curved pieces?

**25. CREATE** Construct a circle and draw a chord. Measure the chord and the distance that the chord is from the center. Find the length of the radius.

**26. ANALYZE** In a circle, $\overline{AB}$ is a diameter and $\overline{HG}$ is a chord that intersects $\overline{AB}$ at point X. Is it *sometimes*, *always*, or *never* true that $HX = GX$? Explain.

# Inscribed Angles

## Explore Angles Inscribed in Circles

 **Online Activity** Use graphing technology to complete the Explore.

> **INQUIRY** What is the relationship between an inscribed angle and the arc it intercepts?

## Learn Inscribed Angles

An **inscribed angle** has its vertex on a circle and sides that contain chords of the circle. In ⊙C, ∠QRS is an inscribed angle. An **intercepted arc** is the part of a circle that lies between the two lines intersecting it. In ⊙C, minor arc $\widehat{QS}$ is intercepted by ∠QRS. There are three ways that an angle can be inscribed in a circle.

| Case 1 | Case 2 | Case 3 |
|---|---|---|
|  |  |  |
| Center P is on the inscribed angle. | Center P is inside the inscribed angle. | The center P is in the exterior of the inscribed angle. |

For each of these cases, the following theorem holds true.

**Theorem 10.7: Inscribed Angle Theorem**

**Words** If an angle is inscribed in a circle, then the measure of the angle equals one half the measure of its intercepted arc.

**Example** $m\angle 1 = \frac{1}{2} m\widehat{AB}$ and $m\widehat{AB} = 2m\angle 1$

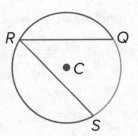

**Theorem 10.8**

**Words** If two inscribed angles of a circle intercept the same arc or congruent arcs, then the angles are congruent.

**Example** ∠B and ∠C both intercept $\widehat{AD}$.

So, ∠B ≅ ∠C.

**Today's Standards**
G.C.2, G.C.3
MP1, MP3, MP5

**Today's Vocabulary**
inscribed angle
intercepted arc

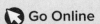 **Go Online**
Proofs of Theorems 10.7 and 10.8 are available.

**Talk About It!**
What is the relationship between an inscribed angle and chords of a circle?

☁ **Think About It!**

If an inscribed angle and a central angle in the same circle intercept the same arc, how are they related?

## Example 1 Use Inscribed Angles to Find Measures

**Find each measure.**

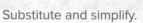

**a.** $m\widehat{CF}$

$m\widehat{CF} = \underline{\quad} \cdot m\angle\underline{\quad}$     Inscribed Angle Theorem

$= 2 \cdot \underline{\quad}$ or $\underline{\quad}°$     Substitute and simplify.

**b.** $m\angle C$

$m\angle C = -\, m\widehat{DE}$     Inscribed Angle Theorem

$= \frac{1}{2}(\underline{\quad})$ or $\underline{\quad}$     Substitute and Simplify

## Example 2 Find Measures of Congruent Inscribed Angles

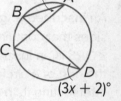

**Find $m\angle A$.**

$\angle A \cong \angle D$     $\angle A$ and $\angle D$ both intercept $\widehat{BC}$.

$m\angle A = \underline{\quad}$     Definition of congruent angles

$5x - 12 = \underline{\quad}$     Substitution

$x = \underline{\quad}$     Solve for $x$.

So, $m\angle A = 5(\underline{\quad}) - 12$ or $\underline{\quad}°$

## Example 3 Use Inscribed Angles in Proofs

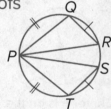

**Write a two-column proof.**

**Given:** $\widehat{QR} \cong \widehat{ST}$; $\widehat{PQ} \cong \widehat{PT}$

**Prove:** $\triangle PQR \cong \triangle PTS$

**Proof:**

| Statements | Reasons |
|---|---|
| 1. $\widehat{QR} \cong \widehat{ST}, \widehat{PQ} \cong \widehat{PT}$ | 1. Given |
| 2. $m\widehat{QR} = m\widehat{ST}, m\widehat{PQ} = m\widehat{PT}$, | 2. _____ |
| 3. $\frac{1}{2}m\widehat{QR} = \frac{1}{2}m\widehat{ST}$, <br> $\frac{1}{2}m\widehat{PQ} = \frac{1}{2}m\widehat{PT}$ | 3. Multiplication Property of Equality |
| 4. $m\angle QPR = \frac{1}{2}m\widehat{QR}, m\angle TPS = \frac{1}{2}m\widehat{ST}, m\angle QRP = \frac{1}{2}m\widehat{PQ}, m\angle TSP = \frac{1}{2}m\widehat{PT}$ | 4. _____ |
| 5. $m\angle QPR = \frac{1}{2}m\widehat{ST}, m\angle QRP = \frac{1}{2}m\widehat{PT}$ | 5. Substitution (Steps 3, 4) |
| 6. $m\angle QPR = m\angle TPS, m\angle QRP = m\angle TSP$, | 6. Substitution (Steps 4, 5) |
| 7. $\angle QPR \cong \angle TPS, \angle QRP \cong \angle TSP$ | 7. Definition of congruent angles |
| 8. $\overline{QR} \cong \overline{ST}$ | 8. $\cong$ arcs have $\cong$ chords. |
| 9. $\triangle PQR \cong \triangle PTS$ | 9. _____ |

🔊 **Go Online** You can complete an Extra Example online.

## Learn Inscribed Polygons

In an **inscribed polygon**, all of the vertices of the polygon lie on a circle. Inscribed triangles and quadrilaterals have special properties.

| Theorem 10.9 | |
|---|---|
| Words | An inscribed angle of a triangle intercepts a diameter or semicircle if and only if the angle is a right angle. |
| Example | If $\overset{\frown}{FJH}$ is a semicircle, then $m\angle G = 90°$. If $m\angle G = 90°$, then $\overset{\frown}{FJH}$ is a semicircle and $\overline{FH}$ is a diameter. 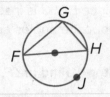 |

While many different types of triangles, including right triangles, can be inscribed in a circle, only certain quadrilaterals can be inscribed in a circle.

| Theorem 10.10 | |
|---|---|
| Words | If a quadrilateral is inscribed in a circle, then its opposite angles are supplementary. |
| Example | If quadrilateral KLMN is inscribed in ⊙A, then ∠L and ∠N are supplementary and ∠K and ∠M are supplementary.  |

**Go Online** Proofs of Theorems 10.9 and 10.10 are available.

## Example 4 Find Angle Measures in Inscribed Triangles

**Find $m\angle K$.**

$\triangle KLM$ is a right triangle because $\angle L$ inscribes a semicircle.

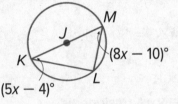

| | |
|---|---|
| $m\angle K + m\angle M = \underline{\quad}$ | Acute ∠s of a right △ are complementary. |
| $5x - 4 + 8x - 10 = \underline{\quad}$ | Substitution |
| $\underline{\quad}x - \underline{\quad} = \underline{\quad}$ | Simplify. |
| $13x = \underline{\quad}$ | Add $\underline{\quad}$ to each side. |
| $x = \underline{\quad}$ | Divide each side by 13. |

So, $m\angle K = 5(8) - 4$ or $\underline{\quad}°$.

## Check

Find $m\angle V$.

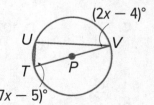

$m\angle V = \underline{\quad}°$

**Go Online** You can complete an Extra Example online.

**Think About It!**

What special quadrilaterals can be inscribed in a circle? Justify your argument.

**Go Online**

You can watch a video to see how to solve problems involving inscribed triangles.

**Think About It!**

If a 45° −45° −90° triangle inscribes a semicircle, what do you know about the minor arcs formed by the vertex of the inscribed angle on the circle? Justify your argument.

### 🌐 Example 5 Find Angle Measures

**STAINED GLASS** Luca is creating a collection of stained glass ornaments. The ornament shown uses a quadrilateral inscribed in a circle. Find m∠F and m∠G.

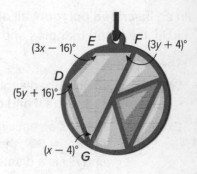

Since *DEFG* is inscribed in a circle, its opposite angles are supplementary.

**Step 1 Find m∠F.**

| | |
|---|---|
| $m\angle D + m\angle F = 180$ | Definition of supplementary |
| $5y + 16 + 3y + 4 = 180$ | Substitution |
| $8y + \underline{\quad} = 180$ | Simplify. |
| $8y = \underline{\quad}$ | Subtract 20 from each side. |
| $y = \underline{\quad}$ | Divide each side by 8. |

So, m∠F = 3(20) + 4 or _____°.

**Step 2 Find m∠G.**

| | |
|---|---|
| $m\angle G + m\angle E = 180$ | Definition of supplementary |
| $x - 4 + 3x - 16 = 180$ | Substitution |
| $\underline{\quad} - \underline{\quad} = 180$ | Simplify. |
| $\underline{\quad}x = \underline{\quad}$ | Add 20 to each side. |
| $x = \underline{\quad}$ | Divide each side by 4. |

So, m∠G = 50 − 4 or _____°.

### Check

**JEWELRY** A designer is making a new line of jewelry with geometric patterns. The ring shown had a quadrilateral gemstone inscribed in a circular piece of metal.

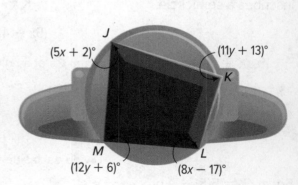

**Part A**

Find m∠J. _____

**A.** 15°          **B.** 42°          **C.** 77°          **D.** 103°

**Part B**

Find m∠K. _____

**A.** 7°          **B.** 46°          **C.** 64°          **D.** 90°

🧭 **Go Online** You can complete an Extra Example online.

# Practice

Go Online You can complete your homework online.

## Example 1
**Find each measure.**

**1.** $m\widehat{AC}$

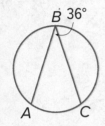

**2.** $m\angle N$

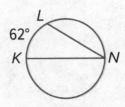

**3.** $m\widehat{QSR}$

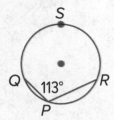

**4.** $m\widehat{XY}$

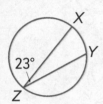

**5.** $m\angle E$

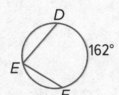

**6.** $m\angle R$

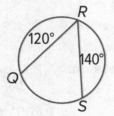

## Example 2
**Find each measure.**

**7.** $m\angle N$

**8.** $m\angle L$

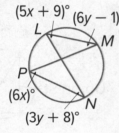

**9.** $m\angle C$

**10.** $m\angle A$

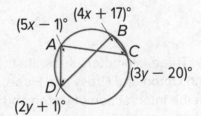

**11.** $m\angle J$

**12.** $m\angle K$

**13.** $m\angle S$

**14.** $m\angle R$

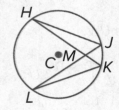

## Example 3
**ARGUMENTS Write the specified type of proof.**

**15.** paragraph proof

Given: $m\angle T = \frac{1}{2} m\angle S$

Prove: $m\widehat{TUR} = 2m\widehat{URS}$

**16.** two-column proof

Given: $\odot C$

Prove: $\triangle KML \sim \triangle JMH$

**Example 4**

**Find each value.**

**17.** x

**18.** m∠W

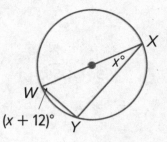
$(x + 12)°$

**19.** x

**20.** m∠T

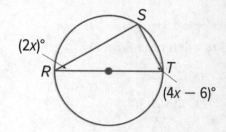
$(2x)°$
$(4x - 6)°$

**21.** m∠J

**22.** m∠K

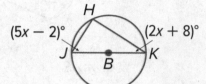
$(5x - 2)°$ H $(2x + 8)°$

**23.** m∠A

**24.** m∠C

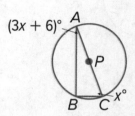
$(3x + 6)°$ A
$x°$

**Example 5**

**Find each measure.**

**25.** m∠R

**26.** m∠S

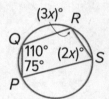
$(3x)°$ R
110° $(2x)°$ S
75°

**27.** m∠W

**28.** m∠X

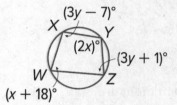
$(3y - 7)°$
$(2x)°$
$(3y + 1)°$
$(x + 18)°$

**29. STREETS** Three kilometers separate the intersections of Cross and Upton and Cross and Hope. What is the distance between the intersection of Upton and Hope and the point midway between the intersections of Upton and Cross and Cross and Hope?

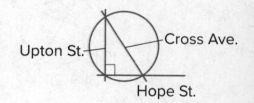
Upton St. — Cross Ave.
Hope St.

**Mixed Exercises**

**Find each measure.**

**30.** m$\widehat{CE}$

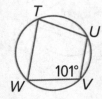
C 62°
D
B
20° E

**31.** m$\widehat{JM}$

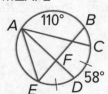
M
J 70° L
N
K 138°

**32.** m∠QPR

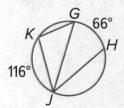
O Q
P
48°
R

**33.** m$\widehat{UW}$

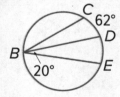
T
U
101°
W V

**34.** m∠AFE

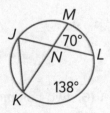
110° B
A
C
F
58°
E D

**35.** m∠KJH

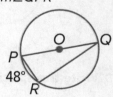
G 66°
K
H
116°
J

**590** **Module 10 ·** Circles

**Find each measure.**

**36.** $m\angle S$

**37.** $m\angle R$

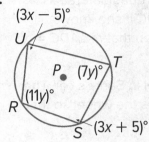

**38.** $m\angle G$

**39.** $m\angle H$

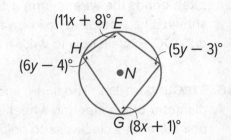

**40. LOGOS** Kendrick is designing a logo for a company that makes equipment for windsurfing, as shown. He knows that the diameter of the circle is 12 centimeters and that $\overarc{BE} \cong \overarc{DE}$. He also knows that $m\overarc{BD} = 136$. What is the length of $\overline{AB}$ to the nearest tenth of a centimeter?

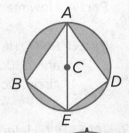

**41. ARGUMENTS** Write a paragraph proof to prove that if $PQRS$ is an inscribed quadrilateral and $\angle Q \cong \angle S$, then $\overline{PR}$ is a diameter of the circle.

**42. ARENA** A concert arena is lit by five lights equally spaced around the perimeter. What is $m\angle 1$?

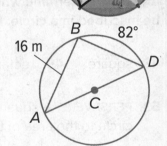

**43. LANDSCAPING** A landscaping crew is installing a circular garden with three paths that form a triangle. The figure shows the plan for the garden and the paths. The leader of the crew wants to determine the radius they should use when they install the circular garden.

**a.** Find the measures of the angles in $\triangle ABD$.

**b. REASONING** Use trigonometry to find the radius of the circular garden. Round to the nearest tenth meter, if necessary.

**c. FIND THE ERROR** A classmate solved the problem and found that the radius of the circular garden was 7.1 meters. Without doing any calculations, how could you convince the classmate that this answer is not reasonable?

**d.** Given that the path will go along the outside of the garden (along the circle), as well as along the three sides of the triangle, find the length of path that will need to be installed.

**44. USE TOOLS** Use dynamic geometry software.

**a.** Construct a circle, and plot four points $J$, $K$, $L$, and $M$ on the circle. Connect the points so that you have an inscribed quadrilateral Drag the points around the circle until you form a parallelogram.

**b. ARGUMENTS** Make a conjecture based on what you notice. Confirm your conjecture by measuring the sides and angles of the quadrilateral.

**c.** Can you construct a counterexample to your conjecture in **part b**? Explain.

**45. JEWELRY** Alyssa makes earrings by bending wire into various shapes. She often bends the wire to form a circle with an inscribed quadrilateral, as shown. She would like to know how she can find $m\overset{\frown}{ADC}$ if she knows $m\angle ADC$. Explain how to write a formula for $m\overset{\frown}{ADC}$ given that $m\angle ADC = x$.

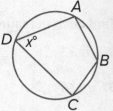

**46. STEERING WHEEL** The figure shows the steering wheel in Sarita's car. The diameter of the steering wheel is 16 inches. Sarita turns the wheel so that point $J$ rotates clockwise to point $K$.

**Part A** How many inches around the circle does point $J$ travel?

**Part B  FIND THE ERROR** Ari said that if a quadrilateral has two 50° angle and a 100° angle, then the quadrilateral cannot be inscribed in a circle. Do you agree? Explain why or why not.

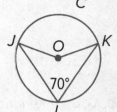

**47. TAPESTRY** Helga is creating a design for a tapestry she is making for her art class. The design shown uses a kite inscribed in a circle. Find $m\angle SRU$ and find $x$.

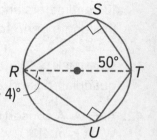

**ANALYZE** Determine whether the quadrilateral can *always*, *sometimes*, or *never* be inscribed in a circle. Explain your reasoning.

**48.** square    **49.** rectangle    **50.** parallelogram    **51.** rhombus    **52.** kite

**53. PERSEVERE** A square is inscribed in a circle. What is the ratio of the area of the circle to the area of the square?

**54. WRITE** a 45° −45° −90° right triangle is inscribed in a circle. If the radius of the circle is given, explain how to find the lengths of the right triangle's legs.

**55. CREATE** Find and sketch a real-world logo with an inscribed polygon.

**56. WRITE** Compare and contrast inscribed angles and central angles of a circle. If they intercept the same arc, how are they related?

**ANALYZE** Determine whether each statement is *always*, *sometimes*, or *never* true. Explain.

**57.** If $\overset{\frown}{PQR}$ is a major arc of a circle, then $\angle PQR$ is obtuse.

**58.** If $\overline{AB}$ is a diameter of circle $O$, and $X$ is any point on circle $O$ other than $A$ or $B$, then $\triangle AXB$ is a right triangle.

**59.** When an equilateral triangle is inscribed in a circle it partitions the circle into three minor arcs that each measure 120°.

# Tangents

## Learn Tangents

A **tangent to a circle** is a line or segment in the plane of a circle that intersects the circle in exactly one point and does not contain any points in the interior of the circle. A tangent can also be a line or segment that intersects a sphere in exactly one point. For a line that intersects a circle in one point, the **point of tangency** is the point at which they intersect.

A **common tangent** is a line or segment that is tangent to two circles in the same plane.

| Theorems: Tangents | |
|---|---|
| 10.11 | In a plane, a line is tangent to a circle if an only if it is perpendicular to a radius drawn to the point of tangency. |
| 10.12: Tangent to a Circle Theorem | If two segments from the same exterior point are tangent to a circle, then they are congruent. |

## Example 1 Identify Common Tangents

**Identify the number of common tangents that exist between each pair of *circles*. If no common tangent exists, state *no common tangent*.**

a.

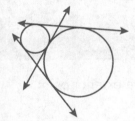

b.

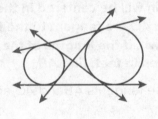

## Example 2 Identify a Tangent

$\overline{AB}$ is a radius of ⊙A. Determine whether $\overline{BC}$ is tangent to ⊙A. Justify your answer.

Test to see if △ABC is a right triangle.

$9^2 + 12^2 \overset{?}{=} (9 + 6)^2$    Pythagorean Theorem

$81 + \underline{\quad} \overset{?}{=} \underline{\quad}^2$    Multiply and add.

$\underline{\quad} = \underline{\quad}$    Simplify.

△ABC is a right triangle with right ∠ABC. So, segment ___ is perpendicular to radius $\overline{AB}$ at point ___. Therefore, by Theorem 10.11, $\overline{BC}$ is tangent to ⊙A.

 **Go Online** You can complete an Extra Example online.

**Today's Standards**
G.C.4, G.CO.13
MP1, MP3, MP6

**Today's Vocabulary**
tangent to a circle
point of tangency
common tangent
circumscribed angle
circumscribed polygon

**Go Online**
You may want to complete the Concept Check to check your understanding.

**Talk About It!**
What is the relationship between a circle and its tangent? When will two circles not have a common tangent? Justify your argument.

**Problem-Solving Tip**
**Solve a Simpler Problem**
You can *solve a simpler problem* by sketching and labeling the right triangle in the example above without the circle. A drawing of the triangle is shown below.

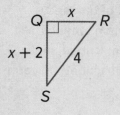

## Example 3 Use a Tangent to Find Missing Values

$\overline{QS}$ **is tangent to $\odot R$ at Q. Find the value of x.**

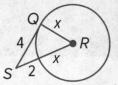

By Theorem 10.11, $\overline{RQ} \perp \overline{QS}$. So, $\triangle RQS$ is a right triangle.

| | |
|---|---|
| $RS^2 + QS^2 = \underline{\quad}^2$ | Pythagorean Theorem |
| $x^2 + \underline{\quad}^2 = (\underline{\quad} + 2)^2$ | $RQ = x$, $QS = 4$, and $RS = \underline{\quad} + 2$ |
| $x^2 + \underline{\quad} = x^2 + \underline{\quad} + \underline{\quad}$ | Multiply. |
| $12 = \underline{\quad}$ | Simplify. |
| $\underline{\quad} = x$ | Divide each side by 4. |

### Check

$\overline{BC}$ is tangent to $\odot A$ at C. Find the value of $x$ to the nearest hundredth.

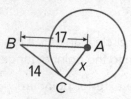

$x = \underline{\quad}$

---

## 🌐 Example 4 Use Congruent Tangents to Find Measures

**PHOTOGRAPHER** A photographer wants to take a picture of a local fountain. She positions herself at point *A* so that the fountain will be centered in the picture. $\overline{AB}$ and $\overline{AC}$ are tangent to the fountain as shown. If the lengths of the tangents are given in feet, find *AB*.

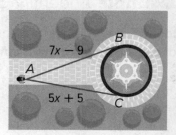

Because tangents $\overline{AB}$ and $\overline{AC}$ are from the same exterior point *A*, $\overline{AB} \cong \overline{AC}$.

| | |
|---|---|
| $AB = AC$ | Definition of congruent segments |
| $\underline{\quad} - \underline{\quad} = 5x + \underline{\quad}$ | Substitution |
| $\underline{\quad} - 9 = \underline{\quad}$ | Subtract 5x from each side. |
| $2x = \underline{\quad}$ | Add 9 to each side. |
| $x = \underline{\quad}$ | Divide each side by 2. |

So, $AB = 7(\underline{\quad}) - 9$ or $\underline{\quad}$ feet.

### Check

**LANDSCAPING** A landscape designer is creating a tiled patio with a circular design pattern. A corner of the patio is shown. $\overline{DE}$ and $\overline{FE}$ are tangent to $\odot G$ as shown, and the lengths of the tangents are given in feet. Find *DE*. $DE = \underline{\quad}$ ft

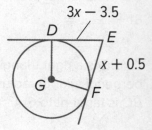

🐾 **Go Online** You can complete an Extra Example online.

## Explore Tangents and Circumscribed Angles

**Online Activity** Use graphing technology to complete the Explore.

> ⊗
>
> @ **INQUIRY** What is the relationship between circumscribed angles, tangents, and the radii of a circle?

## Learn Circumscribed Angles

A **circumscribed angle** is an angle with sides that are tangent to a circle.

| Theorem 10.13 | |
|---|---|
| **Words** | If two segments or lines are tangent to a circle, then the circumscribed angle and the central angle that intercept the arc formed by the points of tangency are supplementary. |
| **Example** | If $\overline{QS}$ and $\overline{RS}$ are tangent to ⊙P, then $m\angle P + m\angle S = 180°$. 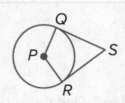 |

**Go Online** A proof of Theorem 10.13 is available.

A **circumscribed polygon** has vertices outside the circle and sides that are tangent to the circle.

| Circumscribed Polygons | Polygons Not Circumscribed |
|---|---|
|  |  |

## Example 5 Use Circumscribed Angles to Find Measures

If $m\angle EGF = 19x + 9$ and $m\angle D = 10x - 3$, find $m\angle D$.

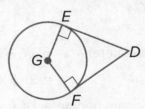

Because $\overline{ED}$ and $\overline{FD}$ are tangent to circle $G$, $\angle EGF$ and $\angle D$ are

_____.

| | |
|---|---|
| $m\angle EGF + m\angle D = 180$ | Definition of _____ angles |
| $19x + 9 + \underline{\ \ }x - \underline{\ \ } = 180$ | Substitution |
| $\underline{\ \ }x + 6 = \underline{\ \ }$ | Simplify. |
| $\underline{\ \ }x = 174$ | Subtract ___ from each side. |
| $x = \underline{\ \ }$ | Divide each side by ___. |

So, $m\angle D = 10(6) - 3$ or ___°.

**Go Online** You can complete an Extra Example online.

### Watch Out!

**Identifying Circumscribed Polygons** Just because a circle is tangent to one or more of the sides of a polygon does not mean that the polygon is circumscribed about the circle, as shown in the second set of figures above.

### 💭 Think About It!

If you do not remember Theorem 10.13, how can you use logic to determine the relationship between a central angle and a circumscribed angle that intercept the same arc?

## Check

If $m\angle XWZ = 7x + 10$ and $m\angle Y = 4x + 5$, find $m\angle Y$.

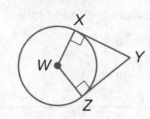

$m\angle Y = \underline{\quad}°$

---

## Example 6 Find Measures in Circumscribed Polygons

$\triangle JKL$ is circumscribed about $\odot Q$.
Find the perimeter of $\triangle JKL$.

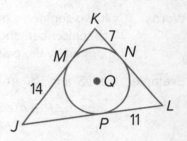

**Step 1 Find the missing measures.**

Because $\triangle JKL$ is circumscribed about $\odot Q$, $\overline{JM}$ and $\overline{JP}$ are tangent to $\odot Q$, as are $\overline{KM}$, $\overline{KN}$, $\overline{LN}$, and $\overline{LP}$. Therefore, $\overline{JM} \cong \overline{JP}$, $\overline{KM} \cong \overline{KN}$, and $\overline{LN} \cong \overline{LP}$.

So, $JM = JP = 14$ feet, $\underline{\quad} = KN = \underline{\quad}$ feet, and $\underline{\quad} = LP = \underline{\quad}$ feet.

**Step 2 Find the perimeter of $\triangle JKL$.**

By Segment Addition, $JK = JM + KM = 14 + 7$ or $\underline{\quad}$ units,

$KL = KN + \underline{\quad} = 7 + \underline{\quad}$ or $\underline{\quad}$ units, and
$JL = \underline{\quad} + LP = \underline{\quad} + 11$ or $\underline{\quad}$ units.

perimeter $\quad= JK + KL + JL$

$\qquad\qquad = 21 + \underline{\quad} + 25$ or $\underline{\quad}$ units

So, the perimeter of $\triangle JKL$ is $\underline{\quad}$ units.

## Check

Quadrilateral $RSTU$ is circumscribed about $\odot J$.
If the perimeter is 18 units, find $x$.

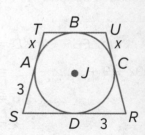

**A.** 1.5 units    **B.** 2 units    **C.** 3 units    **D.** 6 units

> **Go Online** You can complete an Extra Example online.

---

💭 **Think About It!**
How can you check
your answer?

▶ **Go Online**
You can watch a video
to see how to solve
problems involving
circumscribed triangles.

▶ **Go Online**
You may want to
complete the
construction activities
for this lesson.

# Practice

### Example 1

**Identify the number of common tangents that exist between each pair of circles. If no common tangent exists, state *no common tangent*.**

**1.**

**2.**

**3.**

**4.**

### Example 2

**Determine whether each segment is tangent to the given circle. Justify your answer.**

**5.** $\overline{HI}$

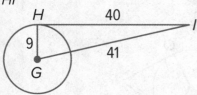

**6.** $\overline{AB}$

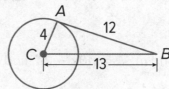

**7.** $\overline{MP}$

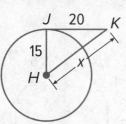

**8.** $\overline{QR}$

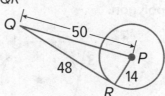

### Example 3

**Find the value of *x*. Assume that segments that appear to be tangent are tangent.**

**9.**

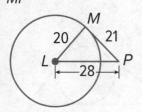

**10.**

**11.**

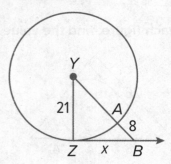

**12.**

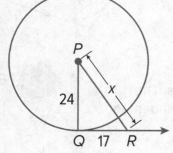

**13.**

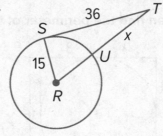

**14.**

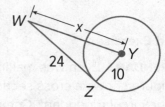

**Example 4**

**Find the value of *x*. Assume that segments that appear to be tangent are tangent.**

**15.**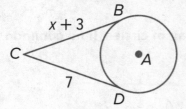

**16.**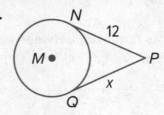

**17.**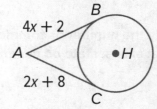

**18.** **CARNIVAL GAMES** Phoebe is playing a carnival game that involves tossing softballs into a wooden basket. She positions herself at point *A* so that the segments $\overline{AB}$ and $\overline{AC}$ are tangent to the basket as shown. If the lengths of the tangents are given in feet, find *AB*.

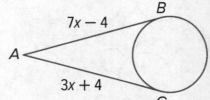

**Example 5**

**19.** If $m\angle BDC = 12x°$ and $m\angle D = (4x + 4)°$, find $m\angle D$.

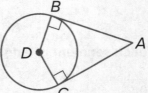

**20.** If $m\angle QPS = (15x + 8)°$ and $m\angle R = (10x - 3)°$, find $m\angle R$.

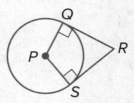

**Example 6**

**Find the perimeter of each polygon.**

**21.**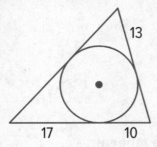

**22.**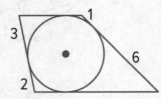

**23.**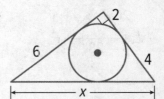

**For each figure, find the value of *x*. Then find the perimeter of the polygon.**

**24.**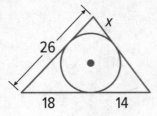

**25.**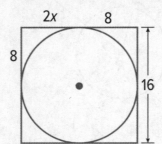

**26.**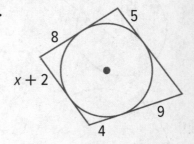

**Mixed Exercises**

**27.** **PENDANT** Evan was making a pendant for his project in metals class. The figure shows a cross section of the pendant. Both $\overline{AV}$ and $\overline{BV}$ and tangent to the circular pendant. Compare the lengths of $\overline{AV}$ and $\overline{BV}$.

Name _____ Period _____ Date _____

**28. CLOCKS** The design shown in the figure is that of a circular clock face inscribed in a triangular base. *AF* and *FC* are equal.

a. Find *AB*.

b. Find the perimeter of the clock.

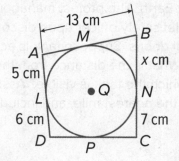

**29. JEWELRY** Juanita is designing a pendant with a circular gem inscribed in a triangle.

a. Find the values of *x*, *y*, and *z*.

b. Find the perimeter of the triangle.

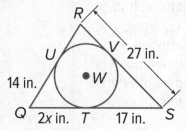

**REASONING Find the value of *x*. Then find the perimeter.**

**30.**

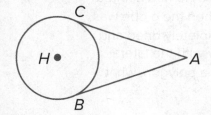

**31.**

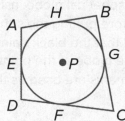

**ARGUMENTS Write the specified type of proof.**

**32.** two-column proof of Theorem 10.12

**Given:** $\overline{AC}$ is tangent to ⊙*H* at *C*.

$\overline{AB}$ is tangent to ⊙*H* at *B*.

**Prove:** $\overline{AC} \cong \overline{AB}$

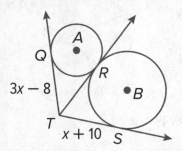

**33.** two-column proof

**Given:** Quadrilateral ABCD is circumscribed about ⊙*P*.

**Prove:** $AB + CD = AD + BC$

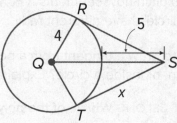

**PRECISION Find *x* to the nearest hundredth. Assume that segments that appear to be tangent are tangent.**

**34.**

**35.**

**36. DESIGN** Amanda wants to make this design of circles inside an equilateral triangle.

a. What is the radius of the large circle to the nearest hundredth inch?

b. What are the radii of the smaller circles to the nearest hundredth inch?

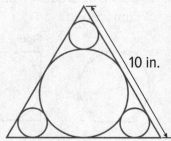

**37. ROLLING** A wheel is rolling down an incline as shown in the figure at the right. Twelve evenly spaced radii form spokes of the wheel. When spoke 2 is vertical, which spoke will be perpendicular to the incline?

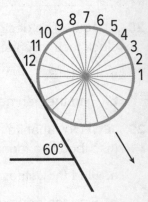

**38. USE TOOLS** Construct a line tangent to a circle through a point on the circle.

Use a compass to draw ⊙A. Choose a point P on the circle and draw $\overleftrightarrow{AP}$. Then construct a segment through point P perpendicular to $\overleftrightarrow{AP}$. Label the tangent line t. Explain and justify each step.

**39. MODELING** NASA has procedures for limiting orbital debris to mitigate the risk to human life and space missions. *Orbital debris* refers to materials from space missions that still orbit Earth. The project manager of a spacecraft in orbit must promptly notify the chief safety officer upon discovering the spacecraft may have generated orbital debris. Suppose tank is accidentally discarded at an altitude of 435 miles. What is the distance from the tank to the farthest point on Earth's surface from which the tank is visible? Assume that the radius of Earth is 4000 miles. Round to the nearest mile, and include a diagram of this situation with your answer.

**40. PERSEVERE** $\overline{PQ}$ is tangent to circles R and S. Find PQ. Explain your reasoning.

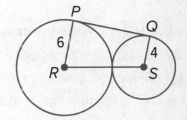

**41. PACKAGING** Taylor packed a spherical globe inside a cubic box. He had painted the sides of the box black before putting the sphere inside. When the globe was later removed, he discovered that the black paint had not completely dried and there were black marks on the globe at the points of tangency with the lateral sides of the box. If the black marks are used as the vertices of a polygon, what kind of polygon results?

**42. CREATE** Draw a circumscribed triangle and an inscribed triangle.

**43. ANALYZE** In the figure, $\overline{XY}$ and $\overline{XZ}$ are tangent to ⊙A. $\overline{XZ}$ and $\overline{XW}$ are tangent to ⊙B. Explain how segments $\overline{XY}$, $\overline{XZ}$, and $\overline{XW}$ can all be congruent if the circles have different radii.

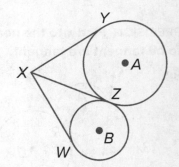

**44. WRITE** Is it possible to draw a tangent from a point that is located anywhere outside, on, or inside a circle? Explain.

**45. WHICH ONE DOESN'T BELONG** Which of the polygons shown below is not circumscribed?

Figure A        Figure B        Figure C        Figure D

 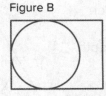 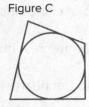 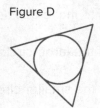

# Tangents, Secants, and Angle Measures

## Explore Relationships Between Tangents and Secants

🔗 **Online Activity** Use graphing technology to complete the Explore.

> ⊗
> @ **INQUIRY** How can you calculate the measure
> of an angle formed when two secants
> intersect or a tangent and a secant intersect?

**Today's Standards**
G.C.2
MP3, MP4, MP7

**Today's Vocabulary**
secant

## Learn Tangents, Secants, and Angle Measures

A **secant** is any line or ray that intersects a circle in
exactly two points. Lines $j$ and $k$ are secants of $\odot C$.

When two secants intersect inside a circle, the
angles formed are related to the arcs they
intercept.

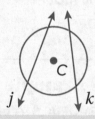

🔗 **Go Online** You
may want to complete
the Concept Check to
check your
understanding.

**Study Tip**
**Absolute Value** In
Theorem 10.16, the
measure of each $\angle A$
can also be expressed
as half the absolute
value of the difference
of the arc measures. In
this way, the order of
the arc measures does
not affect the outcome
of the calculation.

| Theorems | |
|---|---|
| 10.14 | If two secants or chords intersect in the interior of a circle, then the measure of each angle formed is one half the sum of the measures of the arcs intercepted by the angle and its vertical angle. |
| 10.15 | If a secant and a tangent intersect at the point of tangency, then the measure of each angle formed is one half the measure of its intercepted arc. |
| 10.16 | If two secants, a secant and a tangent, or two tangents intersect in the exterior of a circle, then the measure of the angle formed is one half the difference of the measures of the intercepted arcs. |

🔗 **Go Online** Proofs are available for Theorems 10.14 through 10.16.

## Example 1 Intersecting Chords or Secants

**Find the value of x.**
**Step 1 Find $m\angle MKP$.**

$m\angle MKP = \frac{1}{2}(m\widehat{MP} + m\widehat{NQ})$     Theorem 10.14

$= \frac{1}{2}(\underline{\hspace{1cm}} + \underline{\hspace{1cm}})$     Substitution

$= \frac{1}{2}(\underline{\hspace{1cm}})$ or $\underline{\hspace{1cm}}°$     Simplify.

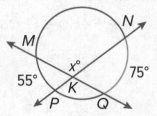

**Step 2 Find the value of x, the measure of $\angle MKN$.**

$\angle MKP$ and $\angle MKN$ are $\underline{\hspace{2cm}}$.

So, $x = \underline{\hspace{1cm}} - 65$ or $\underline{\hspace{1cm}}$.

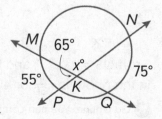

🔗 **Go Online** You can complete an Extra Example online.

## Check
**Find the value of x.**

$x =$ _____

---

## Example 2 Secants and Tangents Intersecting on a Circle

**Find $m\widehat{JLK}$.**

$m\angle HJK = \frac{1}{2}m\widehat{JLK}$      Theorem 10.15

_____ $= \frac{1}{2}m\widehat{JLK}$      Substitution

_____ $= m\widehat{JLK}$      Multiply each side by 2.

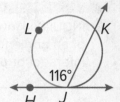

## Check
**Find $m\widehat{DEF}$.**

$m\widehat{DEF} =$ _____ °

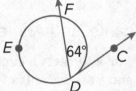

---

### Talk About It!
What is the difference between a tangent and a secant?

## 🌐 Example 3 Tangents Intersecting Outside a Circle

**MEMORIALS A photographer is taking a photo of the Thomas Jefferson Memorial in Washington, D.C., from a boat in the Tidal Basin. The photographer's lines of sight are tangent to the memorial at points Q and S. If the camera's viewing angle measures 36°,**

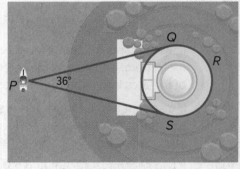

**what portion of the memorial will be visible in the photo?**

Because the Thomas Jefferson Memorial can be modeled by a circle, the arc measure of the memorial is 360°. So, the portion of the memorial

that will be visible in the photo is equal to $\frac{m\widehat{QS}}{360}$.

Let $m\widehat{QS} = x°$. So, $m\widehat{QRS} = 360 - x$.

$m\angle P = \frac{1}{2}m\widehat{QRS} - \widehat{QS}$      Theorem 10.16

_____ $= \frac{1}{2}[(360 -$ _____ $) -$ _____ $]$      Substitution

_____ $= \frac{1}{2}(360 -$ _____ $x)$      Simplify.

_____ $= 360 -$ _____ $x$      Multiply each side by _____.

_____ $=$ _____ $x$      Subtract 360 from each side.

_____ $= x$      Divide each side by _____.

So, $m\widehat{QS} =$ _____ °, and the portion of the memorial that will be visible in the photo is $\frac{144}{360}$ or 40%.

### 🌐 Go Online to practice what you've learned in Lessons 10-1 through 10-6.

🌐 **Go Online** You can complete an Extra Example online.

# Practice

**Go Online** You can complete your homework online.

## Example 1

**REGULARITY Find each value.**

**1.** *x*

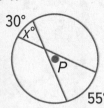

30°
*x*°
*P*
55°

**2.** *m∠1*

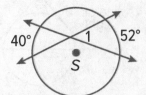

40°  1  52°
*S*

**3.** *m$\widehat{GH}$*

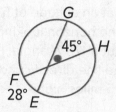

*G*
45°  *H*
*F*
28°  *E*

**4.** *m∠5*

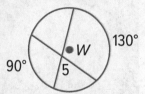

130°
*W*
90°  5

**5.** *m∠1*

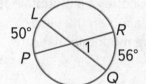

*L*
50°  *R*
*P*  1  56°
*Q*

**6.** *m∠2*

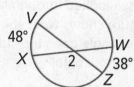

*V*
48°  *W*
*X*  2  38°
*Z*

## Example 2

**Find each value. Assume that segments that appear to be tangent are tangent.**

**7.** *y*

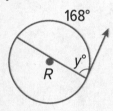

168°
*R*  *y*°

**8.** *m∠3*

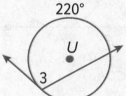

220°
*U*
3

**9.** *m$\widehat{RT}$*

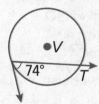

*V*
74°  *T*

**10.** *m∠6*

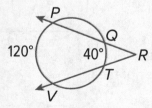

160°  *X*
6

**11.** *m∠3*

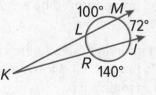

*M*
198°
3
*P*  *Q*

**12.** *m∠4*

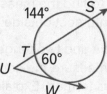

*V*
124°  4  *S*
*R*

## Example 3

**Find each measure. Assume that segments that appear to be tangent are tangent.**

**13.** *m∠R*

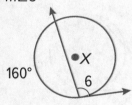

*P*
*Q*
120°  40°  *R*
*T*
*V*

**14.** *m∠K*

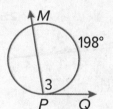

100°  *M*
*L*  72°
*J*
*K*  *R*  140°

**15.** *m∠U*

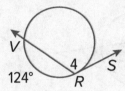

144°  *S*
*T*  60°
*U*
*W*

**16.** *m∠S*

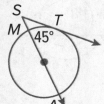

*S*
*M*  *T*
45°

**17.** *m$\widehat{DPA}$*

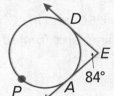

*D*
*E*
84°
*P*  *A*

**18.** *m$\widehat{LJ}$*

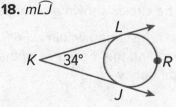

*L*
*K*  34°  *R*
*J*

## Mixed Exercises

19. **REASONING** Francisco places a circular canvas on his A-frame easel and carefully centers it. The apex of the easel is 30° and the measure of arc *BC* is 22°. What is the measure of arc *AB*?

20. **FLIGHT** When flying at an altitude of 5 miles, the lines of sight to the horizon looking north and south make about a 173.7° angle. How much of the longitude line directly under the plane is visible from 5 miles high?

21. **USE TOOLS** Vanessa looked through her telescope at a mountainous landscape. The figure shows what she saw. Based on the view, approximately what angle does the side of the mountain that runs from *A* to *B* with the horizontal?

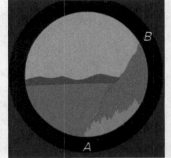

22. **ARGUMENTS** Write a paragraph proof of Theorem 10.15.

   a. **Given:** $\overleftrightarrow{AB}$ is a tangent of ⊙*O*.
      $\overrightarrow{AC}$ is a secant of ⊙*O*.
      ∠*CAE* is acute.
      **Prove:** $m\angle CAE = \frac{1}{2}m\widehat{CA}$

   b. Prove that if ∠*CAB* is obtuse, $m\angle CAB = \frac{1}{2}m\widehat{CDA}$.

**STRUCTURE** Find the value of *x*.

23.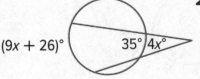
$(9x + 26)°$   $35°$   $4x°$

24.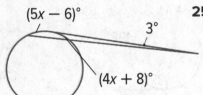
$(5x - 6)°$   $3°$   $(4x + 8)°$

25.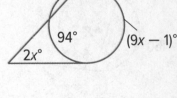
$94°$   $(9x - 1)°$   $2x°$

26. **ANALYZE** Isosceles △*ABC* is inscribed in ⊙*D*. What can you conclude about $m\widehat{AB}$ and $m\widehat{BC}$? Explain.

27. **WRITE** Explain how to find the measure of an angle formed by a secant and a tangent that intersect outside a circle.

28. **CREATE** Draw a circle and two tangents that intersect outside the circle. Use a protractor to measure the angle that is formed. Find the measures of the minor and major arcs formed. Explain your reasoning.

29. **PERSEVERE** The circles shown are concentric. What is *x*?

30. **WRITE** A circle is inscribed within △*PQR*. If $m\angle P = 50°$ and $m\angle Q = 60°$, describe how to find the measures of the three minor arcs formed by the points of tangency.

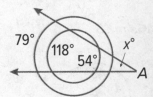

# Equations of Circles

## Explore Deriving Equations of Circles

🧭 **Online Activity** Use guiding exercises to complete the Explore.

> ⊗ **INQUIRY** How can you determine whether a point lies on a circle?

**Today's Standards**
G.GPE.1; G.GPE.4
MP1, MP3, MP6

## Learn Equations of Circles

**Key Concept • Equation of a Circle in Standard Form**

The standard form of the equation of a circle with center at $(h, k)$ and radius $r$ is

$(x - h)^2 + (y - k)^2 = r^2$.

The standard form of the equation of a circle is also called the *center-radius* form.

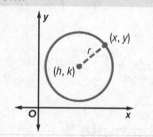

🧭 Watch the online video to see how to write the equation of a circle.

💬 **Talk About It!**

Leonardo says you can also derive the standard form of the equation of a circle by using the Distance Formula. Do you agree or disagree? Justify your reasoning.

## Example 1 Write an Equation Using the Center and Radius

$\overline{AB}$ **is a diameter of the circle. Write the equation of the circle.**

Since $\overline{AB}$ is a diameter of the circle, the center of the circle is the midpoint of $\overline{AB}$.

$M = \left(\dfrac{x_1 + x_2}{2}, \dfrac{y_1 + y_2}{2}\right)$ ⠀⠀Midpoint Formula

$= \left(\dfrac{0 + 8}{2}, \dfrac{-1 + (-1)}{2}\right)$ ⠀⠀Substitution

$= (\underline{\quad}, \underline{\quad})$ ⠀⠀Simplify.

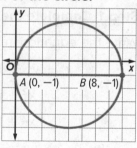

So, the center is at $(\underline{\quad}, \underline{\quad})$ and the radius is 4.

$(x - h)^2 + (y - k)^2 = r^2$ ⠀⠀Equation of a circle

$(x - 4)^2 + [y - (-1)]^2 = 4^2$ ⠀⠀Substitution

$(x - \underline{\quad})^2 + (y + \underline{\quad})^2 = \underline{\quad}$ ⠀⠀Simplify.

## Check

$\overline{LM}$ is a diameter of the circle.

**Part A** Identify the center of the circle and the radius.

center $(\underline{\quad}, \underline{\quad})$; radius $= \underline{\quad}$

**Part B** Write the equation of the circle in standard from. $\underline{\hspace{3cm}}$

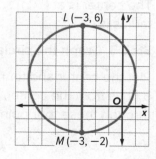

🍄 **Think About It!**

Does the graph of a circle represent a function? Justify your reasoning.

## Example 2 Write an Equation Using the Center and a Point

**Write the equation of the circle with center at (−3, −5) that passes through (0, 0).**

**Step 1 Find the length of the radius.**

Find the distance between the points to determine the radius.

$r = \sqrt{(x_2 - x_1)^2 + (y_2 - y_1)^2}$     Distance Formula

$= \sqrt{[0 - (\underline{\hspace{0.5cm}})]^2 + [0 - (\underline{\hspace{0.5cm}})]^2}$    $(x_1, y_1) = (-3, -5)$ and $(x_2, y_2) = (0, 0)$

$= \sqrt{\underline{\hspace{0.7cm}}}$     Simplify.

**Step 2 Write the equation of the circle.**

Write the equation using $h = -3$, $k = -5$, and $r = \sqrt{34}$.

$$(x - h)^2 + (y - k)^2 = r^2$$     Equation of a circle

$$[x - (\underline{\hspace{0.5cm}})]^2 + [y - (\underline{\hspace{0.5cm}})]^2 = (\sqrt{34})^2$$     Substitution

$$(x + \underline{\hspace{0.5cm}})^2 + (y + \underline{\hspace{0.5cm}})^2 = \underline{\hspace{0.7cm}}$$     Simplify.

## Example 3 Graph a Circle

**The equation of a circle is $x^2 + y^2 + 8x - 14y + 40 = 0$. State the coordinates of the center and the measure of the radius. Then graph the equation.**

**Step 1 Complete the squares.**

Write the equation in standard form by completing the squares.

$$x^2 + y^2 + 8x - 14y + 40 = 0$$     Original equation

$$x^2 + y^2 + 8x - 14y = \underline{\hspace{0.7cm}}$$     Subtract 40 from each side.

$$x^2 + 8x + y^2 - 14y = -40$$     Group terms.

$$x^2 + 8x + \underline{\hspace{0.7cm}} + y^2 - 14y + \underline{\hspace{0.7cm}} = -40 + \underline{\hspace{0.7cm}} + \underline{\hspace{0.7cm}}$$     Complete the squares.

$$(x + \underline{\hspace{0.5cm}})^2 + (y - \underline{\hspace{0.5cm}})^2 = \underline{\hspace{0.7cm}}$$     Factor and simplify.

**Step 2 Identify $h$, $k$, and $r$.**

$h = \underline{\hspace{0.7cm}}$, $k = \underline{\hspace{0.7cm}}$, and $r = \underline{\hspace{0.7cm}}$

The center is at ($\underline{\hspace{0.5cm}}$, $\underline{\hspace{0.5cm}}$), and the radius is $\underline{\hspace{0.5cm}}$.

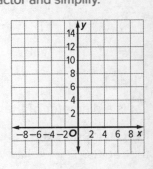

**Step 3 Graph the circle.**

Plot the center and four points that are 5 units from the center. Sketch the circle through these four points.

🍄 **Think About It!**

A circle has the equation $(x - 5)^2 + (y + 7)^2 = 16$. If the center of the circle is shifted 3 units right and 9 units up, what would be the equation of the new circle? Explain your reasoning.

## Example 4 Use a Diameter to Write an Equation

**TRANSPORTATION** The school board is determining the new boundary for Riverdale High School's bus transportation. The high school is at point $H$ and is the center of the circle that represents the new boundary. The students that live on or within the circle will have to walk to school. Students that live at points $J(-5, 2)$, $K(5, 6)$, and $L(5, 2)$ lie on the boundary, and $\overline{JK}$ is a diameter of $\odot H$. Write the equation of $\odot H$ in standard form.

**Understand**

What You Know:

The points $J(\underline{\hspace{1cm}}, 2)$, $K(5, \underline{\hspace{1cm}})$, and $L(\underline{\hspace{1cm}}, \underline{\hspace{1cm}})$ lie on the circle.

$\underline{\hspace{1cm}}$ is a diameter of $\odot H$.

Point $\underline{\hspace{1cm}}$ is the center of the circle.

What You Need to Know: the equation of $\odot H$ in standard form

**Plan**

**Step 1** Find the $\underline{\hspace{1cm}}$ of the circle.

**Step 2** Find the $\underline{\hspace{1cm}}$ of the circle.

**Step 3** Write an equation of the circle in $\underline{\hspace{2cm}}$.

**Solve**

**Step 1** Use the Midpoint Formula to find the coordinates of point $H$ when $J(-5, 2)$ and $K(5, 6)$.

$$H = \left( \frac{x_1 + x_2}{\,}, \frac{y_1 + y_2}{\,} \right) \qquad \text{Midpoint Formula}$$

$$= \left( \frac{-5 + \underline{\;}}{\,}, \frac{2 + \underline{\;}}{\,} \right) \qquad \text{Substitution}$$

$$= (\underline{\hspace{0.7cm}}, \underline{\hspace{0.7cm}}) \qquad \text{Simplify.}$$

**Step 2** To find the radius, use the Distance Formula to find the distance between the center $H(0, 4)$ and point $K(5, 6)$.

$$r = \sqrt{(x_2 - x_1)^2 + (y_2 - y_1)^2} \qquad \text{Distance Formula}$$

$$= \sqrt{(5 - \underline{\;})^2 + (6 - \underline{\;})^2} \qquad \text{Substitution}$$

$$= \sqrt{\underline{\;\;\;} + \underline{\;\;}} \qquad \text{Simplify.}$$

$$= \sqrt{\underline{\;\;}} \qquad \text{Add.}$$

**Step 3** Write the equation of the circle in standard form.

$$(x - h)^2 + (y - k)^2 = r^2 \qquad \text{Equation of a circle.}$$

$$(x - \underline{\;\;\;})^2 + (y - \underline{\;\;\;})^2 = \sqrt{\underline{\;\;}}^2 \qquad (h, k) = (0, 4), r = \sqrt{29}$$

$$(x - \underline{\;\;\;})^2 + (y - \underline{\;\;\;})^2 = \underline{\;\;\;} \qquad \text{Simplify.}$$

😮 **Think About It!**

What is an appropriate unit of measure for the radius of $\odot H$? Justify your reasoning.

## Check

Is your answer reasonable? Explain.

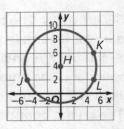

---

## Example 5 Intersections with Circles

Find the point(s) of intersection between $x^2y^2 = 9$ and $y = x - 2$.

**Step 1  Graph the equations on the same coordinate plane.**

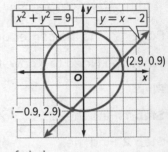

The points of intersection are solutions of both equations, and can be estimated to be at about (_____, _____) and (_____, _____).

**Step 2  Use substitution to find points algebraically.**

$$x^2 + y^2 = 9 \qquad \text{Equation of circle}$$
$$x^2 + (x - 2)^2 = 9 \qquad \text{Substitute}$$
$$x^2 + \underline{\phantom{xx}}^2 - \underline{\phantom{xx}}x + \underline{\phantom{xx}} = 9 \qquad \text{Multiply.}$$
$$\underline{\phantom{xx}}x^2 - \underline{\phantom{xx}}x + \underline{\phantom{xx}} = 9 \qquad \text{Combine like terms.}$$
$$\underline{\phantom{xx}}x^2 - \underline{\phantom{xx}}x - \underline{\phantom{xx}} = 0 \qquad \text{Subtract 9 from each side.}$$

**Step 3  Use the Quadratic Formula.**

Use $\dfrac{-b \pm \sqrt{b^2 - 4ac}}{2a}$ to solve $2x^2 - 4x - 5 = 0$, with $a =$ _____, $b =$ _____, and $c =$ _____.

The solutions are $x = 1 + \dfrac{\sqrt{14}}{2}$ or $x =$ _____

**Step 4  Find the points of intersection.**

Use the equation $y = x - 2$ to find the corresponding $y$-values to be $\left(1 + \dfrac{\sqrt{14}}{2}, -1 + \dfrac{\sqrt{14}}{2}\right)$ and ( _____ , _____ ) or at about ( _____, _____ ) and $(-0.87, -2.87)$.

## Check

Find the point(s) of intersection between $x^2 + y^2 = 8$ and $y = -x$. If there are no points of intersection, select *no intersection points*. _____

A. $(2, -2)$ and $(-2, 2)$

B. $(2\sqrt{2}, -2\sqrt{2})$ and $(-2\sqrt{2}, 2\sqrt{2})$

C. $(2, 2)$ and $(-2, -2)$

D. $(2, -2)$

E. no intersection points.

**Go Online** You can complete an Extra Example online.

**Study Tip:**

**Solving Quadratic Equations** In addition to using the Quadratic Formula, you can also solve quadratic equations by taking square roots, completing the square, and factoring.

**Think About It!**

Will a line always intersect a circle in two points? Justify your reasoning.

Name _____ Period _____ Date _____

# Practice

🔵 **Go Online** You can complete your homework online.

### Examples 1 and 2

**Write the equation of each circle.**

1. center at (0, 0), radius 8

2. center at (−2, 6), diameter 8

3.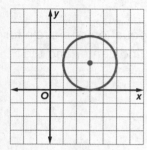

4.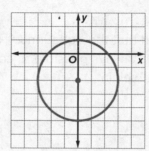

5. center at (3, −4), passes through (−1, −4)

6. center at (0, 3), passes through (2, 0)

7. center at (−4, −1), passes through (−2, 3)

8. center at (5, −2), passes through (4, 0)

### Example 3

**State the coordinates of the center and the measure of the radius of the circle with the given equation. Then graph the equation.**

9. $x^2 + y^2 = 16$

10. $(x - 1)^2 + (y - 4)^2 = 9$

11. $x^2 + y^2 - 4 = 0$

12. $x^2 + y^2 + 6x - 6y + 9 = 0$

### Example 4

13. The Villani family is placing a stake in the ground for which they will tie a dog leash. The proposed location of the stake is at point $Q$ and is the center of the circle that represents the circular boundary in which their dog will be able to roam. The points $R(-4, 1)$ and $S(8, 7)$ lie on the boundary, and $\overline{RS}$ is a diameter of $\odot Q$. Write the equation of $\odot Q$ in standard form.

14. **DELIVERY** A new pizza restaurant is determining the boundary for its delivery service. The pizza restaurant is at point $P$ and is the center of the circle that represents the boundary. Customers who are on or within the circle will be eligible for delivery. Customers who live or work at points $A(-2, 2)$, $B(2, -2)$, and $C(6, 2)$ lie on the boundary, and $\overline{AC}$ is a diameter of $\odot P$. Write the equation of $\odot P$ in standard form.

### Example 5

**Find the point(s) of intersection, if any, between each circle and line with the equations given.**

15. $x^2 + y^2 = 9$; $y = 2x + 3$

16. $(x + 4)^2 + (y - 3)^2 = 25$; $y = x + 2$

17. $(x - 5)^2 + (y - 2)^2 = 100$; $y = x - 1$

18. $x^2 + y^2 = 25$; $y = x$

**Mixed Exercises**

**Write the equation of each circle.**

19. a circle with a diameter having endpoints at (0, 4) and (6, −4)

20. a circle with $d = 22$ and a center translated 13 units left and 6 units up from the origin

21. **REASONING** Adam said that the equation $x^2 + y^2 + 4x − 10y = k$ is the equation of a circle for any value of $k$ since it is always possible to complete the squares to the find center and the radius. Do you agree? Explain.

22. **STRUCTURE** The design of a piece of wallpaper consists of circles that can be modeled by the equation $(x − a)^2 + (y − b)^2 = 4$, for all even integers $b$. Sketch part of the wallpaper on a grid.

23. **REASONING** What is the equation of the circle that is inscribed in the square shown?

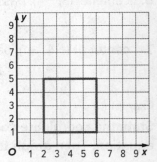

24. **PRECISION** The design for a park is drawn on a coordinate graph. The perimeter of the park is modeled by the equation $(x − 3)^2 + (x − 7)^2 = 225$. Each unit on the graph represents 10 feet. What is the radius of the actual park?

25. **SAFETY RING** A circular safety ring surrounds a fire station. On one map of the fire station grounds, the safety ring is given by the equation $(x − 8)^2 + (y + 2)^2 = 324$. Each unit on the map represents 1 mile. What is the radius of the safety ring?

**Write an equation of a circle that contains each set of points. Then graph the circle.**

26. $A(−2, 3)$, $B(1, 0)$, $C(4, 3)$

27. $F(3, 0)$, $G(5, −2)$, $H(1, −2)$

28. **FIND THE ERROR** An urban planner is designing a new circular road for a housing development. The road will pass through the points $P(−1, 2)$, $Q(5, 2)$, and $R(7, −2)$. The city planner says that the new road will not intersect an existing road that lies along the line $y = 3$. Is the planner correct? Explain.

29. **WRITE** Describe how the equation for a circle changes if the circle is translated $a$ units to the right and $b$ units down.

30. **CREATE** Graph three noncollinear points and connect them to form a triangle. The construct the circle that circumscribes it.

31. **ANALYZE** A circle has center at (2, 3). The point (2, 1) lies on the circle. Find three other points with integer coordinates that lie on the circle. Explain how to do this without finding the equation of the circle.

# Equations of Parabolas

## Explore Focus and Directrix of a Parabola

**Online Activity** Use graphing technology to complete the Explore.

**INQUIRY** How do the focus and directrix affect the shape of a parabola?

## Learn Equations of Parabolas

A **parabola** is the graph of a quadratic function, such as $y = x^2$. Geometrically, a parabola is the set of all points in a plane equidistant from a fixed point, called the **focus**, and a fixed line, called the **directrix**.

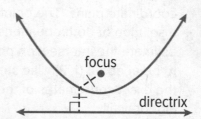

## Example 1 Write an Equation for a Parabola

**Write an equation for the parabola with the focus at (0, −2) and the directrix y = 2.**

**Step 1 Sketch the parabola.**
Graph $F(0, -2)$ and $y = 2$. Sketch a U-shaped curve for the parabola between point $F$ and the directrix as shown. Label a point $P(x, y)$ on the curve.

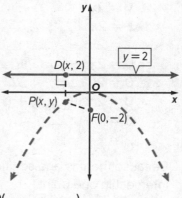

**Step 2 Label point D.**
Label a point $D$ on $y = 2$ such that $\overline{PD}$ is perpendicular to the line $y = 2$. The coordinates of this point must therefore be $D(\underline{\quad}, \underline{\quad})$.

**Step 3 Find PD and PF.**
Use the Distance Formula to find $PD$ and $PF$.

$$PD = \sqrt{(x - \underline{\quad})^2 + (y - \underline{\quad})^2} \qquad P(x, y), D(x, 2)$$
$$= (y - 2)^2 \qquad \text{Simplify.}$$
$$PF = \sqrt{(x - \underline{\quad})^2 + [y - (\underline{\quad})]^2} \qquad P(x, y), D(0, -2)$$
$$= \sqrt{x^2 + (y \underline{\quad})^2} \qquad \text{Simplify.}$$

**Step 4 Write an equation for the parabola.**
Because every point on a parabola is equidistant from the focus and directrix, you can set $PD$ equal to $PF$. Then solve for $y$ to write an equation for the parabola.

$$\sqrt{(y - 2)^2} = \sqrt{x^2 + (y + 2)^2} \qquad PD = PF$$
$$(y - 2)^2 = x^2 + (y + 2)^2 \qquad \text{Square each side.}$$
$$y^2 - \underline{\quad}y + \underline{\quad} = x^2 + y^2 + \underline{\quad}y + \underline{\quad} \qquad \text{Expand } (y - 2)^2 \text{ and } (y + 2)^2$$
$$-\underline{\quad}y = x^2 + \underline{\quad}y \qquad \text{Subtract } y^2 + 4 \text{ from each side.}$$
$$-\underline{\quad}y = x^2 \qquad \text{Subtract } 4y \text{ from each side.}$$
$$y = -x^2 \qquad \text{Multiply each side by } -\tfrac{1}{8}.$$

An equation for the parabola with the focus at (0, −2) and the directrix $y = 2$ is $y = -\frac{1}{8}x^2$.

**Go Online** You can complete an Extra Example online.

---

**Today's Standards**
G.GPE.2
MP4, MP5, MP6

**Today's Vocabulary**
parabola
focus
directrix

**Talk About It!**
How can you find the distance between a line and a point not on the line?

**Watch the online video** to see how to write the equation of a parabola.

## Check

Select the equation of the parabola with the focus at $(0, \frac{1}{2})$ and the directrix $= -\frac{1}{2}$.

**A.** $y = -\frac{1}{2}x^2$    **B.** $y = 2x^2$    **C.** $y = -\frac{1}{2}x^2 - \frac{1}{4}$    **D.** $y = \frac{1}{2}x^2$

---

**Go Online** to see Example 2.

## Example 3 Find Points of Intersection

**Find the points of intersection, if any, between $y = 2x^2$ and $y = 4x - 2$.**

Graph these equations on the same coordinate plane. The point of intersection is a solution of both equations. You can estimate the intersection point on the graph to be at about (1, 2). Use substitution to find the exact coordinates of the point algebraically.

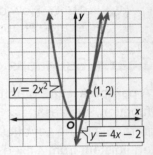

Find the $x$-coordinate of the intersection point.

| | |
|---|---|
| $y = 2x^2$ | Quadratic Equation |
| $4x - 2 = 2x^2$ | Substitute $4x - 2$ for $y$. |
| $-\underline{\phantom{xx}} = 2x^2 - \underline{\phantom{xx}}$ | Subtract $4x$ from each side. |
| $0 = 2x^2 - \underline{\phantom{xx}} + \underline{\phantom{xx}}$ | Add 2 to each side. |
| $0 = 2(\underline{\phantom{xx}}^2 - \underline{\phantom{xx}} + \underline{\phantom{xx}})$ | Factor out the GCF of $2x^2$, $4x$ and 2. |
| $0 = 2(x - \underline{\phantom{xx}}) + (x - \underline{\phantom{xx}})$ | Factor $x^2 - 2x + 1$. |
| $0 = x - \underline{\phantom{xx}}$ | Set the repeated factor equal to zero. |
| $\underline{\phantom{xx}} = x$ | Add 1 to each side. |

Because there is one solution of $4x - 2 = 2x^2$, the parabola and the line intersect in one point.

You can use $y = 4x - 2$ to find the corresponding $y$-value for $x = 1$.

| | |
|---|---|
| $y = 4x - 2$ | Equation of line |
| $y = 4(\underline{\phantom{xx}}) - 2$ | Substitute. |
| $y = \underline{\phantom{xx}}$ | Simplify. |

So, $y = 2x^2$ and $y = 4x - 2$ intersect at $(\underline{\phantom{xx}}, \underline{\phantom{xx}})$.

## Check

Find the point(s) of intersection, if any, between $y = -3x^2$ and $y = 6x$. If there are no points of intersection, select *no intersection points*.

**A.** $(-2, -12)$               **B.** $(0, 0)$ and $(-2, -12)$

**C.** $(2, 12)$                  **D.** no intersection points

**Go Online** You can complete an Extra Example online.

# Practice

**Go Online** You can complete your homework online.

### Example 1

**Find an equation of the parabola with the given focus and directrix.**

**1.** focus $(0, 3)$, directrix $y = -3$

**2.** focus $(8, 0)$, directrix $x = -8$

**3.** focus $(0, -5)$, directrix $y = 5$

**4.** focus $(-11, 0)$, directrix $x = 11$

**5.** focus $(9, 0)$, directrix $x = -9$

**6.** focus $(0, 2)$, directrix $y = -2$

### Example 2

**7. FLASHLIGHTS** The parabolic reflector plate of a flashlight has its bulb located at the focus of the parabola. The distance between the vertex and the focus is 2 centimeters. Write an equation of the cross section of the reflector plate.

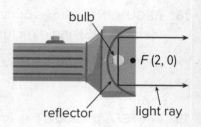

**8. ANTENNAS** A parabolic antenna used at a television station to transmit their signals has a focus that is located the distance shown from the vertex of the parabola. Provided the parabola has a vertical axis of symmetry, write an equation of the cross section of the antenna.

### Example 3

**Find the point(s) of intersection, if any, between each parabola and line with the given equations.**

**9.** $y = x^2$, $y = x + 2$

**10.** $y = 2x^2$, $y = 4x - 2$

**11.** $y = -3x^2$, $y = 6x$

**12.** $y = -(x - 1)^2$, $y = -x$

### Mixed Exercises

**Find an equation of the parabola with the focus and directrix given. Then graph the parabola.**

**13.** focus $(0, -\frac{1}{8})$, directrix $y = \frac{1}{8}$

**14.** Focus $(2, 0)$, directrix $x = 2$

**15. HEADLIGHTS** The mirrored parabolic reflector plate of a car headlight has its bulb located at the focus of the parabola, shown. It has a horizontal axis of symmetry. With a vertex at the origin, the labeled coordinates of the focus, $F$, show the distance between the vertex and the focus. Write an equation of the cross section of the reflector mirror.

16. **SATELLITE DISH** A parabolic satellite dish opens upwards, as shown. The distance between the vertex and the focus is 4.2 feet. Write the equation of the cross section of the dish.

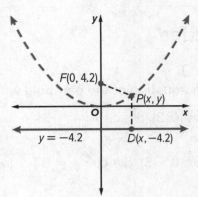

17. **MODELING** The parabolic reflector plate of a flashlight has its bulb located at the focus of the parabola. The distance between the vertex and the focus is 0.9 inch. Provided the parabola has a horizontal axis of symmetry, write an equation of the cross section of the reflector plate.

18. **ARGUMENTS** Prove or disprove that the point $(\sqrt{3}, -4)$ lies on the parabola with focus $(0, -3)$ and directrix $y = 3$.

19. **WRITE** A parabola has focus $\left(-\frac{3}{4}, 0\right)$ and directrix $x = \frac{3}{4}$. Explain how to determine whether the parabola opens upward, downward, left or right.

20. **CREATE** An engineer is using a coordinate plane to design a tunnel in the shape of a parabola. The line shown in the figure represents the top of the wall that will contain the tunnel. This line will be the directrix of the tunnel, and the focus will be point $F$. The base of the tunnel (ground level) is represented by the line $y = -10$.

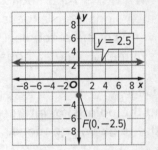

a. Write the equation of the parabola. Then graph the parabola on the coordinate plane.

b. Each unit of the coordinate plane represents one foot. Prove or disprove that the width of the tunnel at a height of 5 feet above the ground is exactly 14 feet.

21. **ANALYZE** Can a parabola have a maximum and a minimum value?

22. **PERSEVERE** The equation of parabola A is $y = \frac{1}{24}x^2$. Parabola B has the same vertex as parabola A and opens in the same direction as parabola A. However, the focus and directrix for parabola B are twice as far apart as they are for parabola A. Write the equation for parabola B. Explain your steps.

## Essential Question

**How can circles and parts of circles be used to model situations in the real world?**

## Module Summary

### Lessons 10-1 and 10-2

Circles, Arcs and Angles

- The circumference, or distance around a circle, $C$ equals diameter times pi ($C = \pi d$).

- All circles are similar. Circles with equal radii are congruent.

- To convert a degree measure to radians, multiply by $\frac{\pi \text{ radians}}{180°}$. To convert a radian measure to degrees, multiply by $\frac{180°}{\pi \text{ radians}}$.

- In the same circle or in congruent circles, two minor arcs are congruent if and only if their corresponding chords are congruent.

### Lessons 10-3 and 10-4

Inscribed Angles, Tangents and Secants

- If an angle is inscribed in a circle, then the measure of the angle equals one half the measure of its intercepted arc.

- If two inscribed angles of a circle intercept the same arc or congruent arcs, then the angles are congruent.

- An inscribed angle of a triangle intercepts a diameter or semicircle if and only if the angle is a right angle.

- If a quadrilateral is inscribed in a circle, then its opposite angles are supplementary.

### Lessons 10-5 and 10-6

Tangents, Secants, and Angle Measures

- If two segments from the same exterior point are tangent to a circle, then they are congruent.

- If two secants or chords intersect in the interior of a circle, then the measure of each angle formed is one half the sum of the measures of the arcs intercepted by the angle and its vertical angle.

- If a secant and a tangent intersect at the point of tangency, then the measure of each angle formed is one half the measure of its intercepted arc.

- If two secants, a secant and a tangent, or two tangents intersect in the exterior of a circle, then the measure of the angle formed is one half the difference of the measures of the intercepted arcs.

### Lessons 10-7 through 10-8

Equations of Circles and Parabolas

- The equation of a circle with center at $(h, k)$ and radius $r$ is $(x - h)^2 + (y - k)^2 = r^2$.

- A parabola is the set of all points in a plane equidistant from a fixed point, called the focus, and a fixed line, called the directrix.

### Study Organizer

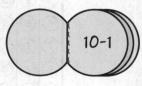

📖 Foldables

Use your Foldable to review this module. Working with a partner can be helpful. Ask for clarification of concepts as needed.

## Test Practice

**1. MULTIPLE CHOICE** Kira is building a fence around the circular field used for barrel racing at the rodeo. If the radius of the field is 45 feet, about how many feet of fencing does she need? (Lesson 10-1)

(A) 71 ft

(B) 90 ft

(C) 142 ft

(D) 283 ft

**2. GRIDDED RESPONSE** The diameter of circle $W$ is 48 centimeters, the diameter of circle $Z$ is 72 centimeters, and $YZ$ is 30 centimeters. What is the length of $\overline{WX}$ in centimeters? (Lesson 10-1)

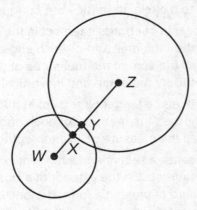

**3. MULTIPLE CHOICE** What is 240° in radians? (Lesson 10-2)

(A) $\frac{8\pi}{3}$

(B) $\frac{4\pi}{3}$

(C) $\frac{8\pi}{9}$

(D) $\frac{2\pi}{3}$

**4. OPEN RESPONSE** $\overline{JK}$ is a diameter of $\odot C$. What is the measure of arc $\widehat{BK}$? (Lesson 10-2)

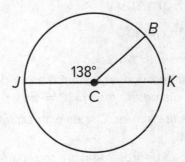

**5. MULTIPLE CHOICE** If the $m\widehat{WXY} = 90°$, then what is the $m\widehat{WX}$? (Lesson 10-3)

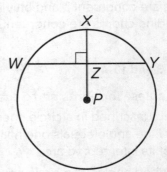

(A) 45°

(B) 90°

(C) 135°

(D) 180°

**6. MULTIPLE CHOICE** Devon is placing mirrors inside a kaleidoscope as shown.

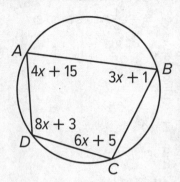

If the four mirrors are placed at the points given, what is the measure of the angle formed at *A*? (Lesson 10-4)

Ⓐ 16°

Ⓑ 23°

Ⓒ 79°

Ⓓ 101°

**7. MULTIPLE CHOICE** Maddie and Selena are visiting a garden. The garden contains a circular area with a fountain at the center and two separate congruent walkways, $\overline{AC}$ and $\overline{DG}$.

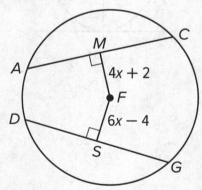

Maddie is at point *M* and Selena is at point *S*. How far is Maddie from the fountain? (Lesson 10-4)

Ⓐ 0.2 unit

Ⓑ 3 units

Ⓒ 6 units

Ⓓ 14 units

**8. OPEN RESPONSE** Use the figure.

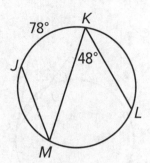

What are $m\widehat{JK}$ and $m\angle LKM$ in degrees? (Lesson 10-4)

**9. TABLE ITEM** Determine whether $\overline{HJ}$ is tangent to each circle. (Lesson 10-5)

**A.**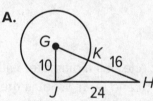

**B.**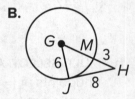

**C.**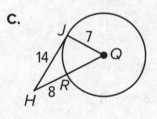

**D.**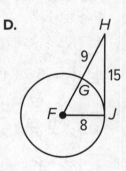

| Circle | Tangent? | |
|--------|-----|-----|
|        | yes | no  |
| A.     |     |     |
| B.     |     |     |
| C.     |     |     |
| D.     |     |     |

**10.** **MULTIPLE CHOICE** What is the measure of $\overset{\frown}{JMK}$? (Lesson 10-6)

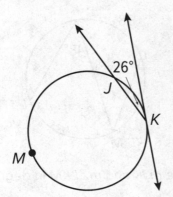

Ⓐ 128°

Ⓑ 308°

Ⓒ 334°

Ⓓ 347°

**11.** **MULTI-SELECT** A circle has a diameter with endpoints at $(-2, 5)$ and $(4, 1)$. What is the equation of the circle? Select all that apply. (Lesson 10-7)

☐ $(x - 1)^2 + (y - 3)^2 = 52$

☐ $(x - 3)^2 + (y - 3)^2 = 52$

☐ $(x - 1)^2 + (y - 3)^2 = 18$

☐ $x^2 + y^2 - 2x - 6y = 8$

☐ $x^2 + y^2 - 6x - 6y = 34$

☐ $x^2 + y^2 - 2x - 6y = 42$

**12.** **OPEN RESPONSE** A meteorologist is tracking a storm and has mapped it to a coordinate plane to make a forecast. At 3 PM, the eye of the storm will be located at $(1, 7)$ and will reach as far as a town located at $(10, -5)$.

What is the equation that could be used to describe this circle? (Lesson 10-7)

**13.** **MULTIPLE CHOICE** Which equation of a parabola has the focus at $\left(0, \frac{1}{5}\right)$ and the directrix $y = -\frac{1}{5}$? (Lesson 10-8)

Ⓐ $y = \frac{5}{8}x^2 - \frac{8}{25}$

Ⓑ $y = \frac{8}{5}x^2$

Ⓒ $y = -\frac{5}{8}x^2$

Ⓓ $y = \frac{5}{8}x^2$

**14.** **OPEN RESPONSE** What are the points of intersection for the parabola and line?

$y = x^2 + 2x - 15$

$y = 12x - 39$ (Lesson 10-8)

**15.** **OPEN RESPONSE** A curve in a highway can be modeled using a parabola where the focus is $(0, 5)$ and the directrix is $y = -5$. Write the equation of the parabola in simplest form. (Lesson 10-8)

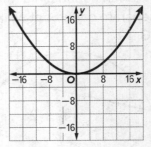

# Measurement

## ⓔ Essential Question

How are measurements of two- and three-dimensional figures useful for modeling situations in the real world?

G.C.5, G.GMD.1, G.GMD.2, G.GMD.3, G.GMD.4, G.MG.2, G.MG.3

## What will you learn?

Place a checkmark (✓) in each row that corresponds with how much you already know about each topic **before** starting this module.

| KEY<br><br>👎 — I don't know.  👍 — I've heard of it.  👍 — I know it! | Before 👎 | 👍 | 👍 | After 👎 | 👍 | 👍 |
|---|---|---|---|---|---|---|
| find areas of quadrilaterals using formulas | | | | | | |
| find areas of regular polygons using formulas | | | | | | |
| find areas of circles and sectors using formulas | | | | | | |
| find surface areas of three-dimensional figures using formulas | | | | | | |
| identify cross sections of three-dimensional solids | | | | | | |
| identify three-dimensional objects generated by rotations of two-dimensional objects | | | | | | |
| find volumes of three-dimensional figures using formulas | | | | | | |
| find measures of similar two- and three-dimensional figures | | | | | | |
| solve real-world problems involving density using area and volume | | | | | | |

📒 **Foldables** Make this Foldable to help you organize your notes about area and volume. Begin with one sheet of notebook paper.

1. **Fold** a sheet of paper in half.

2. **Fold** the paper again, two inches from the top.

3. **Unfold** the paper.

4. **Label** as shown.

# What Vocabulary Will You Learn?

Check the box next to each vocabulary term that you may already know.

- ☐ altitude of a cone
- ☐ altitude of a cylinder
- ☐ altitude of a parallelogram
- ☐ altitude of a prism
- ☐ altitude of a pyramid
- ☐ apothem
- ☐ axis of a cone
- ☐ axis of a cylinder
- ☐ axis symmetry
- ☐ base edge

- ☐ base of a parallelogram
- ☐ center of a regular polygon
- ☐ central angle of a regular polygon
- ☐ chord of a sphere
- ☐ congruent solids
- ☐ composite figure
- ☐ cross section
- ☐ density
- ☐ diameter of a sphere

- ☐ height of a parallelogram
- ☐ height of a solid
- ☐ height of a trapezoid
- ☐ lateral area
- ☐ lateral edges
- ☐ lateral faces
- ☐ lateral surface of a cone
- ☐ lateral surface of a cylinder
- ☐ plane symmetry

- ☐ radius of a regular polygon
- ☐ radius of a sphere
- ☐ regular pyramid
- ☐ sector
- ☐ similar solids
- ☐ slant height
- ☐ slant height of a right cone
- ☐ solid of revolution
- ☐ tangent to a sphere

## Are you ready?

Complete the Quick Review to see if you are ready to start this module.
Then complete the Quick Check.

### Quick Review

**Example 1**

**Find the value of $x$. Round to the nearest tenth if necessary.**

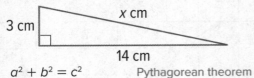

| | |
|---|---|
| $a^2 + b^2 = c^2$ | Pythagorean theorem |
| $3^2 + 14^2 = x^2$ | Substitute the side lengths. |
| $9 + 196 = x^2$ | Evaluate exponents. |
| $205 = x^2$ | Add. |
| $14.3 \approx x$ | Take the square root of each side. |

**Example 2**

**Find the value of $h$.**

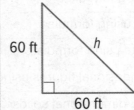

In a 45°-45°-90° triangle, the hypotenuse is $\sqrt{2}$ times the length of a leg.

$h = 60\sqrt{2}$, or about 84.85

So, $h$ is approximately 84.85 feet.

### Quick Check

**Solve for $x$.**

**1.**

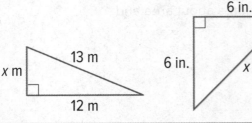

**2.**

**Find the distance between each pair of points. Round to the nearest tenth if necessary.**

**3.** (2, 1) and (4, 10)

**4.** (−3, 7) and (5, 4)

**5.** (−4, −2) and (−1, −2)

**6.** (0, −6) and (3, 1)

**How did you do?**

Which exercises did you answer correctly in the Quick Check? Shade those exercise numbers below.

① ② ③ ④ ⑤ ⑥

# Areas of Quadrilaterals

## Explore  Deriving the Area Formulas

**Online Activity** Use graphing technology to complete the Explore.

> **INQUIRY** How can you use your knowledge of triangles and rectangles to find areas of parallelograms? ×

## Learn  Areas of Parallelograms

A parallelogram is a quadrilateral with both pairs of opposite sides parallel. The **base of a parallelogram** is any side of the parallelogram.
An **altitude of a parallelogram** is defined as a perpendicular segment between any two parallel bases.
The **height of a parallelogram** is the length of an altitude of the parallelogram.

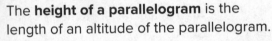
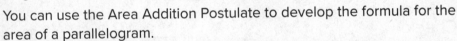

You can use the Area Addition Postulate to develop the formula for the area of a parallelogram.

**Postulate 11.1: Area Addition Postulate**

The area of a region is the sum of the areas of its nonoverlapping parts.

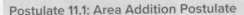

To find the area of a parallelogram, imagine cutting off a right triangle from one side of a parallelogram and translating the triangle to the other side to form a rectangle with the same base and height.

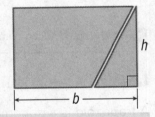

**Key Concept • Area of a Parallelogram**

The area $A$ of a parallelogram is the product of the base $b$ and its corresponding height $h$.

## Example 1  Area of a Parallelogram

**Find the area of the parallelogram.**

$h^2 + 7^2 = 25^2$   Pythagorean Theorem
$h = $ ___   Solve.

$\overline{DC}$ is one base of $\square ABCD$, and $DC = 23$ feet.

$A = bh$   Area of a parallelogram
$= ($___$)($___$)$ or 552 ft$^2$   Solve.

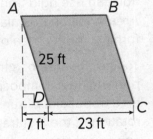

**Go Online** You can complete an Extra Example online.

### Today's Standards
MP1, MP4, MP7

### Today's Vocabulary
base of a parallelogram
altitude of a parallelogram
height of a parallelogram
height of a trapezoid

### Talk About It!
What is the relationship between the area of a parallelogram and the area of a rectangle? Justify your answer.

### Study Tip
**Heights of Figures**
The height of a figure can be measured by extending a base. The height of $\square ABCD$ that corresponds to base $\overline{DC}$ can be measured by extending $\overline{DC}$.

## Check

Find the area of the parallelogram.

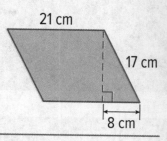

A. 315 cm²

B. 357 cm²

C. 493 cm²

D. 1260 cm²

---

**Watch Out!**

Remember that area is measured in square units such as square feet, square millimeters, and square yards.

🧠 **Think About It!**

Describe two different ways you could use measurement to find the area of parallelogram *PQRS*.

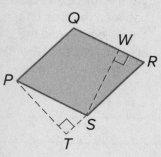

## Example 2 Use Trigonometry to Find the Area of a Parallelogram

You may need to use trigonometry to find the area of a parallelogram.

**Find the area of the parallelogram.**

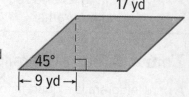

**Step 1  Use special right triangles.**

Use a 45°−45°−90° triangle to find the height *h* of the parallelogram. Recall that if the measure of the leg opposite the 45° angle is *h*, then the measure of the leg adjacent the 45° angle is also *h*.

$h = \underline{\quad}$ yd

**Step 2  Find the area.**

$A = bh$          Area of a parallelogram

$= (\underline{\quad})(9)$ or $\underline{\quad}$ yd²     Solve.

## Check

Find the area of the parallelogram to the nearest tenth if necessary.

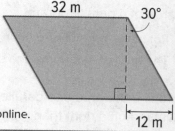

$\underline{\qquad}$ m²

🔎 **Go Online** You can complete an Extra Example online.

## Learn Areas of Trapezoids

A trapezoid is a quadrilateral with exactly one pair of parallel sides.

The parallel sides of a trapezoid are called bases.

The **height of a trapezoid** is the perpendicular distance between the bases of a trapezoid.

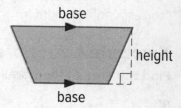

You can use rigid transformations and the Area Addition Postulate to develop the formula for the area of a trapezoid.

To find the area of a trapezoid, imagine performing a composition of transformations on a trapezoid. A translation followed by a rotation of the first trapezoid results in two congruent trapezoids that fit together to form a parallelogram.

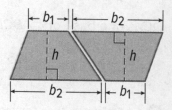

The area of the parallelogram is the product of the height $h$ and the sum of the two bases, $b_1$ and $b_2$. The area of one trapezoid is one half the area of the parallelogram.

**Key Concept • Area of a Trapezoid**

The area $A$ of a trapezoid is one half the product of the height $h$ and the sum of its bases, $b_1$ and $b_2$.

$A = \frac{1}{2} h (b_1 + b_2)$

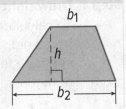

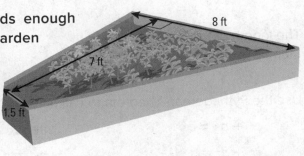

🌐 **Example 3** Area of a Trapezoid

GARDENS Andrea needs enough mulch to cover the garden she planted in a raised bed constructed in the shape of a trapezoid. If one bag of mulch covers 12 square feet at the desired depth, how many bags of mulch does she need to buy?

**Step 1 Find the area of the garden.**

$A = \frac{1}{2} h (b_1 + b_2)$       Area of a trapezoid

$\quad = \frac{1}{2} (7) (8 + 1.5)$       $h = \underline{\quad}$, $b_1 = \underline{\quad}$, and $b_2 = \underline{\quad}$

$\quad = \underline{\quad\quad} ft^2$       Solve.

The area of the garden is _____ square feet.

**Step 2 Calculate the number of bags needed.**

Use unit analysis to determine how many bags of mulch Amanda should buy.

$33.25 \ \cancel{ft^2} \cdot \dfrac{1 \ bag}{12 \ \cancel{ft^2}} = \underline{\quad}$ bags

Round the number of bags up so there is enough mulch. Andrea needs to buy ___ bags of mulch to cover her garden.

Check

ART Miguel wants to cover the top of his desk with butcher paper before working on a project for his art class. If one sheet of butcher paper covers a square meter of work space, how many sheets will Miguel need?

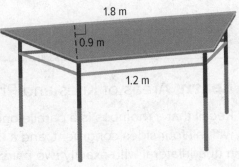

The top of the table has an area of _____ square meters.
Miguel will need ___ sheets of butcher paper.

▶ **Go Online** You can complete an Extra Example online.

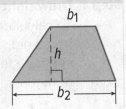

💭 **Think About It!**

What is the relationship between the area of a parallelogram and the area of a trapezoid? Justify your answer.

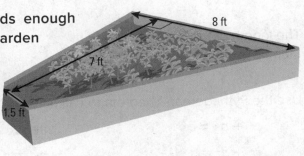

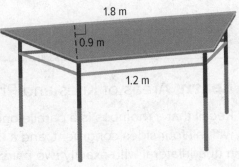

## Example 4 Use Right Triangles to Find the Area of a Trapezoid

**Find the area of the trapezoid.**

**Step 1  Draw right triangles.**

To calculate the height of the trapezoid, draw vertical lines to separate the figure into two right triangles and a rectangle.

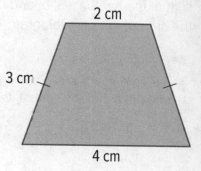

**Study Tip**

**Separating Figures**
To solve some area problems, you may need to draw in parallel and/or perpendicular lines to find information not provided.

**Step 2  Find the height.**

$$x + x + 2 = 4 \qquad \text{Length of } b_2$$
$$x = \underline{\phantom{xx}} \qquad \text{Solve.}$$

The length of the base of each right triangle is 1 centimeter.

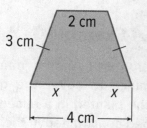

Use the Pythagorean Theorem to calculate the height of the trapezoid.

$$a^2 + b^2 = c^2 \qquad \text{Pythagorean Theorem}$$
$$\underline{\phantom{x}}^2 + h^2 = \underline{\phantom{x}}^2 \qquad \text{Substitute.}$$
$$h = \sqrt{3^2 - 1^2} \text{ or } \underline{\phantom{xx}} \qquad \text{Solve.}$$

**Go Online** An alternate method is available for this example.

**Step 3  Find the area.**

$$A = \tfrac{1}{2} h (b_1 + b_2) \qquad \text{Area of a trapezoid}$$
$$= \tfrac{1}{2} (\sqrt{8}) (4 + 2) \qquad h = \sqrt{8}, b_1 = \underline{\phantom{x}}, \text{ and } b_2 = \underline{\phantom{x}}$$
$$\approx \underline{\phantom{xx}} \text{cm}^2 \qquad \text{Solve.}$$

### Check

Find the area of the trapezoid.

**A.** 4.353 ft$^2$

**B.** 4.375 ft$^2$

**C.** 8.706 ft$^2$

**D.** 8.75 ft$^2$

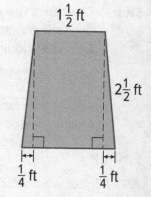

**Go Online** You can complete an Extra Example online.

**Study Tip**

**Review Vocabulary**
A *diagonal* is a segment that connects any two nonconsecutive vertices in a polygon.

## Learn Areas of Kites and Rhombi

Recall that a rhombus is a parallelogram with all four sides congruent, and a kite is a quadrilateral with exactly two pairs of consecutive congruent sides.

The areas of rhombi and kites are related to the lengths of their diagonals.

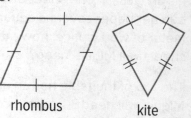

rhombus          kite

## Key Concept • Area of a Rhombus or Kite

The area $A$ of a rhombus or kite is one half the product of the lengths of its diagonals, $d_1$ and $d_2$.

$A = \frac{1}{2} d_1 d_2$

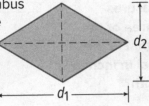

## Example 5  Area of a Rhombus

**Find the area of the rhombus.**

**Step 1  Find the length of each diagonal.**

Because the diagonals of a rhombus bisect each other, the lengths of the diagonals are ___ + ___ or 12 mm and ___ + ___ or 14 mm.

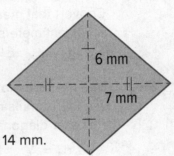

6 mm

7 mm

**Step 2  Find the area of the rhombus.**

$A = \frac{1}{2} d_1 d_2$          Area of a rhombus

$= \frac{1}{2} (12)(14)$          $d_1 = $ ___ and $d_2 = $ ___

$= $ ___ $mm^2$          Solve.

## Check

Find the area of the rhombus.

**A.** 45 $m^2$          **B.** 90 $m^2$

**C.** 180 $m^2$          **D.** 720 $m^2$

10 m

18 m

## Example 6  Area of a Kite

**Find the area of the kite.**

$A = \frac{1}{2} d_1 d_2$          Area of a kite

$= \frac{1}{2} ($___$)(9)$          Substitute.

$= $ ___ $in^2$          Solve.

16 in.

9 in.

## Check

Find the area of the kite.

$A = $ _____ $m^2$

21 m

17 m

🅑 **Go Online** You can complete an Extra Example online.

## Example 7 Use Area to Find Missing Measures

**One diagonal of a kite measures 55.88 centimeters. If the area of the kite is 92 square inches, what is the length of the other diagonal in inches rounded to the nearest tenth?**

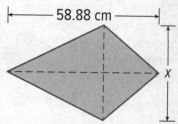

58.88 cm

x

**Watch Out!**

The known length of one of the diagonals is measured in centimeters, which is a metric unit. You are asked to find the measure of the missing diagonal in inches, which is a standard unit. You can use the Internet or other resources to find how to convert from centimeters to inches.

**Step 1  Convert units.**

The length of one of the diagonals is given in centimeters. Convert that measure to inches. Recall that 1 inch equals 2.54 centimeters.

$$(55.88 \text{ cm}) \times \frac{1 \text{ in.}}{\text{cm}} = \underline{\quad} \text{ in.}$$

**Step 2  Use the formula for the area of a kite.**

Use the formula for the area of a kite to find the measure of the other diagonal in inches.

$$\frac{1}{2} d_1 d_2 = A \qquad \text{Area of a kite}$$

$$\frac{1}{2}(\underline{\quad})(x) = \underline{\quad} \qquad A = 92, d_1 = 22, \text{ and } d_2 = x$$

$$x \approx 8.4 \qquad \text{Solve.}$$

The length of the diagonal is about _____ inches.

## Check

In rhombus $ABCD$, $AE = 7.3$ inches. If the area of the rhombus is 96 square inches, find $x$ and the length of each diagonal. Round to the nearest tenth if necessary.

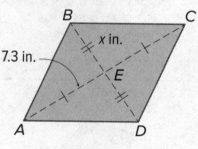

$x \approx$ _____ in.

$AC =$ _____ in.

$BD =$ _____ in.

## Pause and Reflect

Did you struggle with anything in this lesson? If so, how did you deal with it?

Record your observations here

ⓝ **Go Online** You can complete an Extra Example online.

# Practice

🌀 **Go Online** You can complete your homework online.

Examples 1, 2, 4–6

**Find the area of each parallelogram, trapezoid, rhombus, or kite. Round to the nearest tenth if necessary.**

**1.**

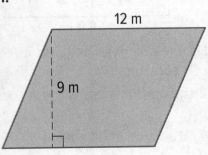

12 m

9 m

**2.**

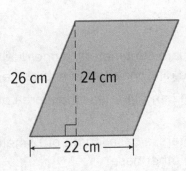

26 cm    24 cm

22 cm

**3.**

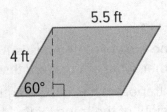

5.5 ft

4 ft

60°

**4.**

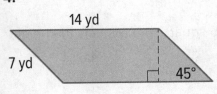

14 yd

7 yd    45°

**5.**

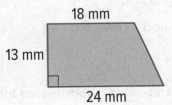

18 mm

13 mm

24 mm

**6.**

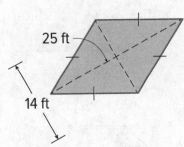

8 in.

10 in.

9 in.    6 in.

**7.**

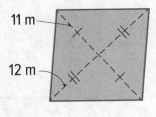

11 m

12 m

**8.**

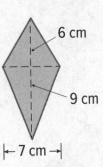

25 ft

14 ft

**9.**

6 cm

9 cm

7 cm

Example 3

**10.** Meghana is making a kite out of nylon. The material is sold for $5.99 per square yard. She needs enough material to cover her kite on both sides.

  **a.** What is the area of Meghana's kite?

  **b.** How many yards of material will Meghana need to cover both sides of her kite? (Hint: watch the units)

  **c.** How much will the material cost before taxes?

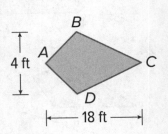

B

A    C

4 ft

D

18 ft

**11.** An architect is planning a building with a glass façade in the shape of a parallelogram with measurements as shown in the diagram. What is the largest amount of glass that would be needed to cover this structure?

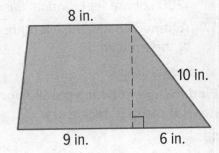

37°

24 m

20 m

**12.** Iker is building an island for his kitchen in the shape of a trapezoid as shown.

   **a.** What is the area of the top surface of the island?

   **b.** The island will be made of granite which costs $65 per square foot. Approximately how much will the material cost?

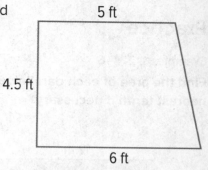

5 ft

4.5 ft

6 ft

**Example 7**

**13.** The area of a rhombus is 168 square centimeters. If one diagonal is three times as long as the other, what are the lengths of the diagonals to the nearest tenth of a centimeter?

**14.** A trapezoid has base lengths of 12 and 14 feet with an area of 322 square feet. What is the height of the trapezoid?

**15.** A trapezoid has a height of 8 meters, a base length of 12 meters, and an area of 64 square meters. What is the length of the other base?

**16.** The height of a parallelogram is 10 feet more than its base. If the area of the parallelogram is 1200 square feet, find its base and height.

**17.** A trapezoid has base lengths of 4 and 19 feet, with an area of 115 square feet. What is the height of the trapezoid?

**Mixed Exercises**

**Find the area of each parallelogram, trapezoid, rhombus, or kite. Round to the nearest tenth if necessary.**

**18.**

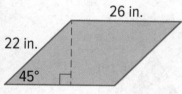

26 in.

22 in.

45°

**19.**

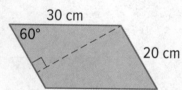

30 cm

60°

20 cm

**20.**

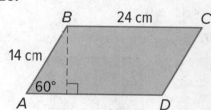

B   24 cm   C

14 cm

60°

A    D

**21.**

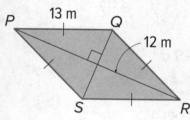

P   13 m   Q

12 m

S    R

**22.**

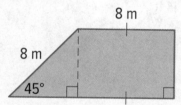

8 m

8 m

45°

**23.**

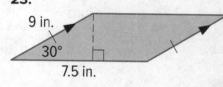

9 in.

30°

7.5 in.

**24.**

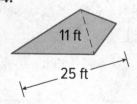

11 ft

25 ft

**25.**

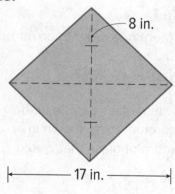

8 in.

17 in.

**26.**

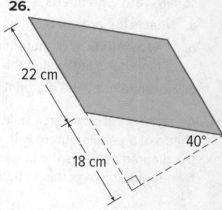

22 cm

18 cm

40°

27. A trapezoid has base lengths of 2.5 inches and 10 inches and an altitude of 8 inches. What is the area of the trapezoid?

28. A parallelogram has a base length of 18.5 kilometers and a height of 9 kilometers. What is the area of the parallelogram?

29. A parallelogram has a base length that is equal to its height. If the base length of the parallelogram is 3.4 meters, what is the area of the parallelogram?

30. A trapezoid has base lengths of 22 feet and 37 feet. If the area of the trapezoid is 678.5 square feet, what is the altitude of the trapezoid?

31. One diagonal of a kite is twice as long as the other diagonal. If the area of the kite is 240 square inches, what are the lengths of the diagonals?

32. A trapezoid has base lengths of 6 and 15 centimeters with an area of 136.5 square centimeters. What is the height of the trapezoid?

33. One diagonal of a kite is four times as long as the other diagonal. If the area of the kite is 72 square meters, what are the lengths of the diagonals?

34. A trapezoid has a height of 24 meters, a base of 4 meters, and an area of 264 square meters. What is the length of the other base?

35. PACKAGING A box with a square opening is reshaped into the rhombus as shown. What is the area of the opening?

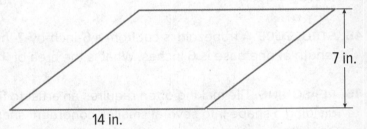

14 in.

7 in.

36. REASONING The diagonal of a square is $6\sqrt{2}$ inches. What is the area of the square?

37. SHADOWS A rectangular billboard casts a shadow on the ground in the shape of a parallelogram. What is the area of the ground covered by the shadow? Round your answer to the nearest tenth.

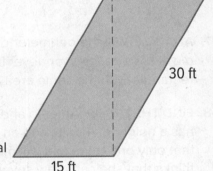

30 ft

38. PATHS A concrete path shown below is made by joining several parallelograms. What is the total area of the path?

15 ft

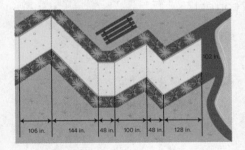

106 in.    144 in.    48 in.  100 in.  48 in.  128 in.

USE TOOLS Find the area of each quadrilateral with the given vertices.

39. $A(-8, 6)$, $B(-5, 8)$, $C(-2, 6)$, and $D(-5, 0)$

40. $W(3, 0)$, $X(0, 3)$, $Y(-3, 0)$, and $Z(0, -3)$

41. **REGULARITY** Given the area of a geometric figure, describe the method for solving for a missing dimension.

42. **PRECISION** Example: Parallelogram *PQRS*.

    a. Explain how a 30° − 60° − 90° triangle is used in finding the area of parallelogram *PQRS*.

    b. Find the area of parallelogram *PQRS* to the nearest tenth.

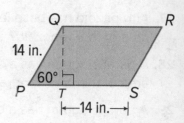

43. **STRUCTURE** For Bruno's birthday, he got a cake shaped like a kite. He cuts the cake along the diagonals into 4 pieces. The diagonals are 6 inches and 10 inches long. Which piece(s) is the largest? What is the area of the top of the cake?

44. **MODELING** The 20-by-20-foot square shows a landscape plan composed of three indoor gardens and one walkway, all congruent in shape. The gardens are centered around a 15-by-15 foot lounging area. What is the area of one of these gardens?

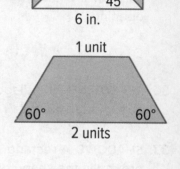

45. **STRUCTURE** A trapezoid is cut from a 6-inch-by-2-inch rectangle. The length of one base is 6 inches. What is the area of the trapezoid?

46. **REASONING** Tile making often requires an artist to find clever ways of dividing a shape into several smaller, congruent shapes. Consider the isosceles trapezoid shown at the right. The trapezoid can be divided into 3 congruent triangles. What is the area of each triangle?

47. **ANALYZE** Will the perimeter of a nonrectangular parallelogram *always*, *sometimes*, or *never* be greater than the perimeter of a rectangle with the same area and the same height? Explain.

48. **FIND THE ERROR** Antonio and Niran want to draw a trapezoid that has a height of 4 units and an area of 18 square units. Antonio says that only one trapezoid will meet the criteria. Niran disagrees and thinks that she can draw several different trapezoids with a height of 4 units and an area of 18 square units. Is either of them correct? Explain your reasoning.

49. **PERSEVERE** Find *x* in parallelogram *ABCD*.

50. **CREATE** Draw a kite and a rhombus with an area of 6 square inches. Label and justify your drawings.

51. **ANALYZE** If the areas of two rhombi are equal, are the perimeters *sometimes*, *always*, or *never* equal? Explain.

52. **WRITE** How can you use trigonometry to find the area of a figure?

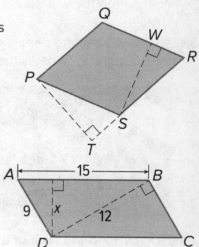

# Areas of Regular Polygons

## Explore Regular Polygons

📲 **Online Activity** Use graphing technology to complete the Explore.

> ⊗
>
> ❓ **INQUIRY** How can the formula for the area of a regular polygon be derived from the formula for the area of a triangle?

## Learn Areas of Regular Polygons

In the figure, a regular pentagon is inscribed in $\odot P$, and $\odot P$ is circumscribed about the pentagon. The **center of a regular polygon** is the center of the circle circumscribed about the polygon. The **radius of a regular polygon** is the radius of the circle circumscribed about the polygon. The **apothem** of a regular polygon is a perpendicular segment between the center of the polygon and a side of the polygon.

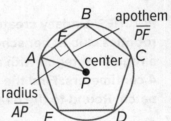

A **central angle of a regular polygon** has its vertex at the center of the polygon and sides that pass through consecutive vertices of the polygon. The measure of each central angle of a regular $n$-gon is $\frac{360}{n}$.

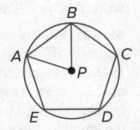

> **Key Concept • Area of a Regular Polygon**
>
> The area $A$ of a regular $n$-gon with side length $s$ is one half the product of the apothem $a$ and the perimeter $P$.
>
> $A = \frac{1}{2}a(ns)$ or $A = \frac{1}{2}aP$

## Example 1 Identify Segments and Angles in Regular Polygons

**In the figure, regular hexagon *JKLMNP* is inscribed in $\odot R$. Identify the center, a radius, an apothem, and a central angle of the polygon. Then find the measure of a central angle.**

The center of the regular hexagon is point _____.
A _____ of the regular hexagon is $\overline{RK}$.
An apothem of the regular hexagon is _____. A central angle of the regular hexagon is $\angle KRL$. A regular hexagon has _____ sides. Thus, the measure of each central angle is $\frac{360}{6}$ or _____°.

📲 **Go Online** You can complete an Extra Example online.

### Today's Standards
G.MG.3

MP3, MP4, MP7

### Today's Vocabulary
center of a regular polygon

radius of a regular polygon

apothem

central angle of a regular polygon

composite figure

## Check

Pentagon *QRTUV* is inscribed in ⊙*P*. Identify the center, a radius, an apothem, and a central angle of the polygon. Then find the measure of a central angle.

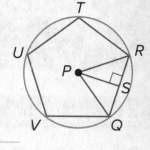

center: _____

radius: segment _____

apothem: segment _____

central angle: angle _____

measure of central angle: _____°

---

## 🌎 Example 2  Area of a Regular Polygon

4 cm

**PATCHES  Lindsay created a patch for the robotics club at her school. The patch is a regular octagon with a side length of 4 centimeters. Find the area covered by the patch. Round to the nearest tenth.**

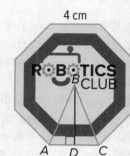

**Step 1  Find the measure of a central angle.**
A regular octagon has 8 congruent central angles, so $m\angle ABC = \frac{360}{\quad}$ or _____°.

**Step 2  Find the length of the apothem.**
Apothem $\overline{BD}$ is the height of isosceles △*ABC*. It bisects ∠*ABC*, so $m\angle DBC = 45 \div 2$ or _____°. It also bisects $\overline{AC}$, so *DC* = _____ centimeters. Use trigonometric ratios to find the length of the apothem.

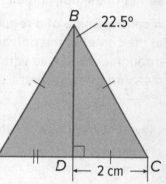

$\tan 22.5° = \dfrac{\quad}{a}$          $\tan B = \dfrac{DC}{BD}$

$a = \dfrac{2}{\underline{\quad\quad}}$          Solve for *a*.

**Step 3  Use the formula for the area of a regular polygon.**

$A = \dfrac{1}{2}\underline{\quad\quad}$          Area of a regular polygon.

$= \dfrac{1}{2}\left(\dfrac{2}{\tan 22.5°}\right)(\underline{\quad})$   $a = \dfrac{2}{\tan 22.5°}$ and $P = 32$

$\approx \underline{\quad\quad}$          Solve.

The patch covers an area of about _____ cm².

## Check

**CERAMICS  Imani is crafting coasters for a local craft fair. Each side measures 4 inches. Find the area of each coaster. Round your answer to the nearest tenth, if necessary.**

$A \approx$ _____ in²

 **Go Online**  You can complete an Extra Example online.

---

🐷 **Think About It!**

James calculated the area of the trampoline. His calculations are shown.

$A = \dfrac{1}{2}aP$

$\approx \dfrac{1}{2}(4)(27.3)$

$\approx 54.6$ ft

Do you agree? Justify your argument.

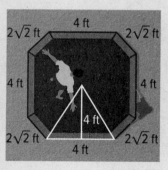

Getty Images

## Learn Areas of Composite Figures

A **composite figure** is a figure that can be separated into regions that are basic figures, such as triangles, rectangles, trapezoids, and circles. To find the area of a composite figure, find the area of each basic figure and then use the Area Addition Postulate.

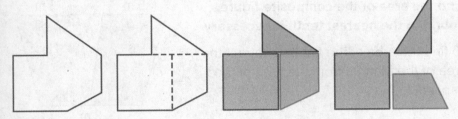

**Go Online**
Go online to watch a video to see how to find the area of a composite figure on the coordinate plane.

## Example 3 Find the Area of a Composite Figure by Adding

**Find the area of the composite figure.**
**Step 1 Separate the composite figure.**

The figure shown is composed of a square, regular hexagon, and trapezoid. So, area of figure = area of square + area of hexagon + area of trapezoid.

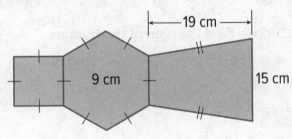

**Step 2 Calculate the area of each basic figure.**

$A = s^2$      Area of square

$= \_\_\_\_^2$      $s = 9$

$= \_\_\_\_ \text{ cm}^2$      Solve.

$A = \frac{1}{2}aP$      Area of regular polygon

$= \frac{1}{2}(\_\_\_\sqrt{3})(\_\_\_)$      $a = 4.5\sqrt{3}$ and $P = 54$

$\approx \_\_\_\_ \text{ cm}^2$      Simplify.

$A = \frac{1}{2}h(b_1 + b_2)$      Area of trapezoid

$= \frac{1}{2}\_\_\_(\_\_\_ + \_\_\_)$      $h = 19$, $b_1 = 15$, and $b_2 = 9$

$= \_\_\_\_ \text{ cm}^2$      Solve.

**Step 3 Calculate the total area of the composite figure.**

area of figure = area of square + area of hexagon + area of trapezoid

$\approx \_\_\_\_ + \_\_\_\_ + \_\_\_\_$

$\approx \_\_\_\_ \text{ cm}^2$

## Check

Find the area of the composite figure.
Round to the nearest tenth if necessary.

$A = \_\_\_\_ \text{ yd}^2$

 **Go Online** You can complete an Extra Example online.

## Example 4 Find the Area of a Composite Figure by Subtracting

The areas of some figures can be found by subtracting the areas of basic figures.

**Find the area of the composite figure. Round to the nearest tenth if necessary.**

To find the area of the figure, subtract the area of the triangle from the area of the trapezoid.

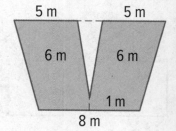

**Step 1  Find the height of the triangle.**

You can use the Pythagorean Theorem and what you know about isosceles triangles to calculate the height of the triangle. The altitude of the isosceles triangle bisects the base of the triangle.

So, $1^2 + h^2 = 6^2$ or $h = \sqrt{\phantom{--}}$.

**Step 2  Find the area of the triangle.**

$$A = \tfrac{1}{2}bh \qquad\qquad \text{Area of a triangle}$$
$$= \tfrac{1}{2}(\underline{\phantom{--}})(\sqrt{\phantom{--}}) \qquad b = 2 \text{ and } h = \sqrt{35}$$
$$= \underline{\phantom{--}} \text{ m}^2 \qquad\qquad \text{Solve.}$$

**Step 3  Find the area of the trapezoid.**

$$A = \tfrac{1}{2}h(b_1 + b_2) \qquad\qquad\qquad \text{Area of trapezoid}$$
$$= \tfrac{1}{2}(\sqrt{35} + \underline{\phantom{--}})(\underline{\phantom{--}} + 8) \qquad b_1 = 12, b_2 = 8, \text{ and } h = \sqrt{35} + 1$$
$$= \underline{\phantom{--}}\sqrt{35} + \underline{\phantom{--}} \text{ m}^2 \qquad\qquad \text{Solve.}$$

**Step 4  Find the area of the figure.**

$$\text{area of figure} = \text{area of trapezoid} - \text{area of triangle}$$
$$= 10\sqrt{35} + 10 - \underline{\phantom{--}}$$
$$\approx \underline{\phantom{--}} \text{ m}^2$$

## Check

Select the area of the composite figure. _____

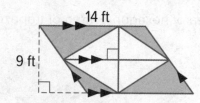

**A.** 31.5 ft² **B.** 63 ft² **C.** 94.5 ft² **D.** 126 ft²

Go Online You can complete an Extra Example online.

### Study Tip

**Drawing Figures**
To solve some area problems by subtracting, you may need to draw figures to represent the basic shapes that are being removed from the composite figure. You can use the figures you draw to help you visualize the situation and calculate missing measures.

# Practice

**Go Online** You can complete your homework online.

## Example 1

In each figure, a regular polygon is inscribed in a circle. Identify the center, a radius, an apothem, and a central angle of each polygon. Then find the measure of a central angle.

**1.**

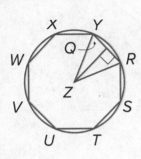

**2.**

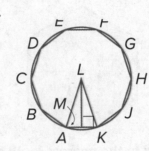

## Example 2

Find the area of each regular polygon. Round to the nearest tenth.

**3.**

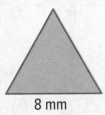

8 mm

**4.**

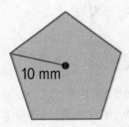

10 mm

**5.**

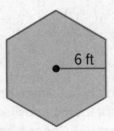

6 ft

**6.** COINS The Susan B. Anthony dollar coin has a hendecagon (11-gon) inscribed in a circle in its design. Each edge of the hendecagon is approximately 7.46 mm. What is the area of this regular polygon? Round to the nearest hundredth.

**7.** An octagonal trampoline has a diameter of 16 feet. What is the area of the surface of the trampoline? Round to the nearest tenth.

## Examples 3 and 4

Find the area of each figure. Round to the nearest tenth if necessary.

**8.**

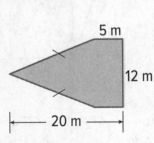

5 m
12 m
20 m

**9.**

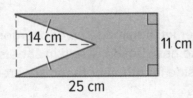

14 cm
11 cm
25 cm

**10.**

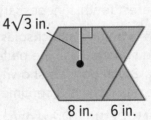

$4\sqrt{3}$ in.
8 in.   6 in.

**11.**

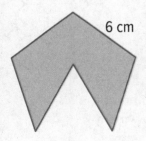

6 cm

**12.**

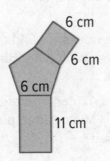

6 cm
6 cm
6 cm
11 cm

**13.**

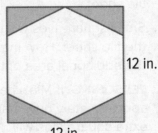

12 in.
12 in.

**14. LAWN** Leila has to buy grass seed for her lawn. Her lawn is in the shape of the composite figure shown. What is the area of the lawn?

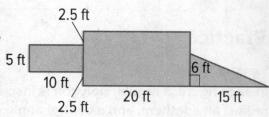

Mixed Exercises

**Find the area of each regular polygon. Round to the nearest tenth.**

15.

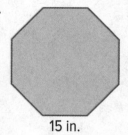

15 in.

16.

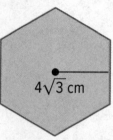

$4\sqrt{3}$ cm

17.

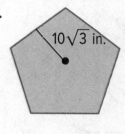

$10\sqrt{3}$ in.

**Find the area of each figure. Round to the nearest tenth if necessary**

18.

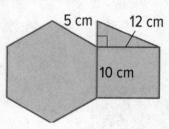

5 cm  12 cm
10 cm

19.

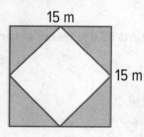

15 m
15 m

20.

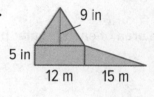

9 in
5 in
12 m   15 m

**21. PATIO** Chenoa is building a patio to surround his fire pit. The patio is in the shape of the composite figure shown. How many square feet of patio pavers will Chenoa need to buy to complete the patio?

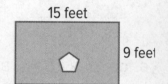

15 feet
9 feet

**22. REASONING** Find the area of a regular hexagon with a perimeter of 72 inches. Round to the nearest square inch.

**23. MODELING** Draw a regular pentagon inscribed in a circle with center of $X$, a radius of $\overline{XV}$, an apothem of $\overline{XY}$, and a central angle of $\angle VXT$ that measures 72°.

**24. STRUCTURE** Find the perimeter and area of a regular hexagon with a side length of 12 centimeters. Round to the nearest tenth, if necessary.

**25. USE TOOLS** Find the total area of the shaded regions. Round to the nearest tenth, if necessary.

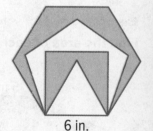

6 in.

**26.** Mia is putting a backsplash in her kitchen made up of octagon and square tiles. The pattern she has chosen is sold in sheets of 4 × 3 octagons as shown. The side length of both the squares and the octagons is 3 centimeters.

**a. MODELING** Draw one octagon and one square. Label the side length and area of each and the apothem of the octagon. Find the area of one tile sheet.

**b.** Small square tiles will be placed to fill the nooks around the edges of the tile sheet. How many complete squares will be needed? What is the additional area of these tiles?

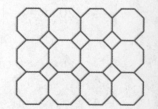

**c. REASONING** If Mia's backsplash measures 42 centimeters by 84 centimeters, approximately how many sheets of tile will Mia need to purchase? How many extra square tiles will need to be purchased?

**27.** Miguel is planning to renovate his living room.

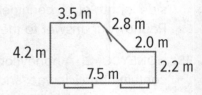

    **a.** STRUCTURE How much finish will he need for the hardwood floor? Assume 1 liter of finish covers 4.5 square meters and round to the nearest tenth.

    **b.** PRECISION The height of the room is 2.6 meters. Approximate how much paint is needed for the walls. Assume that 1 liter of paint covers 7.5 square meters and round to the nearest tenth.

    **c.** ARGUMENTS Why might Miguel adjust your estimates when he purchases materials?

**28.** STATE YOUR ASSUMPTION Chilam is going to build a shelf for his 15 homerun baseballs. He wants the shelf to be in the shape of a regular triangle. Find the area of the regular triangle he must create. State any assumptions you made to support your answer. Round your answer to the nearest tenth.

**29.** USE A SOURCE Research the shape and dimensions of the Pentagon.

    **a.** Draw and label a diagram of the Pentagon including the courtyard.

    **b.** Use your diagram in part a to find the area in square feet of the Pentagon. Round your answer to the nearest tenth.

**30.** REGULARITY Explain how using the formula for the area of a regular polygon is related to finding the area of a triangle.

**31.** SIGNS A stop sign has side lengths of approximately 12.5 inches. What is the area of a regular stop sign? Round your answer to the nearest tenth.

**32.** DIAMONDS Arturo has bought his fiancée an engagement ring with a heptagon diamond as shown. What is the area of the face of the diamond?

**33.** ARCHITECTURE Fort Jefferson in the Florida Keys is the largest brick masonry building in the Americas. The Fort was built in the shape of a hexagon with side lengths of 477 feet. What is the area of the hexagon formed by the Fort's exterior walls? Round your answer to the nearest tenth.

**34.** FRAME Darren is hanging a picture frame that is shaped like a regular pentagon with outside side lengths of 9 inches. What is the area this picture and frame will take up on Darren's wall? Round your answer to the nearest tenth.

**35.** FLOWER Elena photographed the regular triangle shaped flower for a photo contest. The flower had side lengths measuring 4 centimeters. What is the area of the flower? Round your answer to the nearest tenth.

**36.** STAINED GLASS Kenia makes octagonal-shaped stained glass windows. If the apothem of the window is 14 inches, what is the area of the window? Round your answer to the nearest tenth.

**37. GAMING** A 12-sided gaming die is made up of 12 regular pentagons, each with a side length of 1.5 centimeters. What is the area of one face of the gaming die? Round your answer to the nearest hundredth.

**38. HONEYCOMB** A honeycomb is a structure bees make of hexagonal wax cells to contain honey, larvae, and pollen. The height of one hexagon in a honeycomb is approximately 5 millimeters. What is the area of one hexagon? Round your answer to the nearest tenth.

**39. FIND THE ERROR** Chenglei and Flavio want to find the area of the hexagon shown. Is either of them correct? Explain your reasoning.

| Chenglei | Flavio |
|---|---|
| $A = \frac{1}{2}Pa$ | $A = \frac{1}{2}Pa$ |
| $= \frac{1}{2}(66)(9.5)$ | $= \frac{1}{2}(33)(9.5)$ |
| $= 313.5 \ in^2$ | $= 156.8 \ in^2$ |

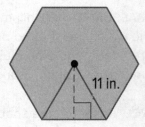

11 in.

**40. ANALYZE** Using the map of Nevada shown, estimate the area of the state. Explain your reasoning.

**41. CREATE** Draw a pair of composite figures that have the same area. Make one composite figure out of a rectangle and a trapezoid, and make the other composite figure out of a triangle and a rectangle. Show the area of each basic figure.

NEVADA

0   140 mi

0.5 in: 140 mi

**42.** Consider the sequence of area diagrams shown.

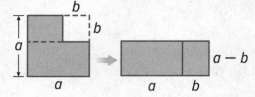

a

b

b

a

a

b

$a - b$

**a. PERSEVERE** What algebraic theorem do the diagrams prove? Explain your reasoning.

**b. CREATE** Create your own sequence of diagrams to prove a different algebraic theorem.

**43. WRITE** How can you find the area of any figure?

**44. WHICH ONE DOESN'T BELONG?** Alden drew the following diagrams to find the area of 4 regular polygons. Which drawing is incorrect and why?

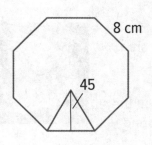

8 cm

45

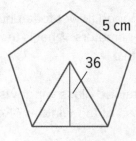

5 cm

36

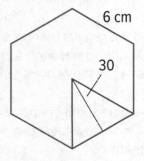

6 cm

30

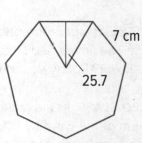

7 cm

25.7

# Areas of Circles and Sectors

## Explore Areas of Circles

**Today's Standards**
G.C.5; G.GMD.1
MP1, MP2, MP5

**Today's Vocabulary**
sector

🡒 **Online Activity** Use graphing technology to complete the Explore.

> ⊘
>
> ② **INQUIRY** How is the formula for the area of a circle related to the formula for the area of a regular polygon?

## Learn Areas of Circles

The formula for the circumference $C$ of a circle with radius $r$ is given by $C = 2\pi r$. You can use this formula to develop the formula for the area of a circle.

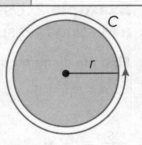

🡒 **Go Online**
to see how the formula for the area of a circle is developed.

| Key Concept • Area of a Circle | |
|---|---|
| **Words** | The area $A$ of a circle is equal to $\pi$ times the square of the radius $r$. |
| **Symbols** | $A = \pi r^2$ |

🡒 **Go Online**
to watch a video to see how to find the area and circumference of a circle.

## 🌐 Example 1 Area of a Circle

PATIO **Keon is building a circular patio in his backyard.**

**Part A** What is the area of the patio to the nearest square foot?

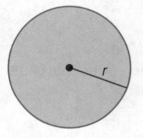

24 ft

The diameter of the circle is 24 feet, so the radius is _____ feet.

$A = \pi r^2$     Area of a circle

$\phantom{A} = \pi(\underline{\phantom{xx}})^2$     $r = 12$

$\phantom{A} \approx 452$     Use a calculator.

So, the area is about _____ square feet.

🡒 **Go Online** You can complete an Extra Example online.

**Study Tip**

**Units of Area**
Remember that when you convert between units of area that you have to multiply or divide by the square of the conversion for the units of length. For example, 1 yd = 3 ft, so 1 yd$^2$ = 1 yd · 1 yd = 3 ft · 3 ft or 9 ft$^2$.

**Part B** Keon can purchase paving stones for $135 per square yard. About how much will he spend on paving stones for his patio?

Keon measured the area of the patio in square feet, but the price for paving stones is given in square yards, so Keon will have to convert from square feet to square yards to find how much he will spend.

$$\left(\underline{\hspace{1cm}} \text{ ft}^2\right)\left(\frac{\text{yd}^2}{9 \text{ ft}^2}\right) \approx \underline{\hspace{1cm}} \text{ yd}^2 \qquad 1 \text{ yd}^2 = 9 \text{ ft}^2$$

$$\left(50.2 \text{ yd}^2\right)\left(\frac{\$}{1 \text{ yd}^2}\right) \approx \$3780 \qquad \text{Solve.}$$

So, Kipton could expect to spend about $\underline{\hspace{1cm}}$ on paving stones.

## Example 2 Use the Area of a Circle to Find a Missing Measure

**Find the diameter of a circle with an area of 196π square yards.**

$$A = \pi r^2 \qquad \text{Area of a circle}$$

$$\underline{\hspace{1cm}} = \pi r^2 \qquad A = 196\pi$$

$$\frac{196\pi}{\underline{\hspace{0.5cm}}} = r^2 \qquad \text{Divide each side by } \pi.$$

$$\underline{\hspace{1cm}} = r \qquad \text{Solve. Take the positive square root.}$$

The radius is $\underline{\hspace{1cm}}$ the diameter, so the diameter of the circle is $\underline{\hspace{1cm}}$ yards.

## Check

Find the diameter of a circle with an area of 915 square feet. Round to the nearest tenth if necessary.

$\underline{\hspace{1cm}}$ ft

## Pause and Reflect

Did you struggle with anything in this lesson? If so, how did you deal with it?

Record your observations here

🔵 **Go Online** You can complete an Extra Example online.

# Learn Areas of Sectors

A **sector** is a region of a circle bounded by a central angle and its intercepted arc. The formula for the area of a sector is similar to the formula for arc length.

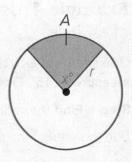

**Go Online**
to see how the formula for the area of a sector is derived.

**Go Online**
to watch a video to see how to find the area of a sector of a circle.

### Key Concept • Area of a Sector

The ratio of the area $A$ of a sector to the area of the whole circle, $\pi r^2$, is equal to the ratio of the degree measure of the intercepted arc $x$ to 360°.

Proportion: $\dfrac{A}{\pi r^2} = \dfrac{x°}{360°}$

Equation: $A = \dfrac{x°}{360°} \cdot \pi r^2$

## 🌐 Example 3 Area of a Sector

**GAMES** Malaya is playing a game where she must track her progress using sectors of a circle and a circular game piece. The game piece has a diameter of 3.5 centimeters and is divided into 6 congruent sectors. What is the area of one sector to the nearest hundredth?

💭 **Think About It!**
What assumptions did you make?

**Step 1 Find the arc measure.**

Because the game piece is equally divided into 6 pieces, each piece will have an arc measure of 360 ÷ _____ = 60°.

**Step 2 Find the area.**

$A = \dfrac{x°}{360°} \cdot \pi r^2$      Area of a sector

$= \dfrac{60°}{360°} \cdot \pi (1.75)^2$      $x =$ _____ and $r =$ _____

$\approx$ _____      Solve.

So, the area of one sector of this game piece is about 1.60 square centimeters.

## Check

**ART** Jorrie is crafting a wall clock using paint and a set of battery-operated clock hands. The clock will have a diameter of 3 feet. What is the area of each sector to the nearest hundredth?

$A =$ _____ $ft^2$

🌐 **Go Online** You can complete an Extra Example online.

## Example 4 Use the Area of a Sector to Find the Area of a Circle

**One sector of a circle has an area of 42 square feet and an arc measure of 45°. Find the area of the circle to the nearest square foot.**

**Step 1  Find the radius.**

Use the area of the sector to find the radius of the circle.

$A = \frac{x°}{360°} \cdot \pi r^2$        Area of a sector

$42 = \frac{45°}{360°} \cdot \pi r^2$        $A =$ _____ and $x =$ _____

$r = \sqrt{\left(\frac{360}{45}\right)\frac{1}{}}$        Solve for $r$.

**Step 2  Find the area.**

Use the radius to calculate the area of the circle.

$A = \pi r^2$        Area of a circle

$= \pi \left( \phantom{xxxxx} \right)^2$        Substitute.

$=$ _____        Solve.

The area of the circle is _____ square feet.

## Check

One sector of a circle has an area of 860 square inches and an arc measure of 60°. Zari found the area of the circle. Provide the justifications for her calculations.

$A = \frac{x°}{360°} \cdot \pi r^2$      _____

$860 = \frac{60°}{360°} \cdot \pi r^2$      _____

$r = \sqrt{860\left(\frac{360}{60}\right)\frac{1}{\pi}}$      _____

$A = \pi r^2$      _____

$= \pi\left(\sqrt{860\left(\frac{360}{60}\right)\frac{1}{\pi}}\right)^2$      _____

$= 5160$ in$^2$      _____

Find each measure for circle $O$ to the nearest tenth.

Area of $O$ (in cm$^2$) _____

Area of sector $AOB$ (in cm$^2$) _____

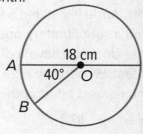

**Go Online**
to practice what you've learned in Lessons 11-1 through 11-3.

**Go Online** You can complete an Extra Example online.

# Practice

**Go Online** You can complete your homework online.

**Example 1**

**Find the area of each circle. Round to the nearest tenth.**

**1.**

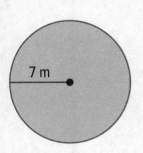

7 m

**2.**

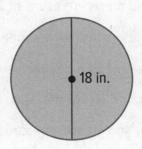

18 in.

**3.**

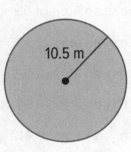

10.5 m

**4.**

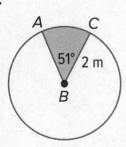

12 in.

**5.**

7.5 in.

**6.**

167 m

**Example 2**

**Find the indicated measure. Round to the nearest tenth.**

**7.** Find the diameter of a circle with an area of 94 square millimeters.

**8.** The area of a circle is 132.7 square centimeters. Find the diameter of the circle.

**9.** The area of a circle is 112 square inches. Find the radius of the circle.

**10.** Find the diameter of a circle with an area of 1134.1 square millimeters.

**11.** The area of a circle is 706.9 square inches. Find the radius of the circle.

**12.** Find the radius of a circle with an area of 2827.4 square feet.

**Example 3**

**Find the area of each shaded sector. Round to the nearest tenth.**

**13.**

A    C
51° 2 m
B

**14.**

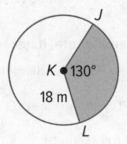

J
K 130°
18 m
L

**15.**

D
12.5 m ● 243°
F    E

**16.** **GAMES** Jason wants to make a spinner for a new board game he invented. The spinner is a circle divided into 8 congruent pieces, what is the area of each piece to the nearest tenth?

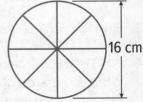

16 cm

(photographer)/Stockbyte/Getty Images

**Example 4**

**Find the area of the circle that contains each sector. Round to the nearest tenth, if necessary.**

**17.** Sector has an area of 210 square centimeters and an arc measure of 30°.

**18.** Sector has an area of 65 square feet and an arc measure of 270°.

**19.** Sector has an area of 325 square millimeters and an arc measure of 72°.

**20.** Sector has an area of 167 square inches and an arc measure of 110°.

**21.** Sector has an area of 98 square meters and an arc measure of 40°.

**22.** Sector has an area of 412 square inches and an arc measure of 82°.

**Mixed Exercises**

**Find the area of each circle. Round to the nearest tenth.**

**23.**

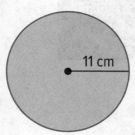

**24.**

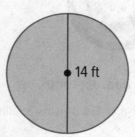

**25.**

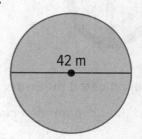

**Find the area of each sector. Round to the nearest tenth.**

**26.**

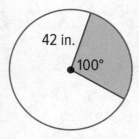

**27.**

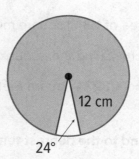

**28.**

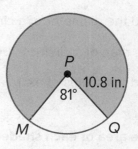

**29.** A sector of a circle has a central angle of 80°. If the circle has a radius of 5.5 inches, what is the area of the sector?

**Find the indicated measure. Round to the nearest tenth, if necessary.**

**30.** The area of a circle is 68 square centimeters. Find the diameter.

**31.** Find the radius of a circle with an area of 206 square feet.

**32.** The area of a circle is 25 square feet. Find the diameter.

**33.** Find the radius of a circle with area of 615.8 square inches.

**34.** PORTHOLES  A circular window on a ship is designed with a radius of 8 inches. What is the area of glass needed for the window? Round your answer to the nearest hundredth.

35. **LOBBY** The lobby of a bank features a large marble circular table for displaying brochures.

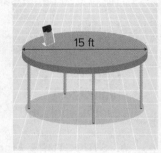

    **a.** The diameter of the circle is 15 feet. What is the area of the circular table? Round your answer to the nearest tenth.

    **b.** If the bank manager adds a floral arrangement with a diameter of 2 feet, how much space remains for brochure displays?

36. **STRUCTURE** A stained-glass artist is making a circle separated into 3 equal sectors with the bottom sector divided equally in two. Suppose the circle has radius *r*. What is the area of each of the larger equal sectors?

37. **SOUP CAN** Jaclynn needs to cover the top and bottom of a can of soup with construction paper to include in her art project. Each circle has a diameter of 7.5 centimeters. What is the total area of the can that Julie must cover?

38. **POOL** A circular pool is surrounded by a circular sidewalk. The circular sidewalk is 3 feet wide. The diameter of the sidewalk and pool is 26 feet.

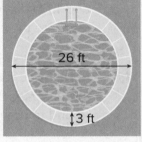

    **a.** What is the diameter of the pool?

    **b.** What is the area of the sidewalk and pool?

    **c.** What is the area of the pool?

39. **REASONING** Explain how to find the area of the shaded region. Then find the area of the shaded region. Round to the nearest tenth if necessary.

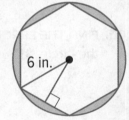

40. **REGULARITY** A sector of a circle has an area *A* and a central angle that measures *x*°. Explain how you can find the area of the whole circle.

41. **STRUCTURE** Given the regular polygon *ABCDEF*. Find the area of the shaded region. Round to the nearest tenth.

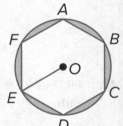

42. **PIE** One sector of an apple pie has an area of 8 square inches and an arc measure of 45°. Find the area of the pie.

43. **SPRINKLER** A lawn sprinkler sprays water in an arc as shown in the picture. The area of the sector covered by the spray is approximately 235.6 square feet. Find the area of the yard that can be covered by the sprinkler if it is set to water a complete circle.

44. **USE TOOLS (ESTIMATION)** Solve the proportion $\frac{m}{360} = \frac{A_s}{\pi r^2}$ to find a formula for the radius of a circle. Then use it to find the radius of a circle that contains a sector with a central angle of 30° that has an area of 100 square inches. Round to the nearest tenth of an inch.

**45.** Lorenzo wants to use sectors of different circles as part of a mural he is painting. The specifications for each sector are shown.

**Red:**
radius: 14 ft
central angle: 60°

**Purple:**
radius: 12 ft
central angle: 75°

**Green:**
radius: 18 ft
central angle: 30°

a. **STRUCTURE** Find the area of each sector to the nearest tenth.

b. **STATE YOUR ASSUMPTION** Was the sector with the largest area the one with the longest radius? Was it the one with the largest central angle? What can you conclude from this?

c. **PRECISION** Lorenzo plans to paint 235 stars on each of the purple sectors and 153 stars on each of the green sectors. To the nearest tenth, how many stars are there per square foot for sectors of each color

d. **REASONING** He also plans to paint stars on the red sectors. He wants there to be twice as many stars per square foot in the red sectors as there are in the green sectors. How many stars should he paint on each red sector? Round to the nearest whole star.

**46.** **FIND THE ERROR** Ketria and Colton want to find the area of a shaded region in the circle shown. Is either of them correct? Explain your reasoning.

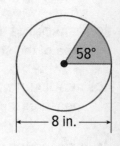

**Ketria**
$$A = \frac{x}{360} \cdot \pi r^2$$
$$= \frac{58}{360} \cdot \pi (8)^2$$
$$= 32.4 \text{ in}^2$$

**Colton**
$$A = \frac{x}{360} \cdot \pi r^2$$
$$= \frac{58}{360} \cdot \pi (4)^2$$
$$= 8.1 \text{ in}^2$$

**47.** **PERSEVERE** Find the area of the shaded region. Round to the nearest tenth.

**48.** **ANALYZE** A **segment of a circle** is the region bounded by an arc and a chord. Is the area of a sector of a circle *sometimes*, *always*, or *never* greater than the area of its corresponding segment?

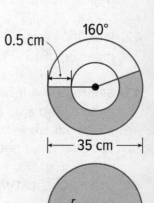

**49.** **WRITE** Describe two methods you could use to find the area of the shaded region of the circle. Which method do you think is more efficient? Explain your reasoning.

**50.** **PERSEVERE** Derive the formula for the area of a sector of a circle using the formula for arc length.

**51.** **WRITE** If the radius of a circle doubles, will the measure of a sector of that circle double? Will it double if the arc measure of that sector doubles?

# Surface Area

**Today's Standards**
G.GMG.3
MP1, MP4, MP8

## Learn Surface Areas of Prisms and Cylinders

The **lateral area** $L$ of a prism is the sum of the areas of the **lateral faces**, which are the faces that join the bases of a solid. The height of a prism or cylinder is the length of the **altitude**, which is the segment perpendicular to the bases that joins the planes of the bases.

**Today's Vocabulary**
lateral area
lateral faces
altitude
regular pyramid
axis
composite solid

| Key Concept • Lateral Area of a Right Prism | |
|---|---|
| Words | The lateral area $L$ of a right prism is $L = Ph$, where $h$ is the height of the prism and $P$ is the perimeter of a base. |
| Symbols | $L = Ph$ |

| Key Concept • Lateral Area of a Right Cylinder | |
|---|---|
| Words | The lateral area $L$ of a right cylinder is $L = 2\pi rh$, where $r$ is the radius of a base and $h$ is the height. |
| Symbols | $L = 2\pi rh$ |

| Key Concept • Surface Area of a Right Prism | |
|---|---|
| Words | The surface area $S$ of a right prism is $S = L + 2B$, where $L$ is the lateral area and $B$ is the area of a base. By substituting $L = Ph$ into the equation, $S = Ph + 2B$, where $P$ is the perimeter of a base and $h$ is the height. |
| Symbols | $S = L + 2B$ or $S = Ph + 2B$ |

| Key Concept • Surface Area of a Right Cylinder | |
|---|---|
| Words | The surface area $S$ of a right cylinder is $S = L + 2B$, where $L$ is the lateral area and $B$ is the area of a base. By substituting $L = 2\pi rh$ and $B = \pi r^2$ into the equation, $S = 2\pi rh + 2\pi r^2$, where $r$ is the radius of a base and $h$ is the height of the cylinder. |
| Symbols | $S = L + 2B$ or $S = 2\pi rh + 2\pi r^2$ |

**Go Online** to learn more about prisms and cylinders.

**Study Tip**

**Right Solids and Oblique Solids**
A solid is a *right solid* if the segment connecting the centers of the bases (for prisms and cylinders) or the center of the base to the vertex (for pyramids and cones) is perpendicular to the base(s). A solid is an *oblique solid* if it is not a right solid.

## Example 1 Lateral Area and Surface Area of a Prism

**Find the lateral area and surface area of the prism. Round to the nearest tenth if necessary.**

**Part A Find the lateral area.**

$L =$ _____          Lateral area of a prism

$= (8 \times 6)11$       The base has 8 sides.

$=$ _____          Solve.

So, the lateral area of the prism is 528 square inches.

*(continued on the next page)*

**Part B Find the surface area.**

**Step 1** Find the area of the base.
A central angle of the octagon is
$\frac{360°}{8}$ or _____ °, so the angle formed in
the triangle is 22.5°.

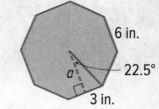

6 in.

22.5°

*a*

3 in.

$\tan 22.5° = \frac{3}{a}$    Write a trigonometric ratio to find *a*.

$a = \frac{3}{\tan 22.5°}$    Solve for *a*.

$\approx 7.2$    Use a calculator.

$A = \frac{1}{2}Pa$    Area of a regular polygon.

$\approx \frac{1}{2}(\underline{\quad})(\underline{\quad})$    $P = 48$ and $a \approx 7.2$

$\approx \underline{\quad}$    Multiply.

So, the area of the base *B* is approximately _____ square inches.

**Step 2** Find the surface area of the prism.

$S = L + 2B$    Surface area of a right prism

$\approx \underline{\quad} + 2(172.8)$    $L = 528$ and $B \approx 172.8$

$\approx 873.6$    Simplify.

The surface area of the prism is about _____ square inches.

**Problem-Solving Tip**

**Use Your Skills** To find the lateral area or surface area of a solid, you may have to use other skills to find missing measures. Remember that you can use trigonometric ratios or the Pythagorean Theorem to solve for missing measures in right triangles.

**Study Tip**

**Estimation** Before finding the lateral area of a cylinder, use mental math to estimate. To estimate, multiply the diameter by 3 (to approximate π) and then by the height of the cylinder.

**Example 2** Lateral Area and Surface Area of a Cylinder

**Find the lateral area and surface area of the cylinder. Round to the nearest tenth.**

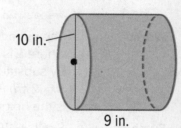

10 in.

9 in.

**Part A Find the lateral area.**

$L = \underline{\quad}$    Lateral area of a cylinder

$= 2\pi(\underline{\quad})(\underline{\quad})$    Substitution

$\approx \underline{\quad}$    Solve.

So, the lateral area of the cylinder is 282.7 square inches.

**Part B Find the surface area.**

$S = 2\pi rh + 2\pi r^2$    Surface area of a cylinder

$\approx 282.7 + 2\pi(\underline{\quad})^2$    Substitute.

$\approx 439.8$    Solve.

The surface area of the cylinder is about _____ square inches.

 **Go Online** You can complete an Extra Example online.

**Go Online** to see Example 3.

# Explore Cone Patterns

 **Online Activity** Use paper folding to complete the Explore.

> ⓠ **INQUIRY** How are the formulas for the circumference and area of a circle related to the formula for the surface area of a right cone?

## Learn Surface Areas of Pyramids and Cones

A **regular pyramid** is a pyramid with a base that is a regular polygon. The height of a pyramid or cone is the length of the **axis**, which is the segment with endpoints that are the centers of the bases.

| Key Concept • Lateral Area of a Regular Pyramid | | |
|---|---|---|
| **Words** | The lateral area $L$ of a regular pyramid is $L = \frac{1}{2}P\ell$, where $\ell$ is the slant height and $P$ is the perimeter of the base. | 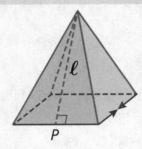 |
| **Symbols** | $L = \frac{1}{2}P\ell$ | |

| Key Concept • Lateral Area of a Right Cone | | |
|---|---|---|
| **Words** | The lateral area $L$ of a right circular cone is $L = \pi r\ell$, where $r$ is the radius of the base and $\ell$ is the slant height. | 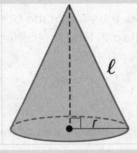 |
| **Symbols** | $L = \pi r\ell$ | |

| Key Concept • Surface Area of a Regular Pyramid | | |
|---|---|---|
| **Words** | The surface area $S$ of a regular pyramid is $S = L + B$, where $L$ is the lateral area and $B$ is the area of the base. By substituting $L = \frac{1}{2}P\ell$, into the equation, $S = \frac{1}{2}P\ell + B$, where $P$ is the perimeter of the base and $\ell$ is the slant height. | 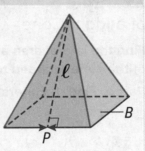 |
| **Symbols** | $S = L + B$ or $S = \frac{1}{2}P\ell + B$ | |

| Key Concept • Surface Area of a Right Cone | | |
|---|---|---|
| **Words** | The surface area $S$ of a right circular cone is $S = L + B$, where $L$ is the lateral area and $B$ is the area of the base. By substituting $L = \pi r\ell$ and $B = \pi r^2$ into the equation, $S = \pi r\ell + \pi r^2$, where $r$ is the radius of the base and $\ell$ is the slant height. | 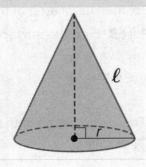 |
| **Symbols** | $S = L + B$ or $S = \pi r\ell + \pi r^2$ | |

ⓔ **Talk About It!**

Describe the similarities and differences between finding the lateral area of a pyramid and the lateral area of a prism.

💭 **Think About It!**

Why does an oblique pyramid or cone not have a slant height?

## Example 4 Lateral Area and Surface Area of a Regular Pyramid

**Find the lateral area and surface area of the pyramid. Round to the nearest tenth if necessary.**

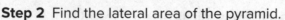

**Part A Find the lateral area.**

**Step 1** Find the perimeter of the base.
The perimeter of the base is ____ × 3 or ____ feet.

**Step 2** Find the lateral area of the pyramid.

$L = \frac{1}{2}P\ell$      Lateral area of a regular pyramid

$= \frac{1}{2}(\underline{\hspace{1cm}})(\underline{\hspace{1cm}})$      $P = 18$ and $\ell = 8$

$= 72$ ft²      Solve.

**Part B Find the surface area.**

**Step 1** Find the length of the apothem and the area of the base.

$a^2 + b^2 = c^2$      Pythagorean Theorem

$a^2 = 6^2 - 3^2$      $c = 6$ and $b = 3$

$a = \sqrt{27}$      Solve.

Calculate the area of the base $B$.

$B = \frac{1}{2}bh$      Area of a triangle

$= \frac{1}{2}(\underline{\hspace{1cm}})\sqrt{27}$      Substitute.

$\approx 15.6$      Solve.

So, the area of the base $B$ is approximately ____ square feet.

**Step 2** Find the surface area of the pyramid

$S = L + B$      Surface area of a regular pyramid

$\approx \underline{\hspace{1cm}} + 15.6$      $L = 72$ and $B \approx 15.6$

$\approx 87.6$      Simplify.

The surface area of the pyramid is about ____ square feet.

## Example 5 Lateral Area and Surface Area of a Right Cone

**Find the lateral area and surface area of the cone rounded to the nearest tenth.**

**Part A Find the lateral area.**

$L = \pi r\ell$      Lateral area of a cone

$= \pi(\underline{\hspace{1cm}})(\underline{\hspace{1cm}})$      Substitution

$\approx \underline{\hspace{1cm}}$      Solve.

So, the lateral area of the cone is 5.5 square millimeters.

**Part B Find the surface area.**

$S = \pi r\ell + \pi r^2$      Surface area of a cone

$\approx 5.5 + \pi(\underline{\hspace{1cm}})^2$      Substitution

$\approx \underline{\hspace{1cm}}$      Solve.

The surface area of the cone is about ____ square millimeters.

**Go Online** You can complete an Extra Example online.

# Example 6 Approximate Surface Areas of Pyramids and Cones

AGRICULTURE Specialty watermelons are carefully cultivated in containers so that the fruits form different geometric shapes. Curt wants to build molds to grow watermelons shaped like triangular pyramids.

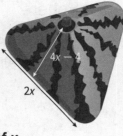

$4x - 4$

$2x$

**Part A Find the surface area of the mold in terms of $x$.**

The watermelon can be modeled by using a regular _____ _____. The perimeter of the base of the pyramid is _____. The slant height of the pyramid is _____.

Calculate the area of the triangular base of the pyramid, $A = \frac{1}{2}bh$. The height of the triangular base is $h = x\sqrt{3}$ and the length is $b =$ _____. Substitute these values into the formula for the area of a triangle to find the area of the base of the pyramid. $B = \frac{1}{2}(2x)(\sqrt{3}x)$ or $\sqrt{3}x^2$

Substitute the values for $P$, $\ell$, and $B$ into the formula for the surface area of a pyramid.

$$S = \frac{1}{2}P\ell + B$$
$$= (12 + \sqrt{3})x^2 - 12x$$

**Part B If $x = 6$, approximate the surface area of the watermelon.**

$$S = (12 + \sqrt{3})x^2 - 12x \qquad \text{Surface area from Part A}$$
$$= (12 + \sqrt{3})6^2 - 12(6) \qquad x = 6$$
$$\approx \underline{\hspace{1cm}} \qquad \text{Solve.}$$

The surface area of the watermelon is about _____ square inches.

# Learn Surface Areas of Spheres

A sphere is the set of all points in space that are a given distance from a given point called the center of the sphere.

| Key Concept • Surface Area of a Sphere | |
|---|---|
| Words | The surface area $S$ of a sphere is $S = 4\pi r^2$, where $r$ is the radius. |
| Symbols | $S = 4\pi r^2$ |

# Example 7 Surface Area of a Sphere

**Find the surface area of the sphere to the nearest tenth.**

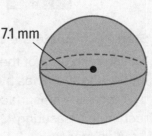

7.1 mm

$$S = 4\pi r^2 \qquad \text{Surface area of a sphere}$$
$$= 4\pi(\underline{\hspace{0.5cm}})^2 \qquad \text{Substitution}$$
$$\approx \underline{\hspace{1cm}} \qquad \text{Solve.}$$

The surface area of the sphere is about 633.5 square millimeters.

## Check

Find the surface area of the sphere to the nearest tenth.

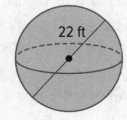

22 ft

$$S = \underline{\hspace{1cm}} \text{ ft}^2$$

 **Go Online** You can complete an Extra Example online.

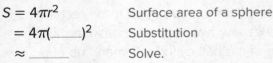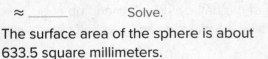

**Go Online** to learn more about the parts spheres.

**Go Online** to watch an animation to learn how you can derive the formula for the surface area of a sphere.

### 🌐 Example 8 Use Formulas to Find the Surface Area of a Sphere

**SPORTS** **If the circumference of the baseball is $C = 3\pi x^2$, find the surface area in terms of $x$ and $\pi$.**

| | | |
|---|---|---|
| $C = 2\pi r$ | Circumference of a sphere |
| _____ $= 2\pi r$ | Substitute. |
| _____ $= r$ | Solve for $r$. |

Use $r$ to find the surface area of the baseball.

| | | |
|---|---|---|
| $S = 4\pi r^2$ | Surface area of a sphere |
| $= 4\pi$ _____ ₂ | Substitute. |
| $=$ _____ | Solve. |

A **composite solid** is a three-dimensional solid that is composed of simpler solids. You can use the formulas you know for calculating the surface areas of prisms, pyramids, cylinders, cones, and spheres to calculate the surface areas of composite solids.

---

### 🌐 Example 9 Surface Area of a Composite Solid

**MANUFACTURING** **A manufacturer wants to know the approximate area of the metal used to create a basic mailbox. Calculate the surface area of the mailbox.**

**Step 1 Choose a model.**

The mailbox can be modeled by a _____ without a top face and one half of a _____.

**Step 2 Draw a net of the composite solid.**

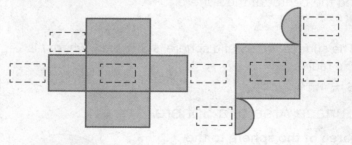

**Step 3 Calculate the surface area.**

The surface area of the square prism is made up of three rectangles measuring 12 inches by 7 inches and two squares measuring 7 inches by 7 inches. The surface area of the half cylinder is made up of a rectangle measuring 12 inches by $3.5\pi$ inches and two half circles with a radius measuring 3.5 inches.

$S = 3(\underline{\quad} \times 7) + 2(7 \times \underline{\quad}) + (12 \times 3.5\pi) + \pi(\underline{\quad})^2$ or about 520.4 square inches.

The surface area of the mailbox is about _____ square inches.

🌐 **Go Online** You can complete an Extra Example online.

Chaikanta/Shutterstock

# Practice

🔵 **Go Online** You can complete your homework online.

**Examples 1, 2, 4, and 5**

**Find the lateral area and surface area of each solid. Round to the nearest tenth if necessary.**

**1.**

12 yd
12 yd  10 yd

**2.**

14 m  5 m

**3.**

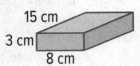

15 cm
3 cm
8 cm

**4.**

12 ft
14 ft

**5.**

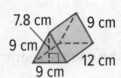

7.8 cm  9 cm
9 cm
9 cm  12 cm

**6.**

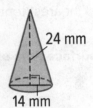

24 mm
14 mm

**7.**

10 in.
12 in.

**8.**

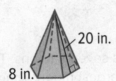

20 in.
8 in.

**9.**

12 yd
6 yd

**10.**

10 ft
25 ft

**11.**

8 in.
12 in.

**12.**

8 in.
6 in.

**Examples 3 and 6**

**13. PAINTING** Greg is painting the four walls of his bedroom and the ceiling.

**Part A** If the height of the walls is $x$ and the edge length of the square ceiling is $2x$, approximate the surface area Greg will be painting in terms of $x$.

**Part B** Approximate the surface area that will be painted to the nearest tenth if $x = 8$ feet.

**14. MANUFACTURING** A food distribution manufacturer is developing a new cylindrical package with a cardboard bottom and sides and a plastic lid. They are evaluating the cost of manufacturing based on the amount of cardboard used.

**Part A** If the radius of the package is $x$ and the height is $x + 4$, approximate the surface area of the new package in terms of $x$ and $\prec$.

**Part B** Approximate the surface area that will be cardboard to the nearest tenth if $x = 6$ centimeters.

**15. CAMPING** A company that manufactures camping gear is designing a new tent shaped like a square pyramid with sidewalls made of a waterproof material.

**Part A** If the base of tent is $x$ units long and the slant height of the walls is 1.5$x$ units, approximate the surface area of the sidewalls in terms of $x$.

**Part B** Approximate the amount of material to manufacture the sidewalls if $x = 9$ feet.

**16. TOPIARY** Davea is planning to prune her landscaping bushes into topiaries the shape of cones.

**Part A** If the radius of the bush is $\frac{1}{2}x$ units and the slant height is 4$x$ units, approximate the surface area of one topiary in terms of $x$ and $\pi$.

**Part B** A frost is expected and Davea is making plastic slipcovers to protect her new topiaries. Approximate the surface area of one slipcover to the nearest tenth if $x = 0.75$ meter.

**Example 7**

**Find the surface area of each sphere to the nearest tenth.**

**17.**

7 in.

**18.**

32 m

**19.**

4.8 mm

**20. MOONS OF SATURN** The planet Saturn has several moons. These can be modeled accurately by spheres. Saturn's largest moon, Titan, has a radius of about 2575 kilometers. What is the approximate surface area of Titan? Round your answer to the nearest tenth.

**Example 8**

**21. AMUSEMENT PARK** Spaceship Earth at Disney's Epcot Center is a sphere with a circumference of 518.1 feet. What is the surface area of Spaceship Earth? Round your answer to the nearest tenth.

**22. BILLIARDS** The eight-ball has a circumference of approximately 7.07 inches. What is the surface area of the eight-ball? Round your answer to the nearest tenth.

**Example 9**

**Find the surface area of each figure to the nearest tenth.**

**23.**

8 cm
4 cm
6 cm

**24.**

24 mm
18 mm

**25.**

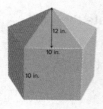

12 in.
10 in.
10 in.

**26.**

3.5 cm
13 cm

## Mixed Exercises

**27.** Find the surface area of the prism.

**28.** **STRUCTURE** A cylinder has a lateral area of $120\pi$ square meters, and a height of 7 meters. Find the radius of the cylinder. Round to the nearest tenth.

**REASONING** Find the lateral area and surface area of each solid. Round to the nearest tenth if necessary.

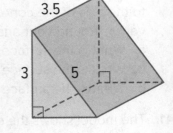

**29.** a rectangular prism with length of 25 centimeters, width of 18 centimeters, and height of 12 centimeters

**30.** a triangular prism with height of 6 inches, right triangle base with legs of 9 inches and 12 inches

**31.** a square pyramid with an altitude of 12 inches and a slant height of 18 inches

**32.** a hexagonal pyramid with a base edge of 6 millimeters and a slant height of 9 millimeters

**33.** a cone with a diameter of 3.4 centimeters and a slant height of 6.5 centimeters

**34.** a cone with an altitude of 5 feet and a slant height of $9\frac{1}{2}$ feet

**35.** The *great circle* of a sphere lies on a plane that passes through the center of the circle. The diameter of a sphere's great circle is the diameter of the sphere.
   **a.** Find the surface area of a sphere with a great circle that has a circumference of $2\pi$ centimeters. Round to the nearest tenth if necessary.
   **b.** Find the surface area of a sphere with a great circle that has an area of about 32 square feet. Round to the nearest tenth if necessary.

**36.** **CLUB HOUSE** Martha's greenhouse is shaped like a square pyramid with four congruent equilateral triangles for its sides. All of the edges are 6 feet long. What is the total surface area of the clubhouse including the floor? Round your answer to the nearest hundredth.

**37.** **PAPER MODELS** Prevan is making a paper model of a castle. Part of the model involves cutting out the net shown and folding it into a pyramid. The pyramid has a square base. What is the surface area of the resulting pyramid?

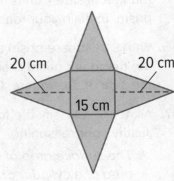

**38.** **CAKES** A cake is a rectangular prism with height 4 inches and base 12 inches by 15 inches. Wallace wants to apply frosting to the sides and the top of the cake. What is the surface area of the part of the cake that will have frosting?

**39.** **CONSTRUCTION** A metal pipe is shaped like a cylinder with a height of 50 inches and a radius of 6 inches. What is the surface area of the pipe? Round your answer to the nearest hundredth.

**40. INSTRUMENTS** A mute for a brass wind instrument is formed by taking a solid cone with a radius of 10 centimeters and an altitude of 20 centimeters and cutting off the tip. The cut is made along a plane that is perpendicular to the axis of the cone and intersects the axis 6 centimeters from the vertex. Round your answers to the nearest hundredth.

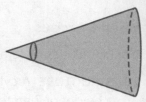

   a. What is the surface area of the original cone?
   b. What is the surface area of the tip that is removed?
   c. What is the surface area of the mute?

**41.** The model shows the dimensions of a sofa.

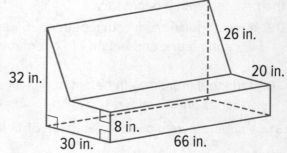

   a. **MODELING** The model shows the dimensions of a sofa. Draw a diagram to show how to calculate the total surface area of the sofa that would be covered by a fitted cover. Explain your technique.
   b. **REASONING** How much fabric is needed for a fitted sofa cover?

**42. STRUCTURE** Marcus builds a sphere inside of a cube. The sphere fits snugly inside the cube so that the sphere touches the cube at one point on each side. The side length of the cube is 2 inches.

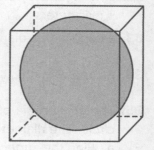

   a. What is the surface area of the cube?
   b. What is the surface area of the sphere? Round your answers to the nearest hundredth.
   c. What is the ratio of the surface area of the cube to the surface area of the sphere? Round your answer to the nearest hundredth.

**43. WRITE** Compare and contrast finding the surface area of a prism and finding the surface area of a cylinder.

**44. CREATE** Given an example of two cylinders that have the same lateral area and different surface areas. Find the lateral area and surface areas of each.

**45. PERSEVERE** A right prism has a height of $h$ units and a base that is an equilateral triangle of side $\ell$ units. Find the general formula for the total surface area of the prism. Explain your reasoning.

**46. WRITE** A square prism and a triangular prism are the same height. The base of the triangular prism is an equilateral triangle, with an altitude equal in length to the side of the square. Compare the lateral areas of the prisms.

**47. ANALYZE** Classify the following statement as *sometimes*, *always*, or *never* true. Justify your reasoning.

   *The surface area of a cone of radius r and height h is less than the surface area of a cylinder of radius r and height h.*

**48. ANALYZE** A cone and a square pyramid have the same surface area. If the areas of their bases are also equal, do they have the same slant height as well? Explain your reasoning.

**49. CREATE** Describe a pyramid that has a total surface area of 100 square units.

# Cross Sections and Solids of Revolution

## Explore Cross Sections

**Online Activity** Use concrete models to complete the Explore.

**INQUIRY** How would you have to cut a polyhedron with *n* faces so that the cross section formed has *n* sides?

## Learn Cross Sections

A **cross section** is the intersection of a solid figure and a plane. The shape of the cross section formed by the intersection of the plane and the figure depends on the angle of the plane.

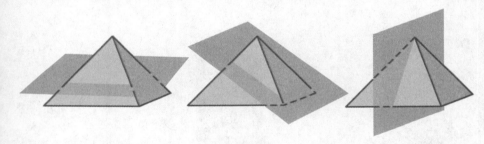

When a plane intersects the face of a solid, one edge of the cross section is formed. When visualizing the shape of the cross section, you can determine the number of sides by counting the number of faces that the plane intersects.

While you must consider a range of possible intersections when determining the cross sections of a solid, one way to start is to determine whether the solid has three-dimensional symmetry.

| Three-Dimensional Symmetry: Plane Symmetry | |
|---|---|
| **Words** | A three-dimensional figure has **plane symmetry** if the figure can be mapped onto itself by a reflection in a plane. |
| **Model** | 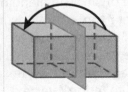 |

### Today's Standards
G.GMD.4
MP3, MP7, MP8

### Today's Vocabulary
cross section

plane symmetry

solid of revolution

axis symmetry

### Go Online
to watch a video to see how to identify the shape of a cross section of a three-dimensional figure.

### Talk About It!
How could you intersect the right prism to form a pentagonal cross section?

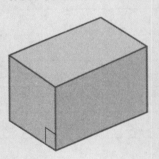

## Example 1 Plane Symmetry

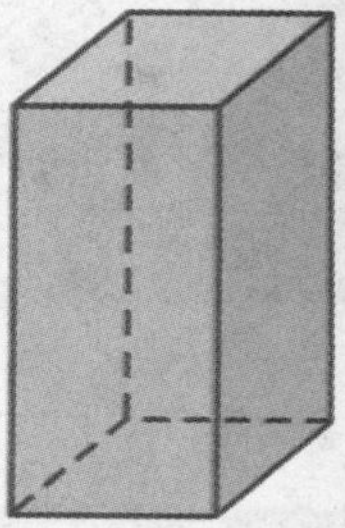

**Describe each plane of symmetry for the rectangular prism.**

The rectangular prism has _____ plane(s) of symmetry.

Choose the correct representation(s) for the plane(s) of symmetry. Select all that apply. _____, _____, _____

**A.**

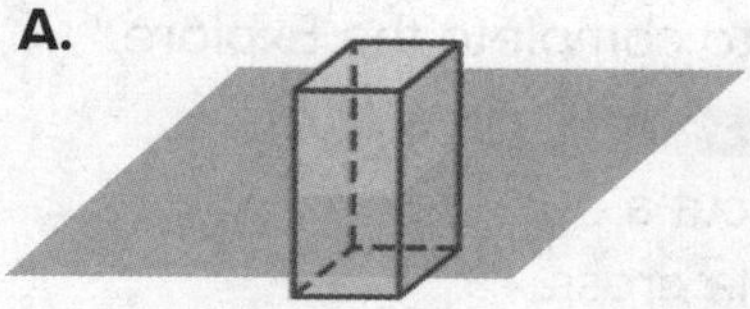

**B.**

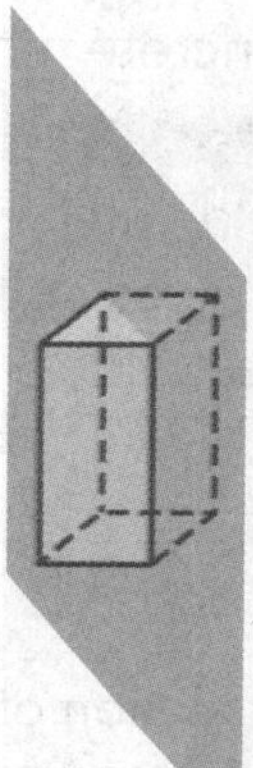

**C.**

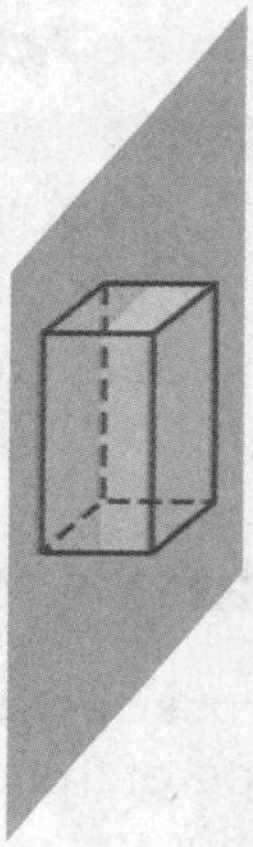

**D.**

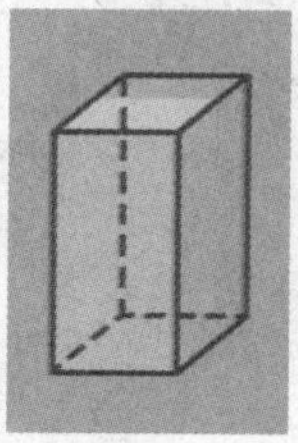

**E.**

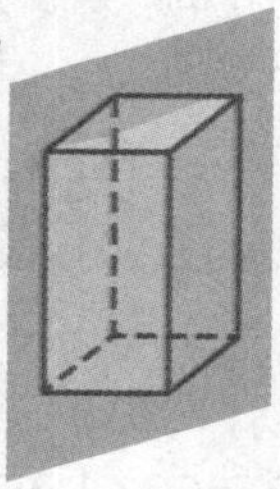

## Check

Describe each plane of symmetry for the regular pentagonal prism.

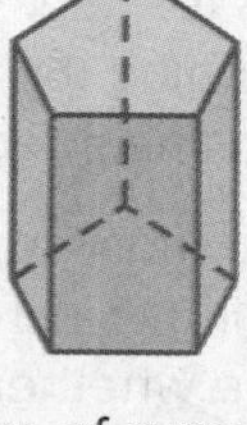

**Part A** The regular pentagonal prism has _____ planes of symmetry.

**Part B** Circle the correct term to complete each sentence.

The regular pentagonal prism has one plane of symmetry that is [**parallel / perpendicular**] to the base of the prism. The plane of symmetry passes through the [**base / center / top**] of the prism. The regular pentagonal prism has five planes of symmetry that are [**parallel / perpendicular**] to the base of the prism. The planes of symmetry each pass through the [**bases / faces**] at one vertex and through the [**edge / midpoint**] of the opposite [**edge / vertex**].

**Go Online** You can complete an Extra Example online.

# Example 2 Identify Cross Sections of Solids

**Identify the shape of each cross section of the cone.**

| | |
|---|---|
| 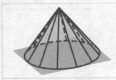 | When a plane parallel to the base of the cone intersects the cone, the cross section is a _____. In fact, this will always be the case unless the plane intersects the _____. |
|  | When a plane perpendicular to the base of the cone intersects the cone, it does so along its curved lateral surface and through its base. This shape has a curved surface and one flat edge. The curve formed has a special name called a _____. |
|  | However, the cross section formed by the intersection of a perpendicular plane and the vertex of the cone is a _____. |
|  | When the plane has the same slope as the slant height of the cone, the intersection is along its curved lateral surface and through its base. This shape has a curved surface and one flat edge. The curve formed has a special name called a _____. |
|  | When the plane has a slope different than the slant height of the cone, it can intersect the cone in an oval, or _____. |

**Go Online**
You may want to complete the Concept Check to check your understanding.

## Learn Solids of Revolution

A **solid of revolution** is a solid figure obtained by rotating a plane figure or curve around an axis. The shape of the solid of revolution depends on the location of the axis and the shape of the plane figure or curve being rotated.

You can determine whether a three-dimensional figure could have been created by rotation by identifying whether the solid figure has axis symmetry.

| Three-Dimensional Symmetry: Axis Symmetry | |
|---|---|
| **Words** | A three-dimensional figure has **axis symmetry** if the figure can be mapped onto itself by a rotation between 0° and 360° in a line. |
| **Model** |  |

**Go Online**
You can watch a video to see how to generate a three-dimensional figure by rotating a two-dimensional figure about a line.

**Go Online** You can complete an Extra Example online.

## Example 3  Identify Solids of Revolution

**Identify the solid formed by rotating the two-dimensional shape about line ℓ.**

Imagine rotating the two-dimensional figure about line ℓ.

The solid of revolution is a _____.

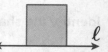

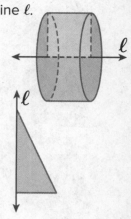

### Check

Identify the solid formed by rotating the two-dimensional shape about line ℓ.

The solid of revolution is a _____.

🔘 **Go Online** You can complete an Extra Example online.

---

## 🌐 Example 4  Axis Symmetry

**Circle each item with axis symmetry.**

### Check

Select all of the images that demonstrate axis symmetry. _____, _____, _____

**A.**   **B.**   **C.**  **D.**  **E.**

**A.** Lightbulb

**B.** Crab apple tree

**C.** Drum

**D.** Armadillo rolled up in a ball

**E.** Hour glass

🔘 **Go Online** You can complete an Extra Example online.

---

## Math History Minute

Russian **Sofia Kovalevskaya** (1850–1891) was a great mathematician, writer, and passionate advocate for women's rights. In 1888, she entered a paper, "On the Rotation of a Solid Body about a Fixed Point," in a competition for the Prix Bordin by the French Academy of Science, and she won. Sofia considered it her greatest personal triumph.

## Study Tip

**Use Tools** You can use many different tools to help you visualize the solids of revolution. Straws or dowel rods could be used to represent the axis of rotations. Card stock or heavy construction paper can be attached to the straws or dowel rods to represent the two-dimensional figures. You could also sketch the figure using dynamic geometry software or graph paper.

Photography/iStockphoto/Getty Images; Don Mason/JUPITERIMAGES/Brand X/Alamy; Lokibaho/iStock/Getty Images; Ingram Publishing; belizar/Shutterstock; Leigh Prather/Shutterstock

# Practice

Go Online You can complete your homework online.

### Example 1

**Describe each plane of symmetry for each figure.**

**1.**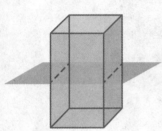

**2.**

**3.**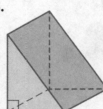

### Example 2

**Identify the shape of each cross section.**

**4.**

**5.**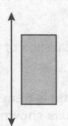

**6.**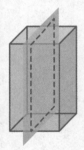

**7.**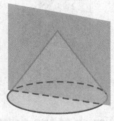

### Example 3

**Identify the solid formed by rotating the two-dimensional shape about the line.**

**8.**

**9.**

**10.**

**11.**

**Example 4**

**Determine whether each item has axis symmetry. Write *yes* or *no*.**

**12.** traffic cone with a base

**13.** one pound weight

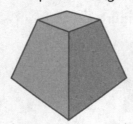

**14.** flower pedal and stem

Mixed Exercises

**Describe each plane of symmetry for each figure.**

**15.**

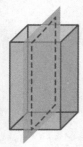

**16.**

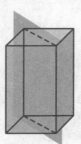

**17.**

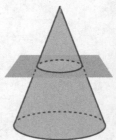

**Identify the shape of each cross section.**

**18.**

**19.**

**20.**

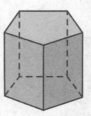

**For Exercises 21 and 22, use the figure at the right.**

**21.** Name the shape of the cross section cut parallel to the base.

**22.** Name the shape of the cross section cut perpendicular to the base.

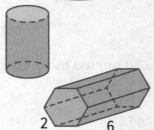

2     6

**23.** Sketch the cross section from a vertical slice of the figure shown at the right.

**Describe the three-dimensional shape that is generated by rotating the two-dimensional shape around the axis shown.**

**24.**

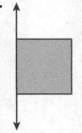

**25.**

**26.**

Name _____ Period _____ Date _____

**Determine if each item has axis symmetry. Write *yes* or *no*.**

**27.** golf ball

**28.** tree

**29.** lamp

**30. BASKETBALL** Reshan rotated a semi-circle about the axis shown to create a basketball, or sphere. Describe another way Reshan could have rotated a semi-circle about an axis to generate the basketball.

**31. WASHER** Consider the washer shown. Describe the two-dimensional shape and axis that could be used to generate the washer by rotating the two-dimensional shape around the axis.

**32. TRIANGLES** Sallie is going to rotate the right triangle around the axis shown. Describe a real-world object that could be generated.

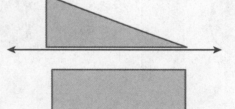

**33. RECTANGLES** Ray is going to rotate the rectangle about the axis shown.

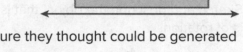

Allen and Patricia selected a real-world figure they thought could be generated by the rotation. Who is correct?

Allen

Patricia

**34.** The rectangle shown will be rotated about the *x*-axis.

**a.** Describe the three-dimensional shape that is generated.

**b. REASONING** What is the length of the radius and height of the figure generated?

**c. STRUCTURE** What is the volume of the figure generated in terms of π?

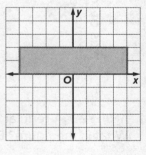

**35. USE TOOLS** Determine whether each cross section can be made from a cube. If so, sketch the cube and its cross section.

**a.** triangle

**b.** square

**c.** rectangle

**d.** pentagon

**e.** hexagon

**f.** octagon

**36. USE A SOURCE** Research the shape and dimensions of the Washington Monument.

 **a.** Describe each plane of symmetry of the Washington Monument.

 **b.** Describe the cross section of the Washington Monument from a vertical slice perpendicular to the base through the vertex.

 **c.** Determine if the Washington Monument has axis symmetry. Write *yes* or *no*.

**MODELING** **Identify and sketch the cross section of an object made by each cut described.**

**37.** square pyramid cut perpendicular to base but not through the vertex

**38.** rectangular prism cut diagonally from a top edge to a bottom edge on the opposite side

**39. ARGUMENTS** You want to cut a geometric object so that the cross section is a circle. There are three objects shown. Give the name for each object. Then describe the cut that results in a circle.

 **a.**

 **b.**

 **c.**

**40. USE TOOLS** Sketch and describe the object that is created by rotating each shape around the indicated axis of rotation.

 **a.**

 **b.**

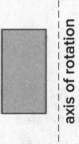

 **c.**

**41. ANALYZE** A regular polyhedron has axis symmetry of order 3, but does not have plane symmetry. What is the figure? Explain.

**42. PERSEVERE** The figure at the right is a cross section of a geometric solid. Describe a solid and how the cross section was made.

**43. WRITE** A hexagonal pyramid is sliced through the vertex and the base so that the prism is separated into two congruent parts. Describe the cross section. Is there more than one way to separate the figure into two congruent parts? Will the shape of the cross section change? Explain.

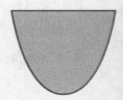

**44. CREATE** Sketch a real-world object that plane symmetry, but not axis symmetry.

# Volumes of Prisms and Pyramids

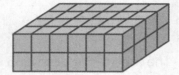

## Learn Volume of Prisms

Recall that the volume of a solid is the measure of the amount of space that the solid encloses. Volume is measured in cubic units.

This rectangular prism has 6 · 4 or 24 cubic units in the bottom layer.

Because there are two layers, the total volume is 24 · 2 or 48 cubic units.

### Key Concept • Volume of a Prism

| | |
|---|---|
| **Words** | The volume $V$ of a prism is $V = Bh$, where $B$ is the area of a base and $h$ is the height of the prism. |
| **Symbols** | $V = Bh$ |
| **Model** | 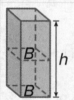 |

### Key Concept • Cavalieri's Principle

| | |
|---|---|
| **Words** | If two solids have the same height $h$ and the same cross-sectional area $B$ at every level, then they have the same volume. |
| **Models** |  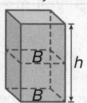 |

## Example 1 Volume of a Prism

**Find the volume of the prism.**

**Step 1 Find the area of the base $B$.**

$B = \frac{1}{2} aP$      Area of a regular polygon

$\quad = \frac{1}{2}(\underline{\quad})(\underline{\quad})$ or 42.5    $a = 3.4$ and $P = 25$

So, the area of the base of the prism is 42.5 mm².

**Step 2 Find the volume of the prism.**

$V = \underline{\quad}$      Volume of a prism

$\quad = 42.5 (8)$ or $\underline{\quad}$    $B = 42.5$ and $h = 8$

The volume of the prism is 340 cubic millimeters.

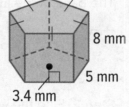

8 mm
5 mm
3.4 mm

## Check

Find the volume of the prism.

$V = \underline{\quad}$ ft³

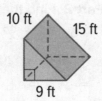

10 ft
15 ft
9 ft

🅑 **Go Online** You can complete an Extra Example online.

### Today's Standards
G.GMD.1, G.GMD.2, G. GMD.3
MP1, MP4, MP7

### Watch Out!

**Cross-Sectional Area**
For solids with the same height to have the same volume, their cross sections must have the same area. The cross sections of the different solids do not have to be congruent polygons.

# Example 2 Volume of an Oblique Prism

**Find the volume of the oblique prism.**

**Part A Find the area of the base.**

The base of the prism is a trapezoid.

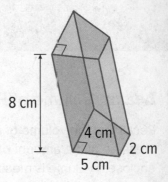

8 cm

4 cm

2 cm

5 cm

$A = \frac{1}{2} h (b_1 + b_2)$     Area of a trapezoid

$= \frac{1}{2} (\underline{\quad}) (4 + \underline{\quad})$ or $\underline{\quad}$     $h = 5, b_1 = 4,$ and $b_2 = 2$

The area of the base is 15 square centimeters.

**Part B Find the volume.**

$V = Bh$     Volume of a prism

$= (\underline{\quad}) (\underline{\quad})$     $B = 15$ and $h = 8$

$= 120$ cm$^3$

The volume of the oblique solid is $\underline{\quad}$ cubic centimeters.

## Check

Find the volume of the oblique rectangular prism.

$V = \underline{\quad}$ m$^3$

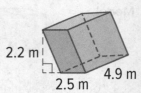

2.2 m

4.9 m

2.5 m

**Watch Out!**

**Area and Volume**

Area is two-dimensional, so it is measured in square units. Volume is three-dimensional, so it is measured in cubic units.

---

# 🌐 Example 3 Volume of a Prism Using Algebraic Expressions

**SNACKS** Gustavo wants to calculate the volume of juice in his juice box.

**Part A Find the volume of the juice box in terms of x.**

PURE Juice

$4x + 3$

$x$    $4x$

**Step 1** Find the area of the base.

The base of the juice box is a rectangle.

$A = \ell w$     Area of a rectangle

$= (\underline{\quad})x$     Substitution

$= \underline{\quad}$     Solve.

**Step 2** Find the volume.

$V = Bh$     Volume of a prism

$= (4x^2)(4x + 3)$     Substitute.

$= \underline{\qquad\qquad}$     Simplify.

**Part B Find the volume of the juice box if x = 2.**

$V = 16x^3 + 12x^2$     Surface area of a cylinder with one base

$= 16(2)^3 + 12(2)^2$     Substitute.

$= \underline{\quad}$ in$^3$     Simplify.

🖱 **Go Online** You can complete an Extra Example online.

## Explore Volumes of Square Pyramids

**Online Activity** Use the guiding exercises to complete the Explore.

> **INQUIRY** How does the formula for the volume of a square pyramid relate to the formula for the volume of a step pyramid with an infinite number of steps?

## Learn Volumes of Pyramids

| Key Concept • Volume of a Pyramid | |
|---|---|
| **Words** | The volume of a pyramid is $V = \frac{1}{3}Bh$, where $B$ is the area of the base and $h$ is the height of the pyramid. |
| **Symbols** | $V = \frac{1}{3}Bh$ |
| **Model** |  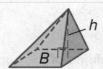 |

**Go Online** to watch a video to learn how you can derive the formula for the volume of a pyramid.

## Example 4 Volume of a Pyramid

**Find the volume of the pyramid.**

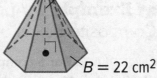

$V =$  \qquad Volume of a pyramid

$\approx 36.7 \text{ cm}^3$  \qquad Substitute and simplify.

The volume of the pyramid is about _____ cubic centimeters.

## Example 5 Volume of a Pyramid Using Algebraic Expressions

**SOUVENIRS Martín bought a bank shaped like a square pyramid, and he wants to calculate its volume.**

**Part A Find the volume of the pyramid in terms of $x$.**

Find the area $B$ of the square base.

$B = (8x)^2$ or _____

Find the height of the pyramid by using the Pythagorean Theorem.

$a^2 + b^2 = c^2$  \qquad Pythagorean Theorem

$(4x)^2 + h^2 = (5x)^2$  \qquad $a = 4x$, $b = h$, and $c = 5x$

$h =$ _____  \qquad Solve for $h$.

Find the volume of the pyramid in terms of $x$.

$V = \frac{1}{3}Bh$  \qquad Volume of a pyramid

$= \frac{1}{3}($ _____ $)($ _____ $)$  \qquad Substitute.

$=$ _____  \qquad Simplify.

**Part B Find the volume in cubic inches if $x = 4$.**

Use the formula in **Part A** to find the volume when $x = 4$.

$V =$ _____ $\text{in}^3$

**Go Online** You can complete an Extra Example online.

## Example 6 Volumes of Composite Solids

**Find the volume of the composite solid.**

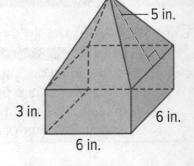

The composite solid is a combination of
a prism and a square pyramid.

**Part A Find the volume of the prism.**

$V = Bh$       Volume of a prism

$= 36(3)$     $B = 36$ and $h = 3$

$= $ _____      Solve.

So, the volume of the prism is _____ cubic inches.

**Part B Find the volume of the pyramid.**

$V = \frac{1}{3}Bh$      Volume of a pyramid

$= \frac{1}{3}(36)(4)$    $B = 36$ and $h = 4$

$= $ _____      Simplify.

The volume of the pyramid is _____ cubic inches.

**Part C Find the volume of the composite solid.**

The volume of the composite solid is _____ cubic inches.

🌎 **Go Online**
to see a common error
to avoid.

## 🌐 Example 7 Approximate Volumes Using Composite Solids

CHOCOLATE **For a competition, a chocolatier created a replica of the Washington Monument made entirely of white chocolate. Approximate the volume of chocolate used to create the sculpture.**

The sculpture can be approximated by a
composite solid made up of a square prism
and a square pyramid.

**Step 1 Find the volume of the square prism.**

$V = Bh$      Volume of a prism

$= 42.25(36)$    $B = 42.25$ and $h = 36$

$= $ _____      Simplify.

The volume of the square prism is
1521 cubic inches.

**Step 2 Find the volume of the square pyramid. Round to the nearest tenth if necessary.**

$V = \frac{1}{3}Bh$      Volume of a pyramid

$= \frac{1}{3}(42.25)(6.5)$    $B = 42.25$ and $h = 6.5$

$\approx$ _____      Simplify.

**Step 3 Find the volume of the sculpture.**

The volume of the sculpture is _____ cubic inches.

🌎 **Go Online** You can complete an Extra Example online.

Name _____ Period _____ Date _____

# Practice

⊘ **Go Online** You can complete your homework online.

Examples 1 and 2

**Find the volume of each prism.Round your answer to the nearest tenth if necessary.**

**1.**

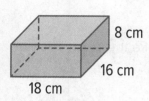
8 cm
16 cm
18 cm

**2.**

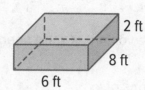
2 ft
8 ft
6 ft

**3.**

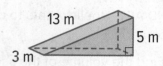
13 m
5 m
3 m

**4.**

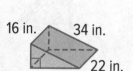
16 in.
34 in.
22 in.

**5.**

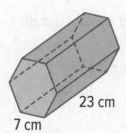
23 cm
7 cm

**6.**

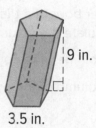
9 in.
3.5 in.

**7.**

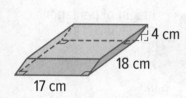
4 cm
18 cm
17 cm

**8.**

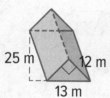
25 m
12 m
13 m

Example 3

**9. CHOCOLATE** Leah wants to calculate the volume of chocolate in her candy bar.

   **a.** Find the volume of the candy bar in terms of x.

   **b.** Find the volume of the candy bar if x = 2 inches.

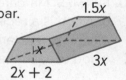
1.5x
x
3x
2x + 2

**10. CANDLE** Benton wants to calculate the volume of wax needed to make a candle.

   **a.** Find the volume of the candle in terms of x.

   **b.** Find the volume of the candle if x = 3 centimeters.

x mm
x m
x + 3

Example 4

**Find the volume of each pyramid. Round to the nearest tenth if necessary.**

**11.**

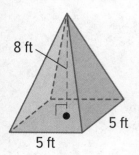
8 ft
5 ft
5 ft

**12.**

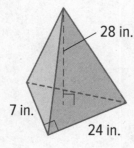
28 in.
7 in.
24 in.

**13.**

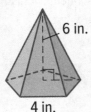
6 in.

4 in.

**14.**

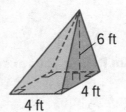
6 ft

4 ft

4 ft

**Example 5**

**15. SAND ART** Noelle wants to calculate the volume of sand needed to make the art decoration.

    **a.** Find the volume of the sand art in terms of $x$.

    **b.** Find the volume of the sand art if $x = 5$ inches.

1.5$x$

$x$

**16. PUZZLE** Roger bought a triangular pyramid shaped puzzle. He wants to calculate the volume of the pyramid.

    **a.** Find the volume of the pyramid in terms of $x$.

    **b.** Find the volume of the pyramid if $x = 10$ centimeters.

$\frac{3}{4}x$  $x$

**Example 6**

**Find the volume of each composite solid.**

**17.**

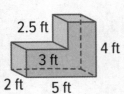
2.5 ft

4 ft

3 ft

2 ft  5 ft

**18.**

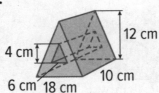
12 cm

4 cm

10 cm

6 cm 18 cm

**19.**

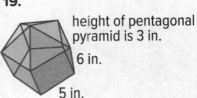
height of pentagonal pyramid is 3 in.

6 in.

5 in.

**Example 7**

**20. FRAMES** Margaret makes a square frame out of four pieces of wood. Each piece of wood is a rectangular prism with a length of 40 centimeters, a height of 4 centimeters, and a depth of 6 centimeters. What is the total volume of the wood used in the frame?

**21. PARKS** Grimby local park is installing new animal-proof trashcans. Approximate the volume of the trashcan.

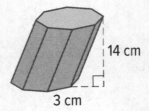
18 in.

18 in.

22 in.

26 in.

**Mixed Exercises**

**Find the volume of each figure. Round to the nearest tenth if necessary.**

**22.**

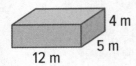
4 m

5 m

12 m

**23.**

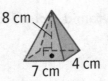
8 cm

7 cm  4 cm

**24.**

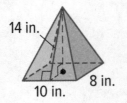
14 in.

10 in.  8 in.

**25.**

14 cm

3 cm

**26.** A rectangular prism has a length of 16 feet, a width of 9 feet, and a height of 8 feet. Find the volume of the prism.

**27.** A pyramid has a height of 18 centimeters and a base with an area of 26 square centimeters. Find the volume.

**28.** **BENCH** Inside a lobby, there is a bench shaped like a simple block with a square base 6 feet on each side and a height of $1\frac{3}{5}$ feet. What is the volume of the bench?

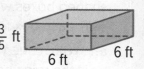

$1\frac{3}{5}$ ft

6 ft        6 ft

**29.** **TUNNELS** Construction workers are digging a tunnel through a mountain. The space inside the tunnel is going to be shaped like a rectangular prism. The mouth of the tunnel will be a rectangle 20 feet high and 50 feet wide and the length of the tunnel will be 900 feet.

   **a.** What volume of rock must be removed to make the tunnel?

   **b.** If instead of a rectangular shape, the tunnel had a semicircular shape with a 50-foot diameter, what would be its volume? Round your answer to the nearest cubic foot.

**30.** **GREENHOUSES** A greenhouse has the shape of a square pyramid. The base has a side length of 30 yards. The height of the greenhouse is 18 yards. What is the volume of the greenhouse available for growing plants?

18 yd

30 yd

**31.** **STAGES** A solid wooden stage is made out of oak, which has a weight of about 45 pounds per cubic foot. The stage has the form of a square pyramid with the top sliced off along a plane parallel to the base. The side length of the top square is 12 feet and the side length of the bottom square is 16 feet. The height of the stage is 3 feet.

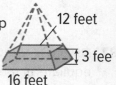

12 feet

3 feet

16 feet

   **a.** What is the volume of the entire square pyramid of which the stage is a part?

   **b.** What is the volume of the top of the pyramid that is removed to create the stage?

   **c.** What is the volume of the stage?

   **d.** What will be the weight of the stage?

**32.** Consider the prisms.

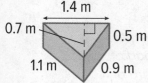

1.4 m

0.7 m

0.5 m

1.1 m   0.9 m

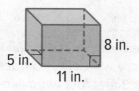

8 in.

5 in.

11 in.

   **a.** What is the volume of the triangular prism?

   **b.** What is the volume of the rectangular prism?

   **c.** **PRECISION** Discuss how the formulas that you used to find the volume of each prism are similar.

**33.** Benjamin finds that baking soda has a mass of 2.2 grams per cubic centimeter and corn flakes have a mass of 0.12 gram per cubic centimeter. A box of baking soda has dimensions of 8 centimeters long by 4 centimeters wide by 12 centimeters high. The dimensions for a box of corn flakes are 30 centimeters long by 6 centimeters wide by 35 centimeters high. Benjamin wants to find the mass of the contents of each box if it is filled to within 2 centimeters of the top.

   **a.** **MODELING** How can Benjamin determine the mass of the contents of each box?

   **b.** **PRECISION** Find the mass of the contents of each box.

**34. STRUCTURE** A model pyramid has a volume of 270 cubic feet and a base area of 90 square feet. What is the height if the pyramid is a right pyramid? What is the height if the pyramid is an oblique pyramid? Explain your reasoning.

**35.** Tristan makes and sells sugar-free candies. She packages them in pyramid-shaped boxes with a 2-inch by 2-inch base and a height of 3 inches. She sells each box for $2.00.

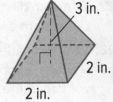

3 in.

2 in.

2 in.

    **a. STURCTURE** What is the volume of sugar-free candies in each box. What is the price per cubic inch?

    **b. REGULARITY** Tristan wants to make a bigger package by doubling the lengths of the sides of the square base. How can she figure out how much to charge if she wants to keep the price per cubic inch the same?

    **c. REASONING** Tristan wants to design a box in the shape of a square-based pyramid that holds between 7 and 8 cubic inches of sugar-free candies. She wants the height to be within $1\frac{1}{4}$ inches of the length of each side of the square. What is one possible set of dimensions that she can use?

**36. ARGUMENTS** Alonzo is building a box in the shape of a right triangular prism for his magic act. Inside the box, he is making a secret compartment. The compartment will be a pyramid with base △*ABE* and vertex at point *C*. After the secret compartment has been made, how is the volume of the space remaining inside the box related to the volume of the secret compartment? Explain your reasoning.

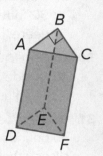

**37. FIND THE ERROR** Francisco and Valerie each calculated the volume of an equilateral triangular prism with an apothem of 4 units and height of 5 units. Is either of them correct? Explain your reasoning.

| Francisco | Valerie |
|---|---|
| $V = Bh$ | $V = Bh$ |
| $= \frac{1}{2}aP \cdot h$ | $= \frac{\sqrt{3}}{2}s^2 \cdot h$ |
| $= \frac{1}{2}(4)(24\sqrt{3}) \cdot 5$ | $= \frac{\sqrt{3}}{2}(4\sqrt{3})^2 \cdot 5$ |
| $= 240\sqrt{3}$ cubic units | $= 120\sqrt{3}$ cubic units |

**38. WRITE** Write a helpful response to the following question posted on an Internet gardening forum. *I am new to gardening. The nursery will deliver a truckload of soil, which they say is 4 yards. I know that a yard is 3 feet, but what is a yard of soil? How do I know what to order?*

**39. CREATE** Draw and label a prism that has a volume of 50 cubic centimeters.

**40. CREATE** Give an example of a pyramid and a prism that have the same base and the same volume. Explain your reasoning.

**41. CREATE** Draw and label the dimensions of a prism and a pyramid that have the same volume.

**42. PERSEVERE** Write an equation to find the dimensions of a composite solid composed of a cube with a square pyramid on top of equal height. If the volume is equal to 36 cubic inches, what are the dimensions of the solid?

**43. ANALYZE** Make a conjecture about how many pentagonal pyramids will fit inside a pentagonal prism of the same height. Justify your answer.

# Volumes of Cylinders, Cones, and Spheres

## Learn Volume of Cylinders

Like a prism, the volume of a cylinder can be thought of as consisting of layers. For a cylinder, these layers are congruent circular discs.

| Key Concept • Volume of a Cylinder | |
|---|---|
| Words | The volume $V$ of a cylinder is $V = Bh$ or $V = \pi r^2 h$, where $B$ is the area of the base, $h$ is the height of the cylinder, and $r$ is the radius of the base. |
| Symbols | $V = Bh$ or $V = \pi r^2 h$ |
| Model |  |

**Today's Standards**
G.GMD.1, G.GMD.2, G.GMD.3
MP3, MP4, MP6

💬 **Talk About It!**
How are the volume formulas for prisms and cylinders similar?

## Example 1 Approximate the Volume of a Cylinder

**POSTAL SERVICE Andrew wants to mail his brother a collection of antique marbles. He needs to calculate the volume of the mail tube before he buys it to ensure that it will hold all of the marbles. Find the volume of the mail tube to the nearest tenth.**

$V =$ _____    Volume of a cylinder

$= \pi(1.25)^2 \cdot 10$    $B = \pi r^2$, $r = 1.25$, and $h = 10$

$\approx$ _____    Simplify.

The volume of the mail tube is about 49.1 cubic inches.

## Example 2 Volume of a Cylinder

**Find the volume of a cylinder with a radius of $x - 5$ centimeters and a height of $x$ centimeters.**

**Part A Find the volume of the cylinder in terms of $x$ and $\pi$.**

Draw and label a diagram to represent the situation.
Find the volume of the cylinder.

$V = Bh$    Volume of a cylinder

$= \pi(x - 5)^2 x$    $B = \pi r^2$, $r = x - 5$, and $h = x$

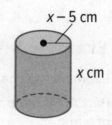
$x - 5$ cm
$x$ cm

▶ **Go Online** You can complete an Extra Example online.

*(continued on the next page)*

**Watch Out!**
**Multiple Expressions**
You could also express the volume of the cylinder in expanded form as $V = \pi x^3 - 10\pi x^2 + 25\pi x$.

So, the volume of the cylinder in terms of $x$ and $\pi$ is $V = \pi(x-5)^2 x$ cubic centimeters.

**Part B Find the volume.**

**Find the volume of the cylinder rounded to the nearest tenth if $x = 8$.**

Substitute the value of $x$ into the expression you found in **Part A**.

$V = \pi(x-5)^2 x$      Expression from Part A

$= \pi(8-5)^2 \cdot 8$      Substitute.

$\approx$ _____      Simplify.

The volume of the cylinder is about 226.2 cubic centimeters.

## **Learn** Volumes of Cones

The pyramid and prism shown have the same base area $B$ and height $h$ as the cylinder and cone. You can use Cavalieri's Principle and similar triangles to show that the volume of the cone is equal to the volume of the pyramid.

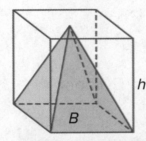

| Key Concept • Volume of a Cone | |
|---|---|
| Words | The volume of a circular cone is $V = \frac{1}{3}Bh$, or $V = \frac{1}{3}\pi r^2 h$, where $B$ is the area of the base, $h$ is the height of the cone, and $r$ is the radius of the base. |
| Symbols | $V = \frac{1}{3}Bh$, or $V = \frac{1}{3}\pi r^2 h$ |
| Model | 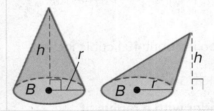 |

## **Example 3** Volume of a Cone

**Examine the cone.**

**Part A Find the volume of a cone in terms of $x$ and $\pi$.**

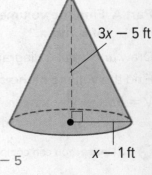

$V = \frac{1}{3}Bh$      Volume of a cone

$= \frac{1}{3}$____$h$      $B = \pi r^2$

$= \frac{1}{3}\pi(\underline{\phantom{xx}})^2(3x-5)$      $r = x-1$ and $h = 3x-5$

 **Go Online** You can complete an Extra Example online.

**Watch Out!**

**Volumes of Cones** The formula for the surface area of a cone only applies to right cones. However, the formula for the volume applies to oblique cones as well as right cones.

 **Think About It!**

Determine whether the following statement is *always*, *sometimes*, or *never* true. Justify your reasoning. The volume of a cone with radius $r$ and height $h$ equals the volume of a prism with height $h$.

So, the volume of the cone in terms of $x$ and $\pi$ is

$V = \frac{1}{3}\pi(x-1)^2(3x-5)$ cubic feet.

**Part B Find the volume of the cone if $x = 4$.**

**Find the volume of the cone to the nearest tenth if $x = 4$.**

Substitute the value of $x$ into the expression you found in **Part A**.

$V = \frac{1}{3}\pi(x-1)^2(3x-5)$      Expression from Part A

$\phantom{V} = \frac{1}{3}\pi(\underline{\hspace{1cm}}-1)^2\,[3(\underline{\hspace{1cm}})-5]$    Substitute.

$\phantom{V} \approx \underline{\hspace{1cm}}$      Simplify.

The volume of the cone is about 66.0 cubic feet.

---

# Example 4 Approximate the Volume of a Cone

TREATS **Alyssa serves ice cream in paper cones with plastic lids. What is the volume of the paper cone?**

10 cm

15 cm

**Step 1 Find the height of the paper cone.**

$a^2 + b^2 = c^2$      Pythagorean Theorem

$5^2 + h^2 = \underline{\hspace{1cm}}$      $a = 5$, $b = h$, and $c = 15$

$\phantom{5^2+} h = \sqrt{15^2 - \phantom{2}^2}$      Solve for $h$.

$\phantom{5^2 + h} = \sqrt{200}$

**Step 2 Find the volume of the paper cone to the nearest tenth.**

$V = \frac{1}{3}Bh$      Volume of a cone

$\phantom{V} = \frac{1}{3}\pi(\underline{\hspace{1cm}})^2\,(\underline{\hspace{1.5cm}})$    $r = 5$ and $h = \sqrt{15^2 - 5^2}$

$\phantom{V} \approx \underline{\hspace{1.5cm}}$      Simplify.

The volume of the paper cone is about 370.2 cubic centimeters.

## Check

DÉCOR A soap dispenser can be modeled by a cone with a height of 15 centimeters and a radius of 4 centimeters. What is the approximate volume of soap that the dispenser can hold? Round your answer to the nearest tenth.

$V \approx \underline{\hspace{1cm}}$ cm$^3$

Go Online You can complete an Extra Example online.

**Watch Out!**

**Rounding** When you find the height of the ice cream container, you may be tempted to round that measure to the nearest tenth. However, rounding at that stage in the calculation will change the final calculation of the volume. Whenever possible, avoiding rounding at intermediate steps. Wait to round until you have calculated the final answer.

## Explore Volume of a Sphere

**Online Activity** Use dynamic geometry software to complete the Explore.

> ✕
>
> @ **INQUIRY** A cone with radius *r* and height *r* is cut out of a cylinder with radius *r* and height *r*. What is the relationship between the volume of the remaining solid and the volume of a sphere with radius *r*?

## Learn Volumes of Spheres

| Key Concept • Volume of a Cone | |
|---|---|
| **Words** | The volume *V* of a sphere is $V = \frac{4}{3}\pi r^3$, where *r* is the radius of the sphere. |
| **Symbols** | $V = \frac{4}{3}\pi r^3$ |
| **Model** | 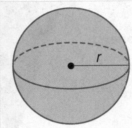 |

**Go Online**
to see how the formula for the volume of a sphere can be derived from the volume of pyramids.

## Example 5 Approximate the Volume of a Sphere

**CANDY** A chocolate company wants to create bite-sized individually wrapped solid chocolate spheres. If the diameter of the sphere is 1.5 inches, find the volume of chocolate used to make the spheres to the nearest hundredth.

$V = \frac{4}{3}\pi r^3$      Volume of a sphere

$= \frac{4}{3}\pi(\underline{\quad})^3$      *r* = 0.75

$\approx \underline{\quad}$      Simplify.

The volume of chocolate used is about 1.77 cubic inches.

## Check

DESIGN Some manufacturers design beverage bottles that are approximately spherical because they have smaller surface areas than cylindrical bottles and require less plastic to produce. If a bottle is approximately spherical and has a radius of 4 centimeters, what is the volume of the bottle? Round your answer to the nearest tenth.

$V \approx \underline{\quad}$ cm$^3$

 **Go Online** You can complete an Extra Example online.

# Example 6 Volume of a Sphere

**Examine the sphere.**

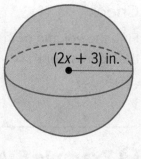

(2x + 3) in.

**Part A** Find the volume of the sphere in terms of $x$ and $\pi$.

$V = \frac{4}{3}\pi r^3$      Volume of a sphere

$= \frac{4}{3}\pi(\underline{\quad\quad})^3$      $r = 2x + 3$

So, the volume of the sphere in terms of $x$ and $\pi$ is $V = \frac{4}{3}\pi(2x + 3)^3$ cubic inches.

**Part B** Find the volume of the sphere if $x = 2.2$.

**Find the volume of the sphere to the nearest tenth if $x = 2.2$.**

Substitute the value of $x$ into the expression you found in **Part A**.

$V = \frac{4}{3}\pi(2x + 3)^3$      Expression from Part A

$= \frac{4}{3}\pi[2(\underline{\quad\quad}) + 3]^3$      Substitute

$\approx \underline{\quad\quad}$      Simplify.

The volume of the sphere is about 1697.4 cubic inches.

## Watch Out!

**Multiple Expressions**
You could also express the volume of the sphere in expanded form as

$V = \frac{32\pi x^3 + 144\pi x^2 + 216\pi x + 108\pi}{3}$

cubic inches.

# Example 7 Volume of a Composite Solid

**Find the volume of the composite solid.**

The composite solid is a combination of a cone and a hemisphere.

**Find the volume of the cone.**

Find the volume of the cone to the nearest tenth.

$V = \frac{1}{3}Bh$      Volume of a cone

$= \frac{1}{3}(\underline{\quad\quad})h$      $B = \pi r^2$

$= \frac{1}{3}\pi(\underline{\quad})^2 (\underline{\quad})$      $r = 8$ and $h = 17$

$\approx \underline{\quad\quad}$      Simplify.

The volume of the paper cone is about 1139.4 mm$^3$.

17 mm

8 mm

**Find the volume of the hemisphere to the nearest tenth.**

$V = \frac{1}{2}\left(\frac{4}{3}\pi r^3\right)$      Volume of a hemisphere

$= \frac{2}{3}\pi(\underline{\quad})^3$      $r = 8$

$\approx \underline{\quad\quad}$      Simplify.

The volume of the hemisphere is about 1072.3 mm$^3$.

**Find the volume of the composite solid.**

The volume of the composite solid is the sum of the volumes of the cone and the hemisphere.

$V \approx \underline{\quad\quad}$ mm$^3$

**Go Online** You can complete an Extra Example online.

## Study Tip

**Hemispheres** A plane can intersect a sphere in a point or in a circle. If the circle contains the center of the sphere, the intersection is called a *great circle*. A great circle separates a sphere into two congruent halves, called *hemispheres*.

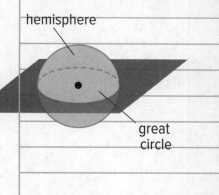

hemisphere

great circle

## Check

Find the volume of the composite solid to the nearest tenth.

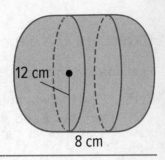

$V \approx$ _____ cm³

---

### 🌐 **Example 8** Approximate the Volume of a Composite Solid

**HOME DECOR  Mia purchased a trash container for her study area. Find the volume of the trash container rounded to the nearest tenth.**

The trash container can be approximated by a composite solid made up of a cylinder and a hemisphere.

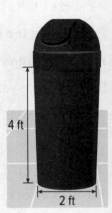

**Step 1  Find the volume of the cylinder in terms of π.**

$V = Bh$ 　　　　　　　Volume of a cylinder

$\quad = \pi(\underline{\quad})^2(\underline{\quad})$ 　　$B = \pi r^2$, $r = 1$, and $h = 4$

$\quad = \underline{\quad}$ 　　　　　Simplify.

**Step 2  Find the volume of the hemisphere in terms of π.**

$V = \frac{1}{2}\left(\frac{4}{3}\pi r^3\right)$ 　　　Volume of a hemisphere

$\quad = \frac{2}{3}\pi(\underline{\quad})^3$ 　　　　$r = 1$

$\quad \approx \underline{\quad}$ 　　　　　Simplify.

The volume of the hemisphere is approximately $\frac{2}{3}\pi$ cubic feet.

**Step 3  Find the volume of the trash container to the nearest tenth.**

The volume of the trash container is _____ cubic feet.

### Check

**RECORDS**  In 2003, a new record was set for the world's largest crayon. The blue crayon in Easton, Pennsylvania, weighs nearly 1500 pounds, and the circular base of the crayon has a diameter of 16 inches.

**Part A** Name the solids that combine to model the composite solid.

_____

**Part B** Find the volume of the crayon if its overall length is 180 inches and the height of the tip is 21 inches. Round your answer to the nearest tenth.

$V \approx$ _____ in³

🌐 **Go Online**  You can complete an Extra Example online.

🌐 **Go Online**
to practice what you've learned in Lessons 11-4 through 11-7.

# Practice

Go Online You can complete your homework online.

### Example 1

**Find the volume of each cylinder. Round to the nearest tenth if necessary.**

**1.**

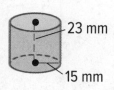
23 mm
15 mm

**2.**

6 yd
10 yd

**3.**

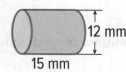
12 mm
15 mm

**4.**

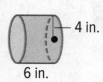
4 in.
6 in.

**5.**

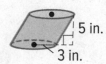
5 in.
3 in.

**6.**

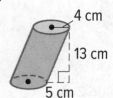
4 cm
13 cm
5 cm

### Example 2

**7. TRASH CANS** The Meyer family uses a kitchen trash can shaped like a cylinder. It has a height of 18 inches and a base diameter of $2x + 4$ inches.

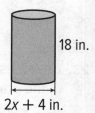
18 in.
$2x + 4$ in.

**Part A** Find the volume of the trash can in terms of $x$ and $\pi$.

**Part B** What is the volume of the trash can if $x = 4$? Round your answer to the nearest tenth of a cubic inch.

**8. COFFEE** A roasting company sells their coffee in canisters shaped like a cylinder. The radius of the cylinder is $x - 7$ inches and the height is $x$ inches.

**Part A** Find the volume of the cylinder in terms of $x$ and $\pi$.

**Part B** Their most popular canister has a height of 7.5 inches. What is its volume rounded to the nearest cubic inch?

### Examples 3 and 4

**Find the volume of each cone. Round to the nearest tenth if necessary.**

**9.**

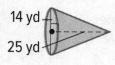
14 ft
8 ft

**10.**

12 m
25 m

**11.**

13 in.
8 in.

**12.**

14 yd
25 yd

**13.**

6 cm
12 cm

**14.**

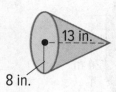
10 mm
6 mm

**15. ICE CREAM DISHES** The part of a dish designed for ice cream is shaped like an upside-down cone. The base of the cone has a radius of 2 inches and the height is 1.2 inches. What is the volume of the cone? Round your answer to the nearest hundredth.

**16. FUNNEL** Matt uses a funnel to pour oil into his car engine. The funnel has a radius of *x* centimeters and a height of *x* + 2 centimeters.

  **Part A** Find the volume of the funnel in terms of *x* and $\pi$.

  **Part B** How much oil, to the nearest cubic centimeter, will the funnel hold if *x* is 6 centimeters?

Example 5

**Find the volume of each sphere. Round to the nearest tenth if necessary.**

17.

18.

19.

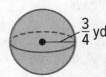
$\frac{3}{4}$ yd

20.

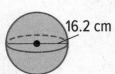

21.

22.

Example 6

**23. ORANGES** Mandy cuts a spherical orange in half along a great circle. The radius of the orange is *x* − 2 inches.

  **Part A** Find the volume of the orange in terms of *x* and $\pi$.

  **Part B** Find the volume, rounded to the nearest tenth, of the orange if *x* = 4 in.

**24. ARCHITECTURE** A scale model of a spherical fountain has a radius of 2*x* − 4.

  **Part A** Find the volume of the sphere in terms of *x* and $\pi$.

  **Part B** Find the volume, rounded to the nearest tenth, of the actual fountain if *x* = 2.75 ft.

Examples 7 and 8

**Find the volume of each composite solid. Round to the nearest tenth if necessary.**

25.

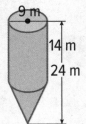

26.

**27. BILLIARDS** A billiard ball set consists of 16 spheres, each $2\frac{1}{4}$ inches in diameter. What is the total volume of a complete set of billiard balls? Round your answer to the nearest thousandth.

**28. SCULPTING** A sculptor wants to remove stone from a cylindrical block 3 feet high and turn it into a cone. The diameter of the base of the cone and cylinder is 2 feet. What is the volume of the stone that the sculptor must remove? Round your answer to the nearest hundredth.

**Consider the composite solid shown at the right.**

**29.** STRUCTURE Write a formula for the volume of this solid in terms of the radius $r$.

**30.** PRECISION Explain how you wrote a formula for the volume of this solid.

**Mixed Exercises**

**31.** A cone has a diameter of 12 centimeters and a height of 9 centimeters. What is the volume of the cone to the nearest tenth?

**32.** A sphere has a radius that is 15.6 inches long. Find the volume of the sphere. Round to the nearest tenth.

**33.** A cylinder has a diameter of 20 inches and a height of 9 inches. Find the volume of the cylinder, round to the nearest tenth.

**34.** Find the volume of the cone. Round to the nearest tenth.

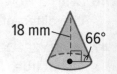

**35.** BARRELS A barrel in the shape of a right cylinder has a diameter of 18 inches and a height of 42 inches. Find the volume of the barrel to the nearest tenth.

**36.** A cone has a radius of 4 meters and a slant height of 5 meters. Find the volume of the cone. Round to the nearest tenth.

**37.** REASONING Find the volume of the cone. Round to the nearest tenth.

**38.** HEMISHPERE A hemisphere has a base with an area that is $25\pi$ square centimeters. Find the volume of the hemisphere. Round to the nearest tenth.

**39.** TEEPEE Caitlyn made a teepee for a class project. Her teepee had a diameter of 6 feet. The angle the side of the teepee made with the ground was 65°. What was the volume of the teepee? Round your answer to the nearest hundredth.

**40.** PENCIL GRIPS A pencil grip is shaped like a triangular prism with a cylinder removed from the middle. The base of the prism is a right isosceles triangle with leg lengths of 2 centimeters. The diameter of the base of the removed cylinder is 1 centimeter. The heights of the prism and the cylinder are the same, and equal to 4 centimeters. What is the exact volume of the pencil grip?

**41.** A wooden sphere is carved from a solid cube of wood so that the least amount of wood is carved away.

   **a.** STRUCTURE If the block of wood had a volume of 729 cubic inches, what is the volume of the sphere? Explain.

   **b.** ARGUMENTS Devon says that he can multiply the volume of any cube by $\frac{\pi}{6}$ to find the volume of the sphere that shares the same diameter as the cube's side. Is he correct?

**42.** REASONING Reginald is creating a scale model of a building using a scale of 4 feet = 3 inches. The building is in the shape of a cube topped with a hemisphere so that the circular base of the hemisphere is inscribed in the square base of the cube. At its highest point, the building has a height of 30 feet and the radius of the hemisphere is shown. Find the volume of his scale model to the nearest cubic inch. Explain.

43. **REGULARITY** A container company manufactures cylindrical containers with a radius of 3 inches and a height of 10 inches. They decided to produce a different cylindrical container with the same volume, but with an 8-inch height. What radius must the new container have for the volumes to be equal? What steps would you use to find the radius of the new cylinder? What is the radius of the new cylinder to the nearest tenth?

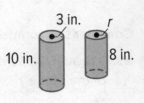

44. **PERSEVERE** The cylindrical can shown is used to fill a container with liquid. It takes three full cans to fill the container. Describe possible dimensions of the container if it is each of the following shapes.

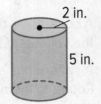

   a. rectangular prism

   b. square prism

   c. triangular prism with a right triangle as the base

45. **ANALYZE** Determine whether the following statement is *true* or *false*. Explain.

   *Two cylinders with the same height and the same lateral area must have the same volume.*

46. **WRITE** How are the volume formulas for prisms and cylinders similar? How are the different?

47. **ANALYZE** Determine whether the following statement is *always*, *sometimes*, or *never* true. Justify your reasoning.

   *The volume of a cone with radius r and height h equals the volume of a prism with height h.*

48. **FIND THE ERROR** Alexandra and Cornelio are calculating the volume of the cone at the right. Is either of them correct? Explain your answer.

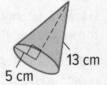

| Alexandra | Cornelio |
|---|---|
| $V = \frac{1}{3} Bh$ | $5^2 + 12^2 = 13^2$ |
| $= \frac{1}{3} \pi (5)^2 (13)$ | $V = \frac{1}{3} Bh$ |
| $= 340.3 \ cm^3$ | $= \frac{1}{3} \pi (5)^2 (12)$ |
| | $= 314.2 \ cm^3$ |

49. **ANALYZE** A cone has a volume of 568 cubic centimeters. What is the volume of a cylinder that has the same radius and height as the cone? Explain your reasoning.

50. **WRITE** Compare and contrast finding volumes of pyramids and cones with finding volumes of prisms and cylinders.

51. **PERSEVERE** A cube has a volume of 216 cubic inches. Find the volume of a sphere that is circumscribed about the circle. Round to the nearest tenth.

52. **ANALYZE** Determine whether the following statement is *true* or *false*. If true, explain your reasoning. If false, provide a counterexample.

   *If a sphere has radius r, there exists a cone with radius r having the same volume.*

53. **CREATE** Sketch a composite solid made of a cylinder and cone that has an approximate volume of 7698.5 cm³.

# Applying Similarity to Solid Figures

## Learn Similar Two-Dimensional Figures

Recall that if two polygons are similar, then their perimeters are proportional to the scale factor between them. The areas of two similar polygons share a different relationship.

| Theorem 11.1: Areas of Similar Polygons | |
|---|---|
| **Words** | In two polygons are similar, then their areas are proportional to the square of the scale factor between them. |
| **Example** | If $ABCD \sim FGHJ$, then $\dfrac{\text{area of } FGHJ}{\text{area of } ABCD} = \left(\dfrac{FG}{AB}\right)^2$. |
| **Model** | 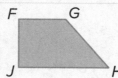 |

**Go Online**
A proof of Theorem 11.1 is available.

**Today's Standards**
G.GMD.3
MP3, MP6, MP7

**Today's Vocabulary**
similar solids
congruent solids

## Example 1 Use Similar Figures to Find Area

□*ABCD* and □*JKLM* are similar rectangles. Find the area of □*JKLM*.

**Step 1** Find the scale factor from □*ABCD* to □*JKLM*.

The scale factor from □*ABCD* to □*JKLM* is —.

**Step 2** Find the ratio of the areas.

If two polygons are similar, then their areas are proportional to the square of the scale factor between them. So, the ratio of their areas is $\left(\frac{5}{8}\right)^2$ or —.

**Step 3** Find the area.

$\dfrac{\text{area of } \square\, JKLM}{\text{area of } \square\, ABCD} = \dfrac{25}{64}$  Write a proportion.

$\text{area of } \square\, JKLM = \dfrac{25}{64}$  Area of □*ABCD* = 32

$area\ of\ \square JKLM = \underline{\qquad}$  Multiply and simplify.

So, the area of □*JKLM* is 12.5 square centimeters.

**Study Tip**

**Ratios** Ratios can be written in different ways. For example, *x* to *y*, *x* : *y*, and $\frac{x}{y}$ are all representations of the ratio of *x* and *y*.

## Example 2 Use Areas of Similar Figures

**Trapezoids *LMNP* and *ABCD* are similar. Find the scale factor of trapezoid *LMNP* to trapezoid *ABCD* and the value of *x*.**

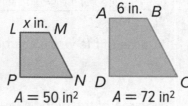

**Go Online** You can complete an Extra Example online.

*(continued on the next page)*

**Part A  Find the scale factor.**

Let $k$ be the scale factor from trapezoid $ABCD$ to trapezoid $LMNP$.

$$\frac{\text{area of } \square LMNP}{\text{area of } \square ABCD} = k^2 \qquad \text{Theorem 11.1}$$

$$\frac{\quad}{72} = k^2 \qquad \text{Substitution}$$

$$\frac{\quad}{\quad} = k^2 \qquad \text{Simplify.}$$

$$\frac{\quad}{6} = k \qquad \text{Take the positive square root.}$$

So, the scale factor from quadrilateral $ABCD$ to $LMNP$ is $\frac{5}{6}$.

**Part B  Find the value of $x$.**

Use the scale factor to find $x$.

$$\frac{LM}{AB} = k \qquad \begin{array}{l}\text{The ratio of corresponding lengths of similar}\\ \text{polygons is equal to the scale factor between}\\ \text{the polygons.}\end{array}$$

$$\frac{x}{\quad} = \frac{5}{6} \qquad \text{Substitution}$$

$$x = \frac{5}{6} \cdot 6 \text{ or } 5 \qquad \text{Multiply each side by 6.}$$

The value of $x$ is 5.

## 🌐 Example 3  Use Similar Figures to Solve Problems

**WORLD RECORDS  An average large pizza has a diameter of 14 inches. The scale factor from an average large pizza to the world's largest pizza is $\frac{786}{7}$. Find the area of the world's largest pizza rounded to the nearest square inch.**

**Step 1  Find the area of an average large pizza.**

$$A = \pi r^2 \qquad \text{Area of a circle}$$

$$= \pi(\underline{\quad})^2 \qquad r = 7$$

$$\approx \underline{\quad\quad} \qquad \text{Simplify.}$$

The area of an average large pizza is about 153.9 square inches.

**Step 2  Find the area of the world's largest pizza.**

Use the scale factor to find the area of the world's largest pizza rounded to the nearest square inch.

$$\frac{\text{area of the world's largest pizza}}{\text{area of an average large pizza}} = 786^2$$

$$\frac{\text{area of the world's largest pizza}}{\quad} \approx 617{,}796$$

$$\text{area of the world's largest pizza} \approx \frac{617{,}796\,(\underline{\quad})}{49}$$

$$\approx \underline{\quad\quad}$$

The area of the world's largest pizza is about 2,034,608 square inches.

🧭 **Go Online** You can complete an Extra Example online.

**Watch Out!**

**Writing Ratios** When finding the ratio of the area of Figure $A$ to the area of Figure $B$, be sure to write your ratio as $\frac{\text{area of Figure } A}{\text{area of Figure } B}$.

## Explore Similar Solids

**▶ Online Activity** Use dynamic geometry software to complete the Explore.

> ✕
>
> **❓ INQUIRY** How are the surface areas and volumes of similar solids related?

## Learn Similar Three-Dimensional Solids

**Similar solids** have exactly the same shape but not necessarily the same size. All spheres are similar, and all cubes are similar.

In similar solids, the corresponding linear measures, such as height and radius, are proportional to the scale factor between them.

> **Similar Solids**
>
> Two solids are similar if and only if they have the same shape and the ratios of their corresponding linear measures are equal.

> **Theorem 11.2**
>
> **Words** If two similar solids have a scale factor of $a:b$, then the surface areas have a ratio of $a^2:b^2$, and the volumes have a ratio of $a^3:b^3$.

> **Key Concept • Characteristics of Congruent Solids**
>
> Two solids are congruent if and only if:
> - Corresponding angles are congruent.
> - Corresponding edges are congruent.
> - Corresponding faces are congruent.
> - Volumes are equal.

**🗨 Talk About It!**

Explain why all spheres are similar.

**▶ Go Online**

A proof of Theorem 11.2 is available.

## Example 4 Use Similar Solids to Find Volume

**The three cones are similar. Find the volume of each cone.**

**Cone 1**   **Cone 2**   **Cone 3**

**Step 1 Find the volume of Cone 1**

$$V = \frac{1}{3}Bh \qquad \text{Volume of a cone}$$

$$= \frac{1}{3}(\pi(\underline{\quad})^2) \cdot \sqrt{\underline{\quad}} \qquad B = \pi r^2, r = 5, \text{ and } h = \sqrt{12^2 - 5^2}$$

$$= \frac{\pi\sqrt{119}}{3} \qquad \text{Simplify.}$$

**Step 2 Find the volume of Cone 2.**

Cone 2 is similar to Cone 1. The scale factor from Cone 2 to Cone 1 is $5:6$. Find the volume of Cone 2.

$$\frac{\text{volume of Cone 1}}{\text{volume of Cone 2}} = \left(\frac{\quad}{\quad}\right)^3 \qquad \text{Theorem 11.2}$$

$$\text{volume of Cone 2} = \left(\frac{6}{5}\right)^3 \times \underline{\qquad} \qquad \text{Substitute, then solve for the volume of Cone 2.}$$

$$= \frac{\pi\sqrt{119}}{5} \qquad \text{Simplify.}$$

The volume of Cone 2 is $\frac{72\pi\sqrt{119}}{5}$ cubic inches.

*(continued on the next page)*

Cone 3 is similar to Cone 1. The scale factor from Cone 3 to Cone 1 is 5:5 or 1:1. Because the two solids are similar and have a scale factor of 1:1, we know that the two solids are congruent. Congruent solids have the same volume, so the volume of Cone 3 is _____ cubic inches.

### Study Tip

**Similar and Congruent Solids** If two solids are similar, then their corresponding linear measures are proportional. If two solids are congruent, then their corresponding linear measures are equal.

## Check

The three prisms are similar. Find the exact volume of each prism.

Prism 1: $V =$ _____ ft$^3$

Prism 2: $V =$ _____ ft$^3$

Prism 3: $V =$ _____ ft$^3$

**Prism 1    Prism 2    Prism 3**

6.2 ft    1.5 ft

2.5 ft

6.2 ft

3 ft

---

## Example 5  Use Similar Solids to Solve Problems

**Two similar rectangular prisms with square bases have surface areas of 98 square centimeters and 18 square centimeters. If one base edge of the larger rectangular prism measures 9 centimeters, what is the perimeter of one base of the smaller prism?**

First, find the scale factor.

$\dfrac{\text{surface area of larger prism}}{\text{surface area of smaller prism}} =$ _____      Substitute.

$= \dfrac{49}{8}$      Simplify.

$= \left(\dfrac{7}{3}\right)^2$      The scale factor is $\dfrac{7}{3}$.

Then, find the length of the base edge of the small prism.

$\dfrac{\text{base edge of larger prism}}{\text{base edge of smaller prism}} = \dfrac{7}{3}$      The scale factor is $\dfrac{7}{3}$.

$\dfrac{\phantom{xxx}}{\text{base edge of smaller prism}} = \dfrac{7}{3}$      Substitute.

base edge of small prism = _____      Use a proportion to find the base edge of the small prism.

The base edge of the smaller prism is $\dfrac{27}{7}$ centimeters.

Find the perimeter of the base of the smaller prism to the nearest tenth.

$P = 4s$      Perimeter of square

$= 4 \cdot \dfrac{27}{7}$      Substitute.

$=$ _____      Simplify.

The perimeter of the base of the smaller prism is about 15.4 centimeters.

## Check

Two similar cylinders have volumes of $270\pi$ and $640\pi$ cubic inches, respectively. If the height of the larger cylinder is 10 inches, what is the area of the base of the smaller cone? _____

**A.** 6 in$^2$      **B.** 7.5 in$^2$      **C.** $36\pi$ in$^2$      **D.** $162\pi$ in$^2$

🔵 **Go Online** You can complete an Extra Example online.

# Practice

🧭 **Go Online** You can complete your homework online.

## Example 1

**For each pair of similar figures, find the area of the shaded figure. Round your answer to the nearest tenth if necessary.**

**1.**

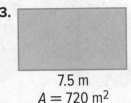

11 m
$A = 20$ m²

44 m

**2.**

8.5 in.    2 in.

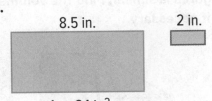

$A = 34$ in²

**3.**

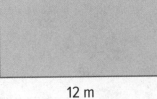

7.5 m
$A = 720$ m²

12 m

**4.**

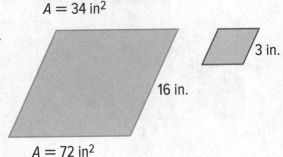

16 in.

$A = 72$ in²

3 in.

## Example 2

**For each pair of similar figures, use the given areas to find the scale factor from the shaded to the unshaded figure. Then find the value of $x$ to the nearest tenth.**

**5.**

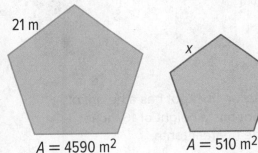

21 m

$x$

$A = 4590$ m²    $A = 510$ m²

**6.**

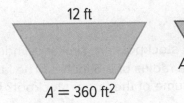

12 ft

$x$

$A = 360$ ft²    $A = 10$ ft²

**7.**

 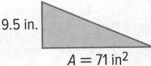

$x$    9.5 in.

$A = 16$ in²    $A = 71$ in²

**8.**

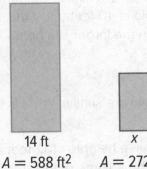

14 ft    $x$
$A = 588$ ft²    $A = 272$ ft²

## Example 3

**9. SCIENCE PROJECT** Matt has two similar posters for his science project. Each poster is a rectangle. The length of the larger poster is 11 inches. The length of the smaller poster is 6 inches. What is the area of the smaller poster if the larger poster is 93.5 square inches?

**10. PINS** Carla has a shirt with decorative pins in the shape of equilateral triangles. The pins come in two sizes. The larger pin has a side length that is three times longer than the smaller pin. If the area of the smaller pin is 6.9 square centimeters, what is the approximate area of the larger pin?

**11. QUILT** A quilt design has one large rectangle surrounded by four congruent rectangles, similar to the large rectangle. If the large rectangle has an area of 45 square inches, what is the area of each small rectangle?

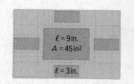

ℓ = 9 in.
A = 45 in²
ℓ = 3 in.

**Example 4**

**Each pair of figures is similar. Find the volume of each figure. Round to the nearest tenth if necessary.**

**12.**

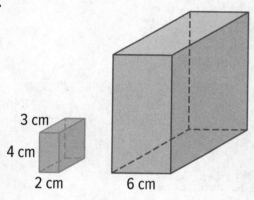

3 cm
4 cm
2 cm
6 cm

**13.**

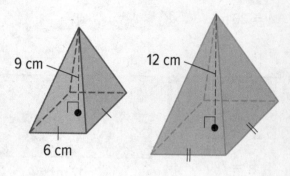

9 cm
12 cm
6 cm

**14.**

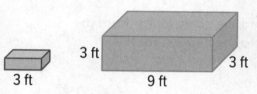

3 ft
3 ft
3 ft
9 ft

**15.**

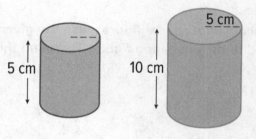

5 cm
5 cm
10 cm

**Example 5**

**16. COOKING** Two stockpots are similar cylinders. The smaller stockpot has a height of 10 inches and a radius of 2.5 inches. The larger stockpot has a height of 16 inches. What is the volume of the larger stockpot? Round to the nearest tenth.

**17. FARMING** A farmer has two similar cylindrical grain silos. The smaller silo is 25 feet tall and the larger silo is 40 feet tall. If the smaller silo can hold 1500 cubic feet of grain, how much can the larger silo hold?

**Mixed Exercises**

**18.** The rectangles shown are similar. What is the area of the smaller rectangle?

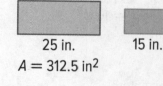

25 in.    15 in.
A = 312.5 in²

**19.** Two similar prisms have heights of 12 feet and 20 feet. What is the ratio of the volume of the small prism to the volume of the large prism?

**20. STUCTURE** Determine whether the pair of solids is *similar*, *congruent*, or *neither*. If the solids are similar, state the scale factor.

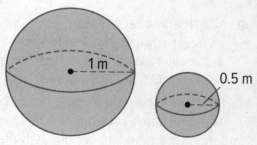

1 m
0.5 m

**21. REASONING** Two cubes have surface areas of 81 square inches and 144 square inches. What is the ratio of the volume of the small cube to the volume of the large cube?

**22.** Two similar pyramids have heights of 4 inches and 7 inches. What is the ratio of the volume of the small pyramid to the volume of the large pyramid?

**23.** Two similar cylinders have surface areas of 40π square feet and 90π square feet. What is the ratio of the height of the large cylinder to the height of the small cylinder?

**24. REGULARITY** A polygon has an area of 225 square meters. If the area is tripled, how does each side length change?

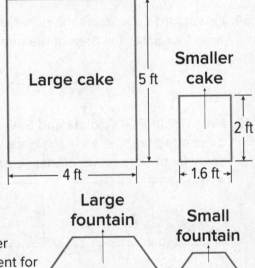

Large cake    5 ft

Smaller cake    2 ft

4 ft    1.6 ft

**25. CAKE** Smith's Bakery is baking several large cakes for a community festival. The cakes consist of two geometrically similar shapes as shown. If 50 pieces of cake can be cut from the smaller cake, how many pieces of the same size can be cut from the larger cake? Round to the nearest piece of cake.

**26. FOUNTAIN** A local park has two fountains in the shape of similar trapezoids as shown. A cement company charges $1000 to pour the cement needed to go under the smaller fountain. How much should the town budget for the cement for both fountains? Explain.

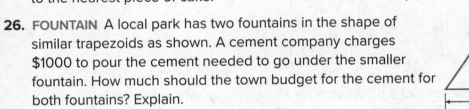

Large fountain    Small fountain

100 ft    40 ft

**27. SCULPTURE** An artist creates metal sculptures in the shapes of regular octagons. The side length of the larger sculpture is 7 inches, and the area of the base of the smaller sculpture is 19.28 square inches.

   **a.** What is the side length of the smaller sculpture?

   **b.** The artist is going to pack the sculptures in a circular box to take them to an art show. Will the larger sculpture fit in a circular box with a 15-inch diameter? Explain your reasoning.

**28. ATMOSPHERE** About 99% of Earth's atmosphere is contained in a 31-kilometer thick layer that surrounds the planet. The Earth itself is almost a sphere with radius 6378 kilometers. What is the ratio of the volume of the atmosphere to the volume of Earth? Round your answer to the nearest thousandth.

**29. COOKING** A cylindrical pot is 4.5 inches tall and has a radius of 4 inches. How tall would a similar pot be if its radius is 6 inches?

**30. MANUFACTURING** Boxes, Inc. wants to make the two boxes at right. How long does the second box need to be so that they are similar?

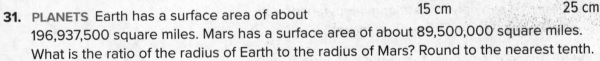

24 cm    25 cm

15 cm

15 cm    25 cm

**31. PLANETS** Earth has a surface area of about 196,937,500 square miles. Mars has a surface area of about 89,500,000 square miles. What is the ratio of the radius of Earth to the radius of Mars? Round to the nearest tenth.

**32. SPORTS** Major League Baseball or MLB, rules state that baseballs must have a circumference of 9 inches. The National Softball Association, or NSA, rules state that softballs must have a circumference not exceeding 12 inches.

   **a.** Find the ratio of the circumference of MLB baseballs to the circumference of NSA softballs.

   **b.** Find the ratio of the volume of MLB baseballs to the volume of NSA softballs. Round to the nearest tenth.

**33. STRUCTURE** At a pet store, toy tennis balls for pets are sold in 3 different sizes. Complete the table by calculating the volume for each size ball. Record the volume of each tennis ball in terms of π. What pattern do you notice as the diameter increases?

| Size | Diameter (cm) | Volume (cm³) |
|---|---|---|
| Small | 3 | |
| Medium | 4.5 | |
| Large | 6.75 | |

**34. REGULARITY** Describe the dimensions of a similar trapezoid that has an area four times the area of the one shown. Explain how you found your answer.

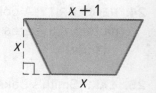

**35. FIND THE ERROR** Violeta and Gavin are trying to come up with a formula that can be used to find the area of a circle with a radius *r* after it has been enlarged by a scale factor *k*. Is either of them correct? Explain your reasoning.

| Violeta |
|---|
| $A = k\pi r^2$ |

| Gavin |
|---|
| $A = \pi(r^2)^k$ |

**36. PERSEVERE** If you want the area of a polygon to be *x*% of its original area, by what scale factor should you multiply each side length?

**37. ANALYZE** A regular *n*-gon is enlarged, and the ratio of the area of the enlarged figure to the area of the original figure is *R*. Write an equation relating the perimeter of the enlarged figure to the perimeter of the original figure *Q*?

**38. WRITE** Explain how the surface areas and volumes of the similar prisms shown at the right are related.

**39. CREATE** Draw a pair of similar figures with areas that have a ratio of 4:1. Explain.

**40. WRITE** Explain how to find the area of an enlarged polygon if you know the area of the original polygon and the scale factor of the enlargement.

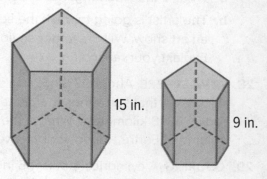

15 in.

9 in.

**41. PERSEVERE** The ratio of the volume of Cylinder A to the volume of Cylinder B is 1:5. Cylinder A is similar to Cylinder C with a scale factor of 1:2 and Cylinder B is similar to Cylinder D with a scale factor of 1:3. What is the ratio of the volume of Cylinder C to the volume of Cylinder D? Explain your reasoning.

**42. CREATE** Describe two nonsimilar triangular pyramids with similar bases.

**43. PERSEVERE** Plane *P* is parallel to the base of cone *C*, and the volume of the cone above the plane is $\frac{1}{8}$ of the volume of cone *C*. Find the height of cone *C*.

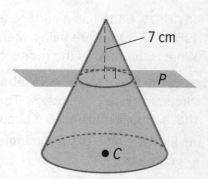

7 cm

P

• C

# Density

**Today's Standards**
G.MG.2

MP4, MP5, MP6

**Today's Vocabulary**
density

## Explore Strategies Based on Density

**Online Activity** Use the video to complete the Explore.

> ⓠ **INQUIRY** How can a knowledge of density
> help you make decisions in video games
> and in real-world situations?

## Learn Density Based on Area

**Density** is a measure of the quantity of some physical property per unit of length, area, or volume. One example of density is population density, which is the measurement of population per unit of area. Population density is calculated for states, major cities, or other areas, based on data collected from the U.S. Census.

| Density Based on Area | |
|---|---|
| **Words** | Density is the ratio of objects to area. |
| **Symbols** | $\text{density} = \dfrac{\text{number of objects}}{\text{area}}$ |

## 🌎 Example 1 Find the Density of an Area

POPULATION **The World Bank reports that the population of Greenland in 2015 was 56,114. If the total land area is about 836,000 square miles, what is the population density of Greenland?**

Calculate population density by adapting the density formula, $\text{population density} = \dfrac{\text{populationland}}{\text{landarea}}$.

The population density of Greenland is $\dfrac{}{836,000}$ or about 0.067 person per square mile.

### Check

**VOLCANOES** The country with the largest number of active volcanoes is Indonesia with 147. The area of Indonesia is 735,400 square miles. What is the density of active volcanoes in Indonesia?

**A.** $\approx 0.0002 \dfrac{\text{volcanoes}}{\text{mi}^2}$

**B.** $\approx 0.0004 \dfrac{\text{volcanoes}}{\text{mi}^2}$

**C.** $\approx 2501.4 \dfrac{\text{volcanoes}}{\text{mi}^2}$

**D.** $\approx 5002.7 \dfrac{\text{volcanoes}}{\text{mi}^2}$

**Go Online** You can complete an Extra Example online.

**Study Tip**

**Dimensional Analysis**
In some cases, it may be necessary to use dimensional analysis to convert between units of measurement. To convert from one unit to another, use a numerical quantity known as a conversion factor. For example, if you are trying to find the number of kilograms in a 16-pound box, you can multiply by the conversion factor $\dfrac{1 \text{ kg}}{2.2 \text{ lbs}}$, because there are about 2.2 pounds in each kilogram.

$16 \text{ lb} \cdot \dfrac{1 \text{ kg}}{2.2 \text{ lb}} \approx 7.3 \text{ kg}$

## 🌐 Example 2 Use the Density of an Area

**GREENHOUSE GASES** Masha has a farm with 220 milking cows that produce 286 pounds per acre for a total of 1,412,840 pounds of milk. Due to recent regulations, Masha must pay a fee if she has more than 30 cows per square mile on her farm. Determine whether Masha will have to pay the fee. (*Hint:* There are 640 acres in 1 square mile.)

**Step 1 Find the area of the farm.**

$$\frac{1,412,840 \text{ lb}}{286 \text{ lb/acre}} = \underline{\hspace{1cm}} \text{ acres on Masha's farm}$$

**Step 2 Convert to square miles.**

Use a conversion factor.

$$4940 \text{ acres} \cdot \frac{1 \text{ mi}^2}{640 \text{ acres}} \approx \underline{\hspace{1cm}} \text{ mi}^2$$

There are about 7.7 square miles on Masha's farm.

**Think About It!**
Why do we multiply 4940 by $\frac{1}{640}$ in Step 2?

**Step 3 Find the density of cows per acre.**

Calculate population density by adapting the density formula.

$$\text{population density} = \frac{\text{population}}{\text{land area}} \qquad \text{Population Density Formula}$$

$$= \frac{}{7.7} \qquad \text{220 cows on 7.7 square miles}$$

$$= \underline{\hspace{1cm}} \qquad \text{Simplify.}$$

Because 28.57 is _____ than 30, Masha _____ have to pay the fee.

## Check

**DUCK POND** For a school carnival game, Adalynn is planning to fill a pool with water and float a layer of numbered rubber ducks on top. She knows that it takes 25 rubber ducks to fill 1 square foot of area. The pool used for the carnival has an area of about 7 square feet. How many rubber ducks should she buy to fill the pool?

Adalynn should buy _____ rubber ducks.

**POPULATION DENSITY** The city of Manila, Philippines, is one of the most densely populated cities on Earth. Its 1,650,000 residents share a space that can be approximated by a rectangle 5.1 miles long by 2.9 miles wide. To the nearest person, what is the approximate density of Manila?

_____ people per square mile

🌐 **Go Online** You can complete an Extra Example online.

## Learn Density Based on Volume

**Density** is the measure of the quantity of some physical property per unit of length, area, or volume. If two objects have the same volume but different masses, the object with the greater mass will be denser.

| Density Based on Volume | |
|---|---|
| **Words** | Density is the ratio of mass (or weight) to volume. |
| **Symbols** | $\text{density} = \dfrac{\text{number of weight}}{\text{volume}}$ |

##  Example 3 Find the Density of a Solid

**ART** **Antonio opens a new brick of clay that weighs 25 pounds.**

**a. What is the density of the brick of clay?**

First find the volume of the clay, which can be approximated by using the formula for the volume of a rectangular prism.

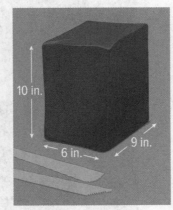

10 in.
6 in.
9 in.

$V = lwh$          Volume of a rectangular prism

$V = 6(\underline{\phantom{xx}})(\underline{\phantom{xx}})$          $l = 6$, $w = 9$, and $h = 10$

$= \underline{\phantom{xx}}$ in$^3$          Simplify.

Next, use the density formula to calculate the density.

$\text{density} = \dfrac{\text{weight}}{\text{volume}}$          Density Formula

$d = \dfrac{25}{540}$          mass = 25 lb, volume = 540 in$^3$

$\approx 0.046$          Simplify.

The density of the clay is about _____ pound per cubic inch.

**b. Antonio uses the same clay to make a foundational cube for a sculpture. If the cube weighs 1.3 pounds, what are the dimensions of the cube?**

Use the density formula to find the volume of the clay Antonio is using given the weight and the density of the clay.

$\text{density} = \dfrac{\text{weight}}{\text{volume}}$          Density Formula

$\underline{\phantom{xxx}} \approx \dfrac{1.3}{V}$          density = 0.046 and weight = 1.3

$V \approx \underline{\phantom{xxx}}$          Simplify.

Because the foundation is a cube, each edge $s$ must be the same length. Therefore, $s^3 = 28.26$ and $s = \underline{\phantom{xx}}$.

Each side of Antonio's cube will be about 5.32 inches long.

**Go Online** You can complete an Extra Example online.

**Think About It!**
If Antonio decides to change the size of the cube so that its weight is greater than 1.3 pounds, how will this change affect the density? Explain.

## Check

**GARDENING** When Kimani filled her planter with soil, the weight of the planter increased by 90 pounds.

15 ft²

3 ft

The density of the soil is _____ pounds per cubic foot.

Kimani uses the same soil to fill another planter, and the weight increased by 154 pounds. The volume of the other planter is _____ cubic feet.

Record your observations here

# Practice

**Go Online** You can complete your homework online.

### Example 1

**Use the data in the table to find the population density of each city.**

1. London, England

2. Paris, France

3. Madrid, Spain

4. Sydney, Australia

| City | Population | Area (mi²) |
|------|-----------|-----------|
| London | 8,674,000 | 607 |
| Paris | 2,224,000 | 40.7 |
| Madrid | 3,165,000 | 234 |
| Sydney | 4,293,000 | 4775 |

### Example 2

**Use the data in the table.**

5. Which city has the greater population density, San Diego, or Los Angeles?

6. What is the population of Sacramento, given that the population density is 4792.1 persons/mi²?

7. What is the area of San Francisco, given that the population density is 17,855.9 persons/mi²?

| California Cities | Population | Area (mi²) |
|------|-----------|-----------|
| San Diego | 1,356,000 | 372.4 |
| Los Angeles | 3,884,000 | 503 |
| Sacramento | -- | 100.1 |
| San Francisco | 837,442 | -- |

### Example 3

8. The mass of the cube is 425 grams. Find the density of the cube to the nearest hundredth.

9. The weight of the rectangular prism is 90 pounds. Find the density of the rectangular prism.

10. A rectangular prism has a length of 25 centimeters, a width of 5.8 centimeters, and a height of 10 centimeters. The mass of the prism is 1650 grams. Find the density of the prism to the nearest tenth.

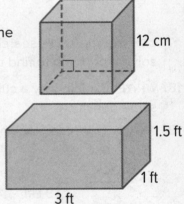

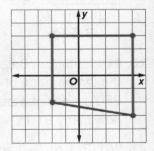

### Mixed Exercises

11. **PARK** A rectangular national sight-seeing park with length of 2 miles and width of 3 miles allows a maximum capacity 250 people. Find the population density of people to the nearest hundredth.

12. **COORDINATE PLANE** Each unit represents 1 mile on the coordinate plane. The population of the region shown is 14,763 people. Find the population density of the region shown. Round to the nearest tenth.

13. **POPULATION** A city is divided by a river. The portion of the city east of the river covers 25% of the city and has a population density of 28 persons/km². The portion on the west side of the river has a population density of 17 persons/km². Find the population density of the city as a whole.

14. **BLACK BEARS** The Great Smoky Mountains National Park runs along the border of Tennessee and North Carolina. The aerial map below shows the outline of the park and the area it covers. It is estimated that there are 1500 black bears that live in the Smoky Mountains National Park. What is the population density of the black bears in the park?

15. **ARGUMENTS** The cargo of a semi-trailer can weigh no more than 34,000 pounds. The interior dimensions of a semi-trailer are shown. Suppose a freight company wants to haul a shipment that will completely fill the entire interior of the trailer, and the freight has a known density of 0.006 pounds/in$^3$. Will this proposed load meet the weight restrictions? Explain.

16. **PAPER WEIGHTS** The cylindrical paper weight with dimensions shown has a mass of 606.7 grams.

   a. Find the density of the paperweight. Use $\pi \approx 3.14$.

   b. Use the table shown to determine which, if any, of the materials may have been used to make the paperweight.

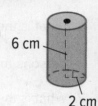

| Material | Density |
|----------|---------|
| Silver | 10.5 g/cm$^3$ |
| Copper | 8.96 g/cm$^3$ |
| Steel | 8.05 g/cm$^3$ |

17. **USE A SOURCE** Research the animal population in a national park. Write and solve a problem to find the population density of the population.

18. **WRITE** Explain why a cubic foot of gas and a cubic foot of gold do not have the same density.

19. **ANALYZE** Which block has a greater density? Explain.

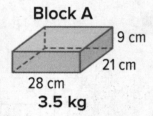

Block A

9 cm
21 cm
28 cm
**3.5 kg**

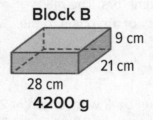

Block B

9 cm
21 cm
28 cm
**4200 g**

20. **PERSEVERE** A British Thermal Unit (BTU), is equal to the amount of energy used to raise the temperature of one pound of water per one degree Fahrenheit. If a water heater holds 250 gallons of water that is currently 58 degrees and needs to be heated to 72 degrees, how many BTU's will be needed? (Hint: 1 gallon of water weighs about 8.34 pounds)

21. **WHICH ONE DOESN'T BELONG?** James is building his own terrarium for 18 plants and he wants 120 cubic inches of space by volume per plant. Which of the following proposed structures will not allow James to meet his requirement?

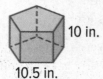

10 in.
10.5 in.

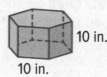

10 in.
10 in.

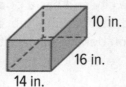

10 in.
16 in.
14 in.

**Essential Question**

How are measurements of two- and three-dimensional figures useful for modeling situations in the real world?

## Module Summary

### Lessons 11-1 through 11-3

Two-Dimensional Areas

- parallelogram: $A = bh$
- trapezoid: $A = \frac{1}{2}h(b_1 + b_2)$
- rhombus or kite: $A = \frac{1}{2}d_1d_2$
- regular $n$-gon:
  $A = \frac{1}{2}aP$
- circle: $A = \pi r^2$
- sector: $A = \frac{x°}{360°} \cdot \pi r^2$

### Lessons 11-4 and 11-7

Surface Area

- right prism: $S = Ph + 2B$, where $P$ is the perimeter of a base and $h$ is the height
- right cylinder: $S = 2\pi rh + 2\pi r^2$, where $r$ is the radius of a base and $h$ is the height
- regular pyramid: $S = \frac{1}{2}P\ell + B$, where $P$ is perimeter of the base and $\ell$ is slant height
- right circular cone: $S = \pi r\ell + \pi r^2$, where $r$ is radius of the base and $\ell$ is slant height
- sphere: $S = 4\pi r^2$, where $r$ is the radius

### Lessons 11-6 and 11-7

Volume

- prism: $V = Bh$, where $B$ is the area of a base and $h$ is the height of the prism
- pyramid: $V = \frac{1}{3}Bh$, where $B$ is the area of the base and $h$ is the height of the pyramid
- cylinder: $V = Bh$ or $V = \pi r^2h$, where $B$ is the area of the base, $h$ is the height of the cylinder, and $r$ is the radius of the base
- circular cone: $V = \frac{1}{3}Bh$ or $V = \frac{1}{3}\pi r^2h$, where $B$ is the area of the base, $h$ is the height of the cone, and $r$ is the radius of the base.
- sphere: is $V = \frac{4}{3}\pi r^3$, where $r$ is the radius

### Lessons 11-5, 11-8, and 11-9

Other Measurement Topics

- A cross section is the intersection of a solid figure and a plane.
- A solid of revolution is obtained by rotating a plane figure or curve around an axis.
- If two similar solids have a scale factor of $a : b$, then the surface areas have a ratio of $a^2 : b^2$, and the volumes have a ratio of $a^3 : b^3$.
- Density is the ratio of objects to area
- Density is the ratio of mass (or weight) to volume.

### Study Organizer

 **Foldables**

Use your Foldable to review this module. Working with a partner can be helpful. Ask for clarification of concepts as needed.

# Test Practice

**1. GRIDDED RESPONSE** What is the area, in square inches, of the parallelogram?
(Lesson 11-1)

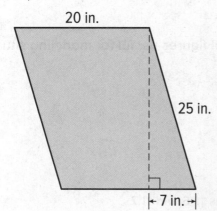

20 in.

25 in.

← 7 in. →

**2. MULTIPLE CHOICE** What is the area of the trapezoid in square centimeters? (Lesson 11-1)

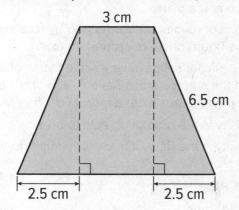

3 cm

6.5 cm

2.5 cm      2.5 cm

- Ⓐ 24 square centimeters
- Ⓑ 33 square centimeters
- Ⓒ 48 square centimeters
- Ⓓ 52 square centimeters

**3. OPEN RESPONSE** Find the area, in square centimeters, of this composite figure.
(Lesson 11-2)

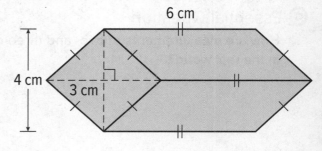

6 cm

4 cm

3 cm

**4. MULTIPLE CHOICE** A stop sign is shaped like a regular octagon. The distance between opposite sides of a stop sign is 30 inches. One side of the stop sign measures approximately 12.4 inches. What is the approximate area of the stop sign to the nearest square inch? (Lesson 11-2)

- Ⓐ 372 square inches
- Ⓑ 588 square inches
- Ⓒ 742 square inches
- Ⓓ 1488 square inches

**5. OPEN RESPONSE** A lawn sprinkler can water a sector with an arc measure of 120° and a radius of 30 feet.

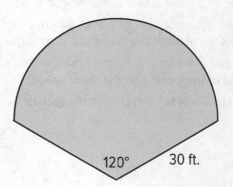

120°      30 ft.

What is the area that can be watered to the nearest square foot? (Lesson 11-3)

**6. MULTIPLE CHOICE** What is the area of a circle with a diameter of 5 inches? Round your answer to the nearest tenth of a square inch. (Lesson 11-3)

(A) 15.7 square inches

(B) 19.6 square inches

(C) 61.7 square inches

(D) 78.5 square inches

**7. OPEN RESPONSE** What are the lateral area and surface area of the cone to the nearest square inch? (Lesson 11-4)

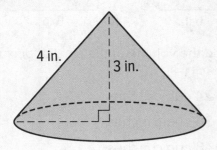

4 in.   3 in.

**8. OPEN RESPONSE** What is the surface area, in square millimeters, of a sphere with a diameter of 10 millimeters? (Lesson 11-4)

**9. MULTIPLE CHOICE** A right square pyramid is intersected by a plane perpendicular to the base that passes through the vertex of the pyramid. What cross section will result? (Lesson 11-5)

(A) isosceles triangle

(B) right triangle

(C) square

(D) trapezoid

**10. MULTI-SELECT** Which shape could be the cross section of a cube? Select all that apply. (Lesson 11-5)

(A) hexagon        (B) octagon

(C) rectangle      (D) triangle

**11. GRIDDED RESPONSE** What is the volume, in cubic centimeters, of this prism? (Lesson 11-6)

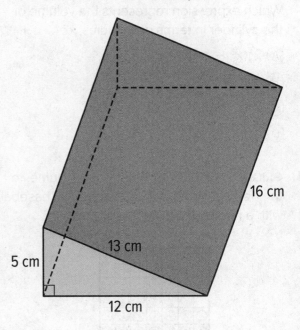

16 cm

13 cm

5 cm

12 cm

**12. OPEN RESPONSE** A square pyramid is constructed so that the height has length $x + 2$ and the sides of the base have length $x$. Write an expression for the volume of the pyramid in terms of $x$. (Lesson 11-6)

**13. MULTIPLE CHOICE** The height of a cylinder is 1 inch less than the diameter of the cylinder. Which expression represents the volume of the cylinder in terms of its radius, $x$? (Lesson 11-7)

Ⓐ $2\pi x^3 - \pi x^2$

Ⓑ $\pi x^3 - \pi x^2$

Ⓒ $2\pi x^3 - 1$

Ⓓ $2\pi x^3 + \pi x^2$

**14. GRIDDED RESPONSE** What is the volume, to the nearest cubic inch, of a spherical baseball with a radius of 2.9 inches? (Lesson 11-7)

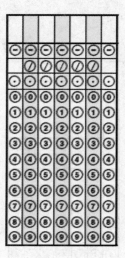

**15. MULTIPLE CHOICE** Triangle *ABC* is similar to triangle *DEF*.

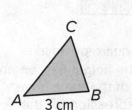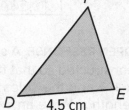

If the area of triangle *ABC* is 4 square centimeters, what is the area of triangle *DEF*? (Lesson 11-8)

Ⓐ 6 square centimeters

Ⓑ 7.5 square centimeters

Ⓒ 9 square centimeters

Ⓓ 10.5 square centimeters

**16. OPEN RESPONSE** A dinner plate has an area of 95 square inches. The matching salad plate has a diameter $\frac{4}{5}$ as long as the diameter of the dinner plate. What is the approximate area of the salad plate to the nearest square inch? (Lesson 11-8)

**17. GRIDDED RESPONSE** These prisms are similar.

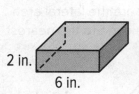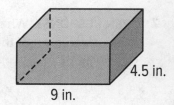

Find the volume of the smaller prism. (Lesson 11-8)

Ⓐ 24 cubic inches

Ⓑ 36 cubic inches

Ⓒ 48 cubic inches

Ⓓ 72 cubic inches

**18. MULTIPLE CHOICE** A pile of sand forms a cone with a diameter of 2 meters and a height of 0.7 meter. The mass of the pile is 1170 kilograms. What is the approximate density of the sand in kilograms per cubic meter? (Lesson 11-9)

Ⓐ 399 kilograms per cubic meter

Ⓑ 532 kilograms per cubic meter

Ⓒ 1140 kilograms per cubic meter

Ⓓ 1596 kilograms per cubic meter

**19. OPEN RESPONSE** Jacksonville, Florida has a land area of 875 square miles and a population of 880,619. What is the approximate population density of Jacksonville to the nearest whole number? (Lesson 11-7)

# Probability

## e Essential Question
How can you use measurements to find probabilities?

S.CP.1, S.CP.2, S.CP.3, S.CP.4, S.CP.5, S.CP.6, S.CP.7, S.CP.8, S.CP.9, S.MD.6, S.MD.7

## What will you learn?
Place a checkmark (✓) in each row that corresponds with how much you already know about each topic **before** starting this module.

KEY

👎 — I don't know.     👉 — I've heard of it.     👍 — I know it!

| | Before | | | After | | |
|---|---|---|---|---|---|---|
| | 👎 | 👉 | 👍 | 👎 | 👉 | 👍 |
| describe events using subsets | | | | | | |
| solve problems involving using the rule for the probability of complementary events | | | | | | |
| find the probability of an event by using lengths of segments and areas | | | | | | |
| solve problems involving probabilities of compound events using permutations and combinations | | | | | | |
| solve problems involving probability of independent events using the Multiplication Rule | | | | | | |
| solve problems involving conditional probability | | | | | | |
| solve problems involving mutually exclusive events using the Addition Rule | | | | | | |
| solve problems involving events that are not mutually exclusive using the Addition Rule | | | | | | |
| solve problems involving conditional probability using the Multiplication Rule | | | | | | |
| decide if events are independent and approximate conditional probabilities using two-way frequency tables | | | | | | |

📖 **Foldables** Make this Foldable to help you organize your notes about probability. Begin with one sheet of notebook paper.

1. **Fold** a sheet of paper lengthwise.

2. **Fold** in half two more times.

3. **Cut** along each fold on the left column.

4. **Label** each section with a topic.

## What Vocabulary Will You Learn?

Check the box next to each vocabulary term that you may already know.

- ☐ combination
- ☐ complement
- ☐ compound event
- ☐ conditional probability
- ☐ dependent events
- ☐ experiment

- ☐ factorial
- ☐ Fundamental Counting Principle
- ☐ independent events
- ☐ joint frequencies
- ☐ marginal frequencies

- ☐ mutually exclusive
- ☐ outcome
- ☐ permutation
- ☐ relative frequency
- ☐ sample space
- ☐ two-way frequency table

## Are You Ready?

Complete the Quick Review to see if you are ready to start this module.
Then complete the Quick Check.

### Quick Review

**Example 1**

**Suppose a die is rolled. What is the probability of rolling less than 5?**

$P(\text{less than } 5) = \frac{\text{number of favorable outcomes}}{\text{number of possible outcomes}}$

$= \frac{4}{6} \text{ or } \frac{2}{3}$

The probability of rolling less than 5 is $\frac{2}{3}$ or 67%.

**Example 2**

**A spinner numbered 1–6 was spun. Find the experimental probability of landing on a 5.**

| Outcome | Tally | Frequency |
|---------|-------|-----------|
| 1 | IIII | 4 |
| 2 | JHT II | 7 |
| 3 | JHT III | 8 |
| 4 | IIII | 4 |
| 5 | II | 2 |
| 6 | JHT | 5 |

$P(5) = \frac{\text{number of times a 5 is spun}}{\text{total number of outcomes}} = \frac{2}{30} \text{ or } \frac{1}{15}$

The experimental probability of landing on a 5 is $\frac{1}{15}$ or 7%.

### Quick Check

**A die is rolled. Find the probability of each outcome**

**1.** $P(\text{greater than } 1)$

**2.** $P(\text{odd})$

**3.** $P(\text{less than } 2)$

**4.** $P(1 \text{ or } 6)$

**The table shows the results of an experiment in which a spinner numbered 1–4 was spun.**

| Outcome | Tally | Frequency |
|---------|-------|-----------|
| 1 | III | 3 |
| 2 | JHT II | 7 |
| 3 | JHT III | 8 |
| 4 | IIII | 4 |

**5.** What is the experimental probability that the spinner will land on a 4?

**6.** What is the experimental probability that the spinner will land on an odd number?

**7.** What is the experimental probability that the spinner will land on an even number?

### How Did You Do?

Which exercises did you answer correctly in the Quick Check? Shade those exercise numbers below.

      ⑦

# Sample Spaces

**Today's Standards**
Prep for S.CP.1
MP5

**Today's Vocabulary**
experiment
outcome
sample space
finite sample space
infinite sample space
event

## Learn Sample Spaces

An **experiment** is a situation involving chance. An **outcome** is the result of a single performance or trial of an experiment.

The set of all possible outcomes make up the **sample space** of an experiment, which may be finite or infinite. A sample space that contains a countable number of outcomes is a **finite sample space**. A sample space with outcomes that cannot be counted is an **infinite sample space**. An **event** is a subset of the sample space.

| Finite | | Infinite Discrete | Infinite Continuous |
|---|---|---|---|
| spinning the spinner {red, blue, purple} | 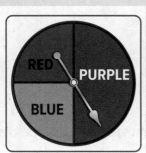 | tossing a coin *until* you get two heads {H, TH, TTH, ...} | all of the diameters in a circle |

Infinite discrete sample spaces have outcomes that can be arranged in a sequence or counted but go on indefinitely. Infinite continuous sample spaces cannot be counted or defined because there are infinite ways to fill the sample space.

You can represent a sample space by using an organized list, a table, or a tree diagram.

## Example 1 Define a Sample Space

**A fair die is tossed once.**

**a. What is the sample space of the experiment?**

The sample space $S$ includes all possible outcomes of rolling a die.

$S = \{1, 2, 3, 4, 5, 6\}$

**b. What is the sample space for the event of rolling a prime number? Write the outcomes to complete the sample space.**

$S$(prime number on a die) $= \{\underline{\quad}, \underline{\quad}, \underline{\quad}\}$

**Go Online** You can complete an Extra Example online.

### 🌐 Example 2 Represent a Sample Space

CLOTHING **Kembe has a black hat and a red hat. He chooses one hat for each day, Saturday and Sunday. Represent the sample space for this experiment by making an organized list, a table, and a tree diagram.**

For each day, Saturday and Sunday, there are two possibilities: the red hat (R) or the black hat (B).

**Organized List**
Pair each possible outcome from the Saturday's hat choice with the possible outcomes from Sunday's hat choice using coordinates.

$S = \{(R, R), (R, B), (B, B), (\underline{\quad}, \underline{\quad})\}$

**Table**
Saturday's hat choices are represented vertically, and Sunday's hat choices are represented horizontally.

|  | Sunday | |
|---|---|---|
| Saturday | R | B |
| R | R, R | R, [B] |
| B | B, [R] | B, B |

**Tree Diagram**
Each event is represented by a different stage of the tree diagram.

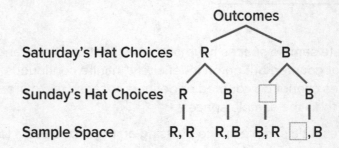

### Check

**GROUP WORK** A geometry teacher always breaks her class up into the red, yellow, and blue groups for class projects. Represent the sample space for the next two class projects by making a tree diagram.

Enter the outcomes to complete the tree diagram.

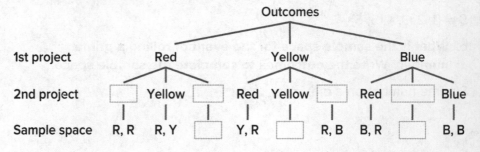

🌐 **Go Online** You can complete an Extra Example online.

💬 **Talk About It!**
Why are R, B and B, R not the same outcome?

**Study Tip**
**Tree Diagram Notation**
Choose notation for outcomes in your tree diagrams that will eliminate confusion. In the example, R stands for red hat and B stands for black hat.

## Example 3 Finite and Infinite Sample Spaces

**Classify each sample space as *finite* or *infinite*. If it is finite, write the sample space. If it is infinite, classify whether it is *discrete* or *continuous*.**

**a. A marble is drawn from a bag that contains 3 orange marbles, 5 green marbles, and 4 blue marbles.**

There are only _____ possible outcomes of this experiment: selecting an orange, green, or blue marble. The sample space $S$ is _____,

$S$ = {orange, green, blue}.

**b. A spinner with four equal parts of green, blue, red, and yellow is spun until it lands on yellow.**

There are a(n) _____ number of possible outcomes of this experiment, so its sample space is _____. Since the experiment ends after a certain number of spins when the spinner lands on yellow, the sample space is _____.

**c. A ball is thrown into the air and its height is recorded in inches.**

There are a(n) _____ number of possible outcomes of this experiment, so its sample space is _____. You can continue to record heights of the thrown ball indefinitely, so the sample space is _____.

## Learn Fundamental Counting Principle

For some large or complicated experiments, listing the entire sample space may not be practical or necessary. To find the *number* of possible outcomes, you can use the Fundamental Counting Principle.

| Key Concepts • Fundamental Counting Principle | |
|---|---|
| **Words** | The number of possible outcomes in a sample space can be found by multiplying the number of possible outcomes from each stage or event. |
| **Symbols** | In a $k$-stage experiment, let<br>$n_1$ = the number of possible outcomes for the first stage.<br>$n_2$ = the number of possible outcomes for the second stage after the first stage has occurred.<br>$n_k$ = the number of possible outcomes for the $k$th stage after the first $k$-1 stages have occurred.<br>Then the total possible outcomes of this $k$-stage experiment is $n_1 \cdot n_2 \cdot n_3 \cdot \ldots \cdot n_k$. |

**Study Tip**

**Multiplication Rule**
The Fundamental Counting Principle is sometimes called the *Multiplication Rule for Counting* or the *Counting Principle*.

**Think About It!**

If each stage of a two-stage experiment has three outcomes, what is the number of total possible outcomes?

## 🌐 **Example 4** Use the Fundamental Counting Principle

**COLLEGE** **Santiago lists the number of sections available for the courses he will take in his first semester at college. How many different schedules could Santiago create for this semester?**

| Course | Sections Offered |
|---|---|
| Art History | 6 |
| French | 5 |
| Mathematics | 9 |
| Art | 4 |
| English | 6 |

You can estimate the total number of different schedules he can make. There are about 10 sections of the _____ course offered. For each of the other four courses, there are about _____ sections offered. Multiply to estimate that Santiago can create about _____ schedules.

Find the number of possible outcomes by using the Fundamental Counting Principle to complete the equation.

Art
History     French     Mathematics     Art     English     Possible Outcomes

$$\boxed{\phantom{xx}} \times \boxed{\phantom{xx}} \times \boxed{\phantom{xxxx}} \times \boxed{\phantom{x}} \times \boxed{\phantom{x}} = 6480$$

Santiago could create 6480 different schedules.

## Check

**CLOTHING** A sneaker company lets you customize your own sneaker on their Web site. Using their most popular sneaker as the base, you have the option to customize the color of each part of the sneaker.

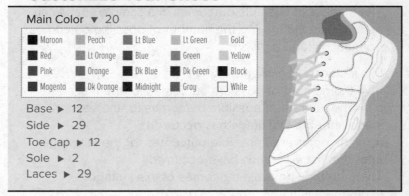

• **Customize Your Shoes**

Main Color ▼ 20

| | | |
|---|---|---|
| ■ Maroon | ■ Peach | ■ Lt Blue | ■ Lt Green | ■ Gold |
| ■ Red | ■ Lt Orange | ■ Blue | ■ Green | ■ Yellow |
| ■ Pink | ■ Orange | ■ Dk Blue | ■ Dk Green | ■ Black |
| ■ Magenta | ■ Dk Orange | ■ Midnight | ■ Gray | ☐ White |

Base ► 12
Side ► 29
Toe Cap ► 12
Sole ► 2
Laces ► 29

a. Which is the best estimate for the number of possible customizations? _____

A. 100                         B. 3,600,000
C. 5,062,500                  D. 36,000,000

b. How many different customizations can be created? _____

🌐 **Go Online** You can complete an Extra Example online.

**Use a Source**

Colleges typically assign general studies courses to freshman undergraduates who haven't yet selected a major. Use available resources to find the freshman curriculum for a college of your choice and determine the number of possible schedules that can be created.

Name _____ Period _____ Date _____

# Practice

Go Online You can complete your homework online.

**Example 1**

1. Define the sample space, *S*, of a fair coin being tossed once.

2. **SPINNERS** The numbered spinner shown is spun once.
   a. What is the sample space of the experiment?

   b. What is the sample space for the event of spinning a prime number?

3. **DODECAGON** The 12-sided dodecagon shown is rolled once.
   a. What is the sample space of the experiment?

   b. What is the sample space for the event of rolling an even number?

4. **SPINNERS** The lettered spinner shown is spun once.
   a. What is the sample space of the experiment?

   b. What is the sample space for spinning a vowel?

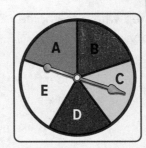

**Example 2**

**Represent the sample space for each experiment by completing the table and tree diagram, and by making an organized list.**

5. The baseball team can wear blue or white shirts with blue or white pants.

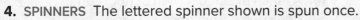

| Outcomes | Blue Pants | White Pants |
|---|---|---|
| Blue Shirts | | |
| White Shirts | | |

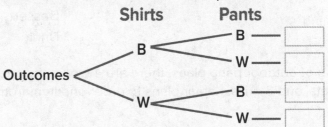

6. The dance club is going to see either *Sleeping Beauty* or *The Nutcracker* at either Symphony Hall or The Center for the Arts.

| Outcomes | Symphony Hall | Center for Arts |
|---|---|---|
| Sleeping Beauty | | |
| Nutcracker | | |

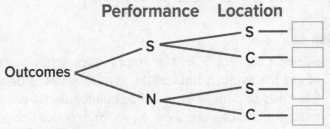

**Represent the sample space for each experiment by making an organized list, a table, and a tree diagram.**

7. Elisa's baby sister can drink either apple juice or milk from a bottle or a toddler cup.

8. Khalid can write his final essay in class or at home on a scientific or a historical topic.

**Example 3**

**Classify each sample space as *finite* or *infinite*. If it is finite, write the sample space. If it is infinite, classify whether it is *discrete* or *continuous*.**

9. A color tile is drawn from a cup that contains 1 yellow, 2 blue, 3 green, and 4 red color tiles.

10. The spinner shown is spun until it lands on 2.

11. An angler casts a fishing line into a body of water and its distance is recorded in centimeters.

12. A letter is randomly chosen from the alphabet.

**Example 4**

**Find the number of possible outcomes for each situation.**

13. A room is decorated with one choice from each category.

| Bedroom Décor | Number of Choices |
|---|---|
| Paint color | 8 |
| Comforter set | 6 |
| Sheet set | 8 |
| Throw rug | 5 |
| Lamp | 3 |
| Wall hanging | 5 |

14. A lunch at Lincoln High School contains one choice from each category.

| Cafeteria Meal | Number of Choices |
|---|---|
| Main dish | 3 |
| Side dish | 4 |
| Vegetable | 2 |
| Salad | 2 |
| Salad Dressing | 3 |
| Dessert | 2 |
| Drink | 3 |

15. In a catalog of outdoor patio plans, there are 4 types of stone, 3 types of edging, 5 dining sets, and 6 grills. Kamar plans to order one item from each category.

16. The drama club held tryouts for 6 roles in a one-act play. Five people auditioned for lead female, 3 for lead male, 8 for the best friend, 4 for the mother, 2 for the father, and 3 for the humorous aunt.

**Mixed Exercises**

17. **SCHOOL SUPPLIES** Eva is shopping for school supplies. She chooses one of each of the following: 6 backpacks, 8 notebooks, 3 pencil cases, 3 brands of pencils, 8 brands of pens, 4 types of calculator, and 4 colors of highlighter. How many different choices does she have for school supplies?

18. **LAPTOPS** Catalina is buying a laptop. She has a choice of 3 hard drive sizes, 3 processor speeds, 4 colors, 2 screen sizes, 2 warranty options, and 4 cases. She knows she wants a blue laptop with the longest warranty. How many choices does she have for laptops?

19. **BOARD GAMES** The spinner shown is used in a board game. If the spinner is spun 4 times, how many different possible outcomes are there?

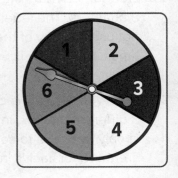

20. **BASKETBALL** In the NBA there must be a minimum of 14 players on a team's roster. A team has the minimum number of players where 3 are centers, 4 are power forwards, 2 are small forwards, 3 are shooting guards, and the rest are point guards. For this team, if one player is picked from each position to be a starter, how many different possible outcomes are there?

21. **VACATION RENTAL** A website describes available vacation rentals in Colorado and Florida. In Colorado, you can choose a 1 or 2 week stay in a 1- or 2-bedroom suite. In Florida, you can choose a 1, 2, or 3 week stay in a 2- or 3-bedroom suite, on the beach or not.

   **a.** How many outcomes are available in Colorado?

   **b.** How many outcomes are available in Florida?

   **c.** How many total outcomes are available?

> **Maurice's Packing List**
> 1. Suits: Gray, black, khaki
> 2. Shirts: White, light blue
> 3. Ties: Striped (But optional)

22. **MODELING** Maurice packs suits, shirts, and ties that can be mixed and matched. Using the packing list at the right, draw a tree diagram to represent the sample space for business suit combinations.

23. For an art class assignment, Mr. Green gives students their choice of red or blue paint to use as a base. Then the students can use glitter, beads, or fasters to decorate the project. Represent the sample space by making an organized list, a table, and a tree diagram.

**Find the number of possible outcomes for each situation.**

24. Valentine gift sets come with a choice of 4 different teddy bears, 8 types of candy, 5 balloon designs, and 3 colors of roses.

25. Tala wears a school uniform that consists of a skirt or pants, a white shirt, a blue jacket or sweater, white socks and black shoes. She has 3 pairs of pants, 3 skirts, 6 white shirts, 2 jackets, 2 sweaters, 6 pairs of white socks and 3 pairs of black shoes.

| Bread | Meats | Cheeses |
|---|---|---|
| White | Turkey | American |
| Wheat | Ham | Swiss |
| Whole Grain | Roast Beef | Provolone |
| | Chicken | Colby-Jack |
| | | Muenster |

26. **DELICATESSEN** A sandwich shop offers the table of options to its customers when making a sandwich. Provided one item from each category is selected, how many different sandwiches can be made?

27. Carlito is calculating the area of the composite figure at the right. List six different expressions he can evaluate to do this.

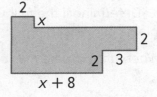

**28. LICENSE PLATES** The pattern for a certain license plate is 3 letters followed by 3 numbers. The letter "O" is not used as any of the letters and the number "0" is not used as any of the numbers. Any other letter or number can be used multiple times. How many license plates can be created with this pattern?

**29. INTERNSHIP** Jack is choosing an internship program that could take place in 3 different months, in 4 different departments, and in 3 different companies. Jack is only available to complete his internship in July. How many different outcomes are there for Jack's internship?

**30. COMBINATIONS** Talula got a new bicycle lock that has a four-number combination. Each number in the combination is from 0 to 9.

  **a.** How many combinations are possible if there are no restriction on the number of times Talula can use each number?

  **b.** How many combinations are possible if Miranda can use each number only once? Explain.

**31. BOARD GAMES** Hugo and Monette are playing a board game in which you roll two dice per turn.

  **a.** In one turn, how many outcomes result in a sum of 8?

  **b.** How many outcomes in one turn result in an odd sum?

**32. ARGUMENTS** Explain when it is necessary to show all the possible outcomes of an experiment by using a tree diagram and when using the Fundamental Counting Principle is sufficient.

**33. REASONING** A multistage experiment has $n$ possible outcomes at each stage. If the experiment is performed with $k$ stages, write an equation for the total number of possible outcomes $P$. Explain.

**34. PERSEVERE** A box contains $n$ different objects. If you remove three objects from the box, one at a time, without putting the previous object back, how many possible outcomes exist? Explain your reasoning.

**35. CREATE** Sometimes a tree diagram for an experiment is not symmetrical. Describe a two-stage experiment where the tree diagram is asymmetrical. Include a sketch of the tree diagram. Explain.

**36. WRITE** Explain why it is not possible to represent the sample space for a multi-stage experiment by using a table.

**37. ANALYZE** Determine if the following statement is *sometimes*, *always*, or *never* true. Explain your reasoning.

  *When an outcome falls outside the sample space, it is a failure.*

# Probability and Counting

## Explore Venn Diagrams

 **Online Activity** Use the interactive tool to complete the Explore.

> ⑦ **INQUIRY** How can you identify the objects in
> the sample space of two sets? ✕

## Learn Intersections and Unions

When two events *A* and *B* occur, the **intersection of *A* and *B*** is the set of all outcomes in the sample space of event *A* that are also in the sample space of event *B*. In the Venn diagram, the shaded portion represents the intersection.

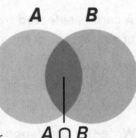

To determine the probability of the intersection of two or more evens, find the ratio of the number of outcomes in both evens to the total number of possible outcomes.

### Key Concept • Probability Rule for Intersections

The probability of the intersection of two events *A* and *B* occurring is the ratio of the number of outcomes in both *A* and *B* to the total number of possible outcomes.

$$P(A \cap B) = \frac{\text{number of outcomes in } A \text{ and } B}{\text{total number of possible outcomes}}$$

When two events *A* and *B* occur, the **union of *A* and *B*** is the set of all outcomes in the sample space of event *A* combined with all outcomes in the sample space of event *B*. In the Venn diagram, the shaded portion represents the union.

### Key Concept • Union of Two Events

The number of elements in the union of two events *A* and *B* is the number of outcomes in both event *A* and *B* minus the number of outcomes in their intersection.

$$n(A \cup B) = n(A) + n(B) - n(A \cap B)$$

## Example 1 Find Intersections

**A fair die is rolled once. Let *A* be the event of rolling an odd number, and let *B* be the event of rolling a number greater than 3. Find *A* ∩ *B*.**

The possible outcomes for event ____ are all the numbers on a die that are odd, or {1, 3, 5}.

The possible outcomes for event ____ are all the numbers on a die that are greater than 3, or {4, 5, 6}.

*A* ∩ *B* contains ____ of the outcomes that are in both sample space *A* and *B*.

*A* ∩ *B* = ____

 **Go Online** You can complete an Extra Example online.

**Today's Standards**
S.CP.1
MP3, MP4

**Today's Vocabulary**
intersection of *A* and *B*
union of *A* and *B*
complement of *A*

**Study Tip**

**Intersection** The symbol for intersection is ∩, and it is associated with the word *and*.

*P*(*A* ∩ *B*) is read as *the probability of A and B*.

**Study Tip**

**Union** The symbol for union is ∪, and it is associated with the word *or*.

*n*(*A* ∪ *B*) is read as, *the number of elements in A or B*.

**Go Online**

You may want to complete the Concept Check to check your understanding.

**⨀ Think About It!**

Is the sample space for *A* ∩ *B* finite or infinite? Justify your reasoning.

## Check

Let $A$ be the event of the spinner landing on a blue section, and let $B$ be the event of the spinner landing on a section with a number divisible by 3. What are the possible outcomes of each event?

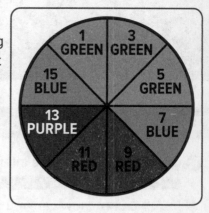

$A = \{7, \underline{\quad}\}$

$B = \{3, \underline{\quad}, 15\}$

$A \cap B = \{\underline{\quad}\}$

---

## Example 2  Probability of Intersections

**A card is selected from a standard deck of cards. What is the probability that the card is a queen and is red?**

Let $A$ be the event of choosing a queen, and let $B$ be the event of choosing a red card. The total number of outcomes is the total number of cards in a deck, or 52.

**Write the corresponding number of each card in its correct place in the Venn diagram.**

Card 1        Card 2        Card 3        Card 4

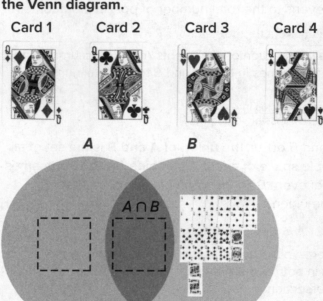

From the diagram, there are only 2 red cards that are also Queens.

$P(A \cap B) = \dfrac{\text{number of outcomes in } A \text{ and } B}{\text{total number of possible outcomes}}$   Probability Rule for Intersections

$= \dfrac{\quad}{52}$   Substitution

$= \underline{\quad}$   Simplify.

The probability that the card is both a queen and is red is $\frac{1}{26}$, or about

$\underline{\qquad}$.

🌐 **Go Online** You can complete an Extra Example online.

## Math History Minute

Scottish physician **John Arbuthnot (1667–1735)** published *Of the Laws of Chance* anonymously in 1692, the first work on probability published in English. This appears to be the first time the word "probability" is used in print. In a 1710 paper, Arbuthnot discusses the first application of probability to social statistics.

## Check

Two fair dice are rolled one time. What is the probability that the same number is rolled on both dice and that the sum of the numbers on the two dice is 10 or greater? Enter the probability as a ratio.

$P(A \cup B) = $ _____

---

## Example 3  Find Unions

**A fair die is rolled once. Let _A_ be the event of rolling a number less than 5, and let _B_ be the event of rolling a multiple of 2. Find _A_ ∪ _B_.**

The possible outcomes for event _A_ are all the numbers on a die that are less than 5, or { ____, ____, ____, ____ }.

The possible outcomes for event _B_ are all the numbers on a die that are multiples of ____, or {2, 4, 6}.

_A_ ∪ _B_ contains all of the outcomes that are in _____ sample space(s) _A_ or _B_.

$A \cup B = \{$ ____, ____, ____, ____, ____, ____ $\}$

## Check

Let _A_ be the event of the spinner landing on a blue section, and let _B_ be the event of the spinner landing on a section with a number divisible by 3. What are the possible outcomes of each event?

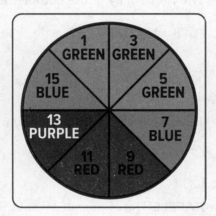

$A = \{7, $ ____ $\}$

$B = \{3, $ ____ $, 15\}$

$A \cup B = \{3, 7, 9, $ ____ $\}$

## Learn  Complements

The **complement of _A_** consists of all the outcomes in the sample space that are not included as outcomes of event _A_. The complement of event _A_ can be noted as _A'_. In the Venn diagram, the shaded portion represents the complement of _A_.

The probability of rolling a die and getting a 3 is $\frac{1}{6}$. What is the probability of _not_ getting a 3?

There are 5 possible outcomes for this event: 1, 2, 4, 5, or 6. So, _P_(not 3) = $\frac{5}{6}$. Notice that this probability is also $1 - \frac{1}{6}$ or $1 - P(3)$.

**Think About It!**

Why are 2 and 4 only listed once in _A_ ∪ _B_?

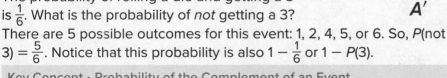

| Key Concept • Probability of the Complement of an Event | |
|---|---|
| **Words** | The probability that an event will not occur is equal to 1 minus the probability that the event will occur. |
| **Symbols** | For an event _A_, $P(A') = 1 - P(A)$. |

### 🌐 Example 4 Complementary Events

**DIGITAL MEDIA Panju subscribes to a movie streaming service. For movie night, he is going to let the program randomly pick a movie from his list of favorites. What is the probability that a comedy movie will not be chosen?**

| MY MOVIE QUEUE | |
|---|---|
| **GENRES** | |
| ▶ ACTION | 44 |
| ▶ ANIME | 109 |
| ▶ CHILDREN'S | 8 |
| ▶ COMEDIES | 112 |
| ▼ DOCUMENTARIES | 13 |
| Worms in the Garden | 1990 - Directed by Nigh Crowder - 96 minutes |
| Department of Redundancy, Department | 2004 - directed by William Williams - 118 minutes |
| Matter of Fact - What We're Made Of | 2010 - Narrated by D. Cy Edu.k - 54 minutes |
| ▶ DRAMAS | 30 |
| ▶ FOREIGN | 5 |
| ▶ HORROR | 29 |

Let event *A* represent selecting a comedy movie from Panju's favorites. Then find the probability of the complement of *A*.

There are ____ comedy movies in Panju's favorites list.
There are ____ total movies in Panju's favorites list.

The probability of the complement of *A* is $P(A') = $ ____ $- P($____$)$.

$P(A') = 1 - P(A)$      Probability of a complement

$= 1 - \dfrac{}{350}$      Substitution

$= \dfrac{}{350}$ or $\dfrac{17}{}$      Subtract and Simplify.

The probability that a comedy movie will not be chosen is $\dfrac{17}{25}$ or ____%.

## Check

**RAFFLE** The Harvest Fair sold 967 raffle tickets for a change to win a new TV. Match each probability of not winning the TV with the given number of tickets.

| | | |
|---|---|---|
| 20 tickets | ↔ | |
| 200 tickets | ↔ | |
| 100 tickets | ↔ | |
| 1 ticket | ↔ | |

| | |
|---|---|
| 0.02% | 0.1 |
| 99.8% | 0.79 |
| 0.90 | 98% |
| 21% | 2% |

🌐 **Go Online** You can complete an Extra Example online.

💬 **Talk About It!**

Why do you think the probability of the complement of an event is found by subtracting from 1?

Name _____ Period _____ Date _____

# Practice

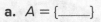

icon **Go Online** You can complete your homework online.

### Example 1

1. A fair die is rolled once. Let $A$ be the event of rolling an even number, and let $B$ be the event of rolling a number greater than 4. Find $A \cap B$.

2. A fair die is rolled once. Let $A$ be the event of rolling an even number, and let $B$ be the event of rolling an odd number. Find $A \cap B$.

**NUMBERED SPINNERS** **For Exercises 3 and 4, use the spinner shown.**

3. Let $A$ be the event of the spinner landing on a blue section, and let $B$ be the event of the spinner landing on a section with a number divisible by 4. What are the possible outcomes of each event?

   a. $A = \{$\_\_\_\_$\}$

   b. $B = \{$\_\_\_\_$\}$

   c. $A \cap B = \{$\_\_\_\_$\}$

4. Let $A$ be the event of the spinner landing on a red section, and let $B$ be the event of the spinner landing on a section with a number that is a multiple of 3. What are the possible outcomes of each event?

   a. $A = \{$\_\_\_\_$\}$

   b. $B = \{$\_\_\_\_$\}$

   c. $A \cap B = \{$\_\_\_\_$\}$

### Example 2

**Consider the experiment of selecting one of the letters of the alphabet at random. Find the probability of picking each of the following. Round to the nearest hundredth, if necessary.**

5. a vowel (not including the letter $y$)

6. one of the letters in the word *Mississippi*

**Consider the experiment of drawing a card from a standard deck of playing cards at random. Find the probability of selecting each of the following. Round to the nearest hundredth, if necessary.**

7. a 2 or an 8

8. a club

**Consider the experiment of picking one of the twelve months of the year at random. Find the probability of picking each of the following. Round to the nearest hundredth, if necessary.**

9. a month of the year that begins with the letter J

10. a month of the year that begins with the letter A

11. a month of the year that ends with the letters *er*

12. a month of the year that starts with the letter F

Example 3

LETTERED SPINNERS  **For Exercises 13 and 14, use the spinner shown.**

13. Let *A* be the event that the spinner lands on a vowel. Let *B* be the event that it lands on the letter J. What are the possible outcomes of each event?

   a. $A = \{$_____$\}$

   b. $B = \{$_____$\}$

   c. $A \cup B = \{$_____$\}$

14. Let *A* be the event that the spinner lands on a consonant. Let *B* be the event that it lands on the letter A. What are the possible outcomes of each event?

   a. $A = \{$_____$\}$

   b. $B = \{$_____$\}$

   c. $A \cup B = \{$_____$\}$

15. A random number generator is used to generate one integer between 1 and 20. Let *A* be the event of generating a multiple of 5, and let *B* be the event of generating a number less than 12. What are the possible outcomes of each event?

   a. $A = \{$_____$\}$

   b. $B = \{$_____$\}$

   c. $A \cup B = \{$_____$\}$

16. A random number generator is used to generate one integer between 1 and 100. Let *A* be the event of generating a multiple of 10, and let *B* be the event of generating a factor of 30. What are the possible outcomes of each event?

   a. $A = \{$_____$\}$

   b. $B = \{$_____$\}$

   c. $A \cup B = \{$_____$\}$

**Example 4**

**Determine the probability of each event. Round to the nearest hundredth, if necessary.**

17. If there is a 4 in 5 chance that Jeremias's mom will tell him to clean his room today after school, what is the probability that she won't tell Jeremias to clean his room?

18. What is the probability of drawing a card from a standard deck and not getting a spade?

19. What is the probability of flipping a coin and not landing on tails?

20. Jimena purchased 10 raffle tickets. If 250 were sold, what is the probability that one of Jimena's tickets will not be drawn?

21. If the chance of being selected for the student bailiff program is 10 in 200, what is the probability of not being chose?

22. What is the probability of spinning a spinner numbered 1 to 6, and not landing on 5?

**Mixed Exercises**

23. **SURVEYS** A survey found that about 90% of the junior class is right handed. If 1 junior is chosen at random out of 100 juniors, what is the probability that he or she is not right handed?

24. **RAFFLE** Miguel bought 50 raffle tickets. If 1000 were sold, what is the probability that Miguel will not win the raffle, assuming that only 1 ticket is chosen.

25. **MASCOT** At Riverview High School, 120 students were asked whether they prefer a lion or a timber wolf as the new school mascot. The table shows the results of the survey.

| | Upperclassmen | Lowerclassmen | Totals |
|---|---|---|---|
| Lion | 38 | 40 | 78 |
| Timber Wolf | 19 | 23 | 42 |
| Totals | 57 | 63 | 120 |

   a. What is the probability of selecting an upperclassman who prefers the timber wolf?

   b. What is the probability of not selecting a lowerclassman who prefers the lion?

26. **COLLEGE** In Evan's senior class of 240 students, 85% are planning to attend college after graduation. What is the probability that a senior chosen at random is not planning to attend college after graduation?

**27. STUDENT ACTIVITIES** The Venn diagram shows the cast members of two school musicals who also participate in the local children's theater. One of the students will be chosen at random to attend a statewide performing arts conference. Let *A* be the event that a student is a cast member of *Suessical* and let *B* be the event that the student is a cast member of *Wizard of Oz*.

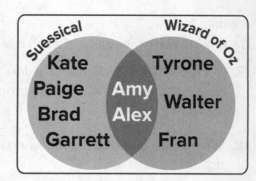

a. Find $A \cap B$.

b. What is the probability that the student who is chosen to attend the conference is a participant in the local children's theater who is a cast member of only one of the musicals?

**28. MODELING** Emma was playing a game that involved the spinner shown. What is the probability of each of Emma spinning each of the following in her next turn?

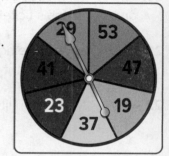

a. a prime number

b. a number with a "4" in the tens place

c. a number that has a "3" in either its tens or ones place

**29. MALL** Outside a shopping mall, 20 people were asked whether they had made a purchase at one of the mall's stores, eaten at one of the mall's restaurants, done both, or done neither. The Venn diagram shows the results of the survey. One person will be chosen at random to be interviewed on the local evening news. Find the probability that the person chosen will be someone who made a purchase and ate at one of the restaurants?

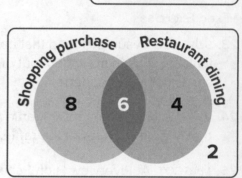

**30. PERSEVERE** Let *A* be the possible side measures of the rectangle with perimeter shown. Let *B* represent the possible measures of $\overline{XY}$ in $\triangle XYZ$.

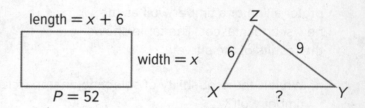

a. Find $A \cap B$.

b. Find $A \cup B$.

**31. CREATE** Make a Venn diagram to display the following: let *A* be the months of the year with 31 days and let *B* be the months of the year that begin with the letter J.

**32. WRITE** Suppose you need to explain the concept of *intersections* and *unions* to someone with no knowledge of the topic. Write a brief description of your explanation.

**33. ANALYZE** Determine if the following statement is *sometimes, always,* or *never* true. Explain your reasoning.
*The union of two sets has more elements than the intersection of two sets.*

**34. FIND THE ERROR** Let *A* be the event that the spinner lands on a vowel. Let *B* be the event that it lands on the letter J. Truc says $A \cup B$ is {A, A, E, O, U, J, J}, and Alan says $A \cup B$ is ø. Who is correct? Explain.

# Geometric Probability

## Explore Probability Using Lengths of Segments

 **Online Activity** Use dynamic geometry software to complete the Explore.

**⊘ INQUIRY** How can lengths of segments be used to determine probability?

**Today's Standards**
S.MD.6, SMD.7
MP4, MP5

**Today's Vocabulary**
geometric probability

## Learn Probability with Length

Probability that involves a geometric measure such as length or area is called **geometric probability**.

| Key Concept · Length Probability Ratio | |
|---|---|
| **Words** | If a line segment (1) contains another segment (2) and a point on segment (1) is chosen at random, then the probability that the point is on segment (2) is $\frac{\text{length of segment (2)}}{\text{length of segment (1)}}$. |
| **Example** | If a point $E$ on $\overline{AD}$ is chosen at random, then $P(E \text{ is on } \overline{BC}) = \frac{BC}{AD}$. |

When determining geometric probabilities, we assume
- that the object lands within the target area, and
- it is equally likely that the object will land anywhere in the region.

## Example 1 Use Length to Find Geometric Probability

**Point $X$ is chosen at random on $\overline{PS}$. Find the probability that $X$ is on $\overline{RS}$.**

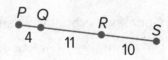

What is the length of $\overline{PS}$? $PS =$ _____
What is the length of $\overline{RS}$? $PS =$ _____

$$P(X \text{ is on } \overline{RS}) = \frac{RS}{PS} \qquad \text{Length probability ratio}$$

$$= \frac{10}{25} \qquad RS = \underline{\quad}; PS = \underline{\quad}$$

$$= \frac{2}{5} \text{ or } 0.4 \qquad \text{Simplify.}$$

The probability that 3 is on $\overline{RS}$ is _____%.

### Check

Point $A$ is chosen at random on $\overline{WZ}$. Find the probability to the nearest percent that $A$ in not on $\overline{YZ}$.

What is the length of $\overline{WZ}$? $WZ =$ _____
What is the length of $\overline{YZ}$? $YZ =$ _____
To the nearest percent, $P(A \text{ is not on } \overline{YZ})$ is about _____%.

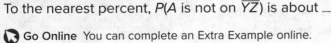 **Go Online** You can complete an Extra Example online.

> 🍩 **Think About It!**
> What is the probability that $X$ is not on $\overline{RS}$?
> _____ %

## 🌐 Example 2 Model Real-World Probabilities

**IMAGE SHARING  A Web site that hosts an image sharing gallery updates its front page for new content every 6 minutes. Assuming that you open the gallery at a random time, what is the probability that you will have to wait 5 or more minutes for the content to refresh?**

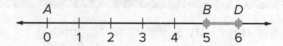

You can use a number line to model the situation. Since the page is updated every 6 minutes, the next update will be in 6 minutes or less.

The event of waiting 5 or more minutes is modeled by $\overline{BD}$ on the number line.

Find $P(\text{waiting 5 minutes or more}) = \dfrac{BD}{AD}$        Length probability ratio

$= \underline{\hspace{1cm}}$        $BD = 1$ and $AD = 6$

So, the probability of waiting 5 or more minutes for the gallery to refresh is $\dfrac{1}{6}$ or about $\underline{\hspace{1cm}}$%.

### 🔖 Go Online
You can complete an Extra Example online.

## Explore Probability and Decision Making

🔖 **Online Activity** Use the guiding exercises to complete the Explore.

> ❓ **INQUIRY** How can you use geometric probability to make decisions?

## Learn Probability with Area

Geometric probability can also involve area.

| Key Concept • Area Probability Ratio | |
|---|---|
| **Words** | If region $A$ contains a region $B$ and a point $E$ in region $A$ is chosen at random, then the probability that point $E$ is in region $B$ is $\dfrac{\text{area of region } B}{\text{area of region } A}$. |
| **Example** | If point $E$ is chosen at random in rectangle $A$, then $P(\text{point } E \text{ is in circle } B) = \dfrac{\text{area of region } B}{\text{area of region } A}$.  |

## 🌐 Example 3 Use Area to Find Geometric Probabilities

**LAWN GAMES  Haruko Games is designing a new lawn game where each player will attempt to hit a circular target by tossing a beanbag onto a larger, circular board.**

**a. What is the probability that a toss will land on the target?**

You need to find the ratio of the area of the circular target to the area of the board. Change feet to inches to compare the measurements.

$P$(toss lands on target) $= \dfrac{\text{area of target}}{\text{area of game board}}$    Area probability ratio

$$= \dfrac{\frac{\pi}{4}(4)^2}{\frac{\pi}{4}(\underline{\hspace{1cm}})^2}$$    Area $= \frac{1}{4}\pi d^2$

$$= \dfrac{16}{\underline{\hspace{1cm}}} \text{ or } \dfrac{1}{\underline{\hspace{1cm}}}$$    Simplify.

The probability that the toss lands on the target is $\frac{1}{36}$ or about _____%.

**b. During the testing phase of the game, one common complaint received is that the target is too difficult to hit. In order to make the game more enjoyable, the company wants to increase the probability of hitting the target. What diameter should they use for the circle so that the probability of a toss landing on the target is 10%? Round to the nearest hundredth of a inch.**

You need to find the diameter of a circle so that the geometric probability of a toss landing in the circle is 10% or 0.1.

$P$(toss lands on target) $= \dfrac{\text{area of target}}{\text{area of game board}}$    Area probability ratio

$$0.1 = \dfrac{\frac{\pi}{4}(d)^2}{\frac{\pi}{4}(\underline{\hspace{1cm}})^2}$$    Substitution

$$0.1 = \dfrac{d^2}{\underline{\hspace{1cm}}}$$    Simplify.

$$\underline{\hspace{1cm}}(0.1) = d^2$$    Multiply each side by 576.

$$\underline{\hspace{1cm}} = d^2$$    Simplify.

$$\underline{\hspace{1cm}} \approx d$$    Take the square root of each side.

To increase the probability of a toss landing on the target, the diameter of the target should be about 7.59 inches.

---

## Example 4 Use Angle Measures to Find Geometric Probability

**Use the spinner to find the probability of landing in each section.**

**a.** $P$(pointer landing on purple): The angle measure of the purple region is _____°.

$P$(point landing on purple) $= \dfrac{30}{360}$ or _____%.

(continued on the next page)

**Study Tip**

**Assumptions** When determining the probability that a toss will land on the target, we are assuming that the toss must land on the game board, and not on the ground surrounding the board.

**Study Tip**

**Units of Measure** Notice that the diameter of the target is given in inches, while the diameter of the game board is given in feet. When finding geometric probabilities, be sure to check that all measurements are in the same unit, in order to avoid miscalculations.

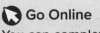

**Go Online**
You can complete an Extra Example online.

**Problem-Solving Tip**

**Use Estimation** In part b, the area of the green sector is a little more than $\frac{1}{3}$ of 33% of the spinner. Therefore, an answer of 34.7% is reasonable.

**Go Online**
to learn how to use probability tools to make fair decisions in Expand 10-2.

**Think About It!**

Jayla can wait no more than 20 minutes without risking her favorite exhibit closing. If Cody's tour should depart first, should she wait for Demarcus' tour or go on a tour with Cody. Explain your reasoning.

**Go Online**
You can complete an Extra Example online.

**b.** $P$(pointer landing on green):
The angle measure of the green region is _____°.
$P$(pointer landing on green) $= \frac{125}{360}$ or _____%.

**c.** $P$(pointer landing on neither yellow nor red):
Combine the angle measures of the red and yellow regions: $55 + 90$.
$P$(pointer landing on neither yellow nor red) $= \frac{360 - 125}{360}$ or $\frac{\quad}{360}$
or _____%

**Go Online** You can complete an Extra Example online.

## Example 5 Use Probability to Make Decisions

**DECISION MAKING Jayla is visiting the museum and wants to take a guided tour. A friend suggested that she do the tour with Demarcus, rather than Cody, because Demarcus' tour was more informative. Tours with Demarcus depart every 45 minutes, while tours with Cody depart every 30 minutes.**

**a. What is the probability that Jayla will have to wait 20 minutes or less for each tour guide? Explain your reasoning.**

The region below $y = 45$ represents the possible wait time for Demarcus' tour. The region to the left of $x = 30$ represents the possible wait time for Cody's tour. The area formed by the intersection is $45 \cdot 30$ or _____ units$^2$.

The region to the left of $x = 20$ and below $y = 20$ represents the possible waiting times of 20 minutes or less for both tour guides. The area of the square is _____ units$^2$.

The geometric probability is $\frac{400}{1350}$ or about _____%.

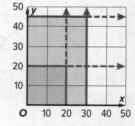

**b. What is the probability that Jayla will have to wait 20 minutes or less for one tour guide? Explain your reasoning.**

The region representing the possible wait times for Demarcus' and Cody's tour is the same as in part **a**.

The region bounded by the lines $x = 30$ and $y = 20$ represents the possibility of waiting 20 minutes or less for Cody's tour. The area of this rectangle is _____ units$^2$.

The region bounded by the lines $x = 20$ and $y = 45$ represents the possibility of waiting 20 minutes or less for Demarcus' tour. The area of this rectangle is _____ units$^2$.

Because the rectangles that describe Cody and Demarcus' tour times overlap, the waiting time of 20 minutes or less is counted twice. So, the geometric probability is $\frac{600}{1350} + \frac{900}{1350} - \frac{400}{1350} = \frac{1100}{1350}$, or about _____%.

Name _____ Period _____ Date _____

# Practice

Go Online You can complete your homework online.

### Example 1

REGULARITY **Point *M* is chosen at random on $\overline{ZP}$. Find the probability of each event.**

1. $P(M$ is on $\overline{ZQ})$

2. $P(M$ is on $\overline{QR})$

3. $P(M$ is on $\overline{RP})$

4. $P(M$ is on $\overline{QP})$

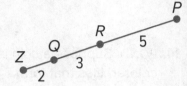

**Point *X* is chosen at random on $\overline{LP}$. Find the probability of each event.**

5. $P(X$ is on $\overline{LN})$

6. $P(X$ is on $\overline{MO})$

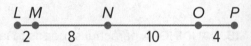

7. **FROGS** Three frogs are sitting on a 15-foot log. The first two are spaced 5 feet apart and the third frog is 10 feet away from the second one. What is the probability that when a fourth frog hops onto the log it lands between the first two?

8. **PIGS** Four pigs are lined up at the feeding trough as shown in the picture. What is the probability that when a fifth pig comes to eat it lines up between the second and third pig?

### Example 2

9. **TRAFFIC LIGHT** In a 5-minute traffic cycle, a traffic light is green for 2 minutes 27 seconds, yellow for 6 seconds, and red for 2 minutes 27 seconds. What is the probability that when you get to the light it is green?

10. **GASOLINE** Your minivan has a 24-gallon gasoline tank. What is the probability that, when the engine is turned on, the needle on the gas gauge is pointing between $\frac{1}{4}$ and $\frac{1}{2}$ full?

11. **PRECISION** A certain company plays Mozart's *Eine Kleine Nachtmusik* when its customers are on hold on the telephone. If the length of the complete recording is 2 hours long, what is the probability a customer put on hold will hear the Allegro movement, which is 6 minutes, 31 seconds long?

12. **RADIO CONTEST** A radio station is running a contest in which listeners call in when they hear a certain song. The song is 2 minutes 40 seconds long. The radio station promised to play it sometime between noon and 4 P.M. If you tune in to that radio station during that time period, what is the probability the song is playing?

Example 3

13. **DESKTOP** The diagram shows the top of a student's desk at home. A pen is dropped on the desk. What is the probability the tip of the pen lands on the book report?

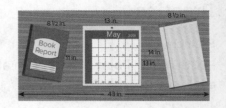

14. **POND** Suppose a coin is flipped into a reflection pond designed with colored tiles that form 3 concentric circles on the bottom. The diameter of the center circle is 4 feet and the circles are spaced 2 feet apart. What is the probability the coin lands in the center circle?

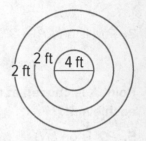

15. **LANDING** A parachutist needs to land in the center of a target on a rectangular field that is 120 yards by 30 yards. The target is a circular design with a 10-yard radius. What is the probability the parachutist lands somewhere in the target?

16. **CLOCKS** Jonas watches the second hand on an analog clock as it moves past the numbers. What is the probability that at any given time the second hand on a clock is between the 2- and the 3-hour numbers?

Example 4

**Use the spinner to find each probability. If the spinner lands on a line it is spun again.**

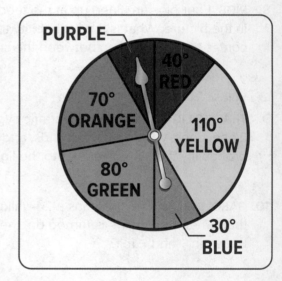

17. *P*(pointer landing on red)

18. *P*(pointer landing on blue)

19. *P*(pointer landing on green)

20 *P*(pointer landing on either green or blue)

21. *P*(pointer landing on neither red nor yellow)

**Describe an event with a 33% probability for each model.**

22.

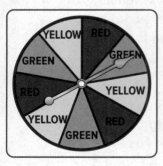

23.

24.

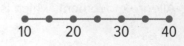

**Example 5**

**25. REASONING** Terryl is trying to plan a bus ride to the airport. He can choose between the Crimson bus, which arrives at the bus stop every 8 minutes, or the Gold bus, which arrives at the bus stop every 15 minutes. He would prefer to take the Gold bus because the bus has Wi-fi available for its passengers. In order to get to the airport by a certain time, Terryl cannot wait more than 5 minutes without risking being late for his flight.

    **a. STATE YOUR ASSUMPTION** What assumption is made in this scenario?

    **b.** To the nearest percent, what is the probability that Terryl will have to wait 5 minutes or less to see both buses?

    **c.** To the nearest percent, what is the probability that Terryl will have to wait 5 minutes or less to see only one of the buses?

**Mixed Exercises**

**Find the probability that a point chosen at random lies in the shaded region.**

**26.**

**27.**

**28.**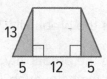

**29. DARTS** A dart is thrown at the dartboard shown. Each sector has the same central angle. The dart has equal probability of hitting any point on the dartboard. What is the probability that the dart will land in a shaded sector?

**30. MODELING** Jamie, Noa, and Pat celebrate the end of each work week by ordering spring rolls from a Chinese restaurant. The order comes with 4 spring rolls so somebody gets an extra roll. Because Jamie works full time and Noa and Pat work half time, they decide who gets the extra roll by using a spinner that has a 50% chance of coming up Jamie, and 25% chances of coming up either Noa or Pat. Design such a spinner.

**31. ELECTRON MICROSCOPES** Prima places a 7-millimeter by 10-millimeter rectangular plate into the sample chamber of an electron microscope. A black and white checkerboard pattern of 1-millimeter squares was painted over the plate to identify different treatments of the material. When she turns on the monitor, she has no idea at what point on the plate she is looking because the white and black contrast does not show up on the screen. If there are 2 more black squares than white squares, what is the probability that she is looking at a white square?

**32. OUTDOOR GAMES** Washers is a popular outdoor game where players attempt to throw washers into the 17-inch by 17-inch square pit as shown. Generally, when a player throws a washer into the square box he or she scores 1 point; however, if their washer lands within the centered circular region that has a 4-inch diameter, then they score 5 points.

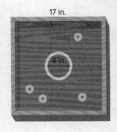

**a. STATE YOUR ASSUMPTION** What is an assumption you have to make in this scenario?

**b.** What is the probability a washer is tossed into the circular region? Round to the nearest percent. Use $\pi \approx 3.14$.

**33. ENTERTAINMENT** A rectangular dance stage is lit by two lights that light up circular regions of the stage. The circles have radii of the same length and each circle passes through the center of the other. The stage perfectly circumscribes the two circles. A spectator throws a bouquet of flowers onto the stage. Assume the bouquet has an equal chance of landing anywhere on the stage. (*Hint:* Use inscribed equilateral triangles.)

**a.** What is the probability that the flowers land on a lit part of the stage?

**b.** What is the probability that the flowers land on the part of the stage where the spotlights overlap?

**34. ARGUMENTS** Prove that the probability that a randomly chosen point in the circle will lie in the shaded region is equal to $\frac{x}{360}$.

**35. PERSEVERE** Find the probability that a point chosen at random would lie in the shaded area of the figure. Round to the nearest tenth of a percent.

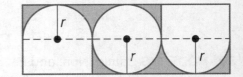

**36. ANALYZE** An isosceles triangle has a perimeter of 32 centimeters. If the lengths of the sides of the triangle are integers, what is the probability that the area of the triangle is exactly 48 square centimeters? Explain.

**37. WRITE** Can athletic events be considered random events? Explain.

**38. CREATE** Represent a probability of 20% using three different geometric figures.

**39. WRITE** Explain why the probability of a randomly chosen point falling in the shaded region of either of the squares shown is the same.

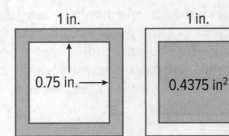

# Probability with Permutations and Combinations

**Today's Standards**
S.CP.9

MP5, MP8

## Explore Permutations and Combinations

**Online Activity** Use dynamic geometry software to complete the Explore.

> **INQUIRY** How does the order in which objects are arranged affect the sample space of events? ✕

**Today's Vocabulary**
permutation
factorial of $n$
combination

## Learn Probability with Permutations

A **permutation** is an arrangement of objects in which order is important.

| Key Concept • Factorial | |
|---|---|
| **Words** | The **factorial** of a positive integer $n$ is the product of the positive integers less than or equal to $n$, and is written as $n!$. |
| **Symbols** | $n! = n \cdot (n-1) \cdot (n-2) \cdot \ldots \cdot 2 \cdot 1$, where $0! = 1$ |

| Key Concept • Permutations | |
|---|---|
| **Words** | The number of permutations of $n$ distinct objects taken $r$ at a time is denoted by $_nP_r$ and given by $_nP_r = \dfrac{n!}{(n-r)!}$ |
| **Symbols** | The number of permutations of 6 objects taken 4 at a time is $_6P_4 = \dfrac{6!}{(6-4)!} = \dfrac{6 \cdot 5 \cdot 4 \cdot 3 \cdot 2 \cdot 1}{4 \cdot 3 \cdot 2 \cdot 1}$ or 30. |

| Key Concept • Permutations with Repetition |
|---|
| The number of distinguishable permutations of $n$ objects in which one object is repeated $r_1$ times, another is repeated $r_2$ times, and so on, is $\dfrac{n!}{r_1! \cdot r_2! \cdot \ldots \cdot r_k!}$ |

**Go Online**
to watch a video to see how to find a probability using permutations.

## ⊕ Example 1 Probability and Permutations of $n$ Objects

**PERFORMING ARTS Tyesha and Liam sign up for an open mic night with 32 available slots that are filled at random. What is the probability that Tyesha will perform first and Liam will perform second?**

**Step 1 Find the number of possible outcomes.**

The number of possible outcomes in the sample space is the number of permutations of the _____ performers' order, or 32!.

**Step 2 Find the number of favorable outcomes.**

The number of favorable outcomes is the number of permutations of the other performers' order given that Tyesha performs first and Liam performs second: $(32 - 2)!$ or _____!.

*(continued on the next page)*

💭 **Think About It!**
How would the probability differ, if at all, of Liam performing first and Tyesha performing second?

**Step 3 Calculate the probability.**

$P(\text{Tyesha 1, Liam 2}) = \dfrac{30!}{32!}$    ←number of favorable outcomes
←number of possible outcomes

$= \dfrac{\overset{1}{\cancel{30!}}}{32 \cdot 31 \cdot \underset{1}{\cancel{30!}}}$    Expand 32! and divide out common factors

$= \dfrac{1}{992}$    Simplify.

The probability that Tyesha will perform first and Liam will perform second is $\dfrac{1}{992}$, or about _____%.

## Check

**GEOMETRY** Five geometry students are asked to randomly choose a polygon and describe its properties. What is the probability that the first three students choose the hexagon, the pentagon, and the triangle, in that order? _____

🌐 **Example 2** Probability and Permutations with No Repetition

**SLIDESHOW** For a project, Rami selects 12 family photographs that will randomly play in a slideshow. The slideshow will not show repeat photos until all 12 photos have been shown. Three photos are of Rami's entire family, two photos are of his brother, three photos are just of him, and four photos are of his sister. What is the probability that the first four pictures in the slideshow will be of Rami's sister?

**Step 1 Find the number of possible outcomes.**
Because the photos will not repeat, order in this situation is important. The number of possible outcomes in the sample space is the number of permutations of _____ photos taken 4 at a time, $_{12}P_4$.

$_{12}P_4 = \dfrac{(12!)}{(12-4)!} = \dfrac{12 \cdot 11 \cdot 10 \cdot 9 \cdot \cancel{8!}}{\cancel{8!}}$ or 11,880

**Step 2 Find the number of favorable outcomes.**
The number of favorable outcomes is the number of permutations of the _____ photos of Rami's sister, or 4!.

**Step 3 Calculate the probability.**
The probability of the four photos of Rami's sister appearing as the first four in the slideshow is $\dfrac{4!}{11,880}$, or _____.

## Check

**SCHOOL** Mr. Jardin's music class consists of 15 girls and 23 boys. If Mr. Jardin pulls the names of 4 students to be team leaders, what is the probability that Jay will lead team 1, Lorenzo will lead team 2, Kellie will lead team 3, and Patricia will lead team 4?

_____

🔎 **Go Online** You can complete an Extra Example online.

# Example 3 Probability and Permutations with Repetition

**GAME SHOW** One game show has contestants reach into a bag of 13 number tiles. Tiles are randomly selected from the bag by the contestant. Each number tile can be used or discarded as the contestant sees fit, but the order of the numbers cannot change. If you select a permutation of four numbers at random, what is the probability that they will result in the correct price of $1500?

**Step 1** There is a total of 13 numbers. Of these numbers, 0 occurs 2 times, 1 occurs 3 times, 2 occurs 2 times, 5 occurs 2 times, and 7 occurs 2 times. So, the number of distinguishable permutations of these letters is

$$\frac{13!}{3! \cdot 2! \cdot 2! \cdot 2! \cdot 2!} = \frac{6,227,020,800}{\rule{1.5cm}{0.4pt}}, \text{ or } \underline{\hspace{2cm}}$$   Use a calculator.

**Step 2** There is only 1 favorable arrangement, the actual price, _____.

**Step 3** The probability that a permutation of these numbers selected at random results in the correct price of 1500, is _____.

## Check

**GAMES** The physics team is holding a game night fundraiser. To win a grand prize in a particular game, you must spin the spinner four times and land on blue, red, green, and yellow, in that order. What is the probability that you will spin the winning sequence? _____

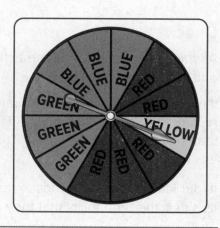

## Learn Probability Using Combinations

A **combination** is a selection of objects in which order is *not* important.

A combination of *n* objects taken *r* at a time, or $_nC_r$, is calculated by dividing the number of permutations $_nP_r$ by the number of arrangements containing the same elements, *r*!.

| Key Concept • Combinations | |
|---|---|
| **Symbols** | The number of combinations of *n* distinct objects taken *r* at a time is denoted by $_nC_r$ and is given by $_nC_r = \frac{n!}{(n-r)!r!}$. |
| **Example** | The number of combinations of 7 objects taken 3 at a time is $_7C_3 = \frac{7!}{(7-3)!3!} = \frac{7!}{4!3!} = \frac{7 \cdot 6 \cdot 5 \cdot 4!}{4! \cdot 6}$ or 35. |

**Go Online** You can complete an Extra Example online.

**Study Tip**

**RANDOMNESS** When outcomes are decided at random, they are equally likely to occur and their probabilities can be calculated using permutations and combinations.

**Think About It!**

What is the probability that the numbers selected will not result in the correct price of $1500? Justify your reasoning.

**Study Tip**

**Permutations and Combinations**
Use permutations when the order of an arrangement is important, and use combinations when the order is not important.

**Go Online**

Go online to watch a video to see how to find a probability using combinations.

## Example 4 Probability and Combinations

**GEOMETRY** **If three points are randomly chosen from those named on pentagon *ACEGJ*, what is the probability that they all lie on the same line segment?**

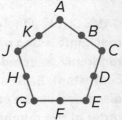

**Step 1 Find the number of possible outcomes.**

Since the order in which the points are chosen does not matter, the number of possible outcomes in the sample space is the number of combinations of _____ points taken _____ at a time, $_{10}C_3$.

$$_{10}C_3 = \frac{10!}{(10-3)!3!} = \frac{10 \cdot 9 \cdot 8 \cdot 7!}{7! \cdot 3!} \text{ or } \underline{\quad}$$

**Step 2 Find the number of favorable outcomes.**

There are _____ favorable outcomes – the points could lie on $\overline{AC}$, $\overline{CE}$, $\overline{EG}$, $\overline{GJ}$, or $\overline{JA}$.

**Step 3 Calculate the probability.**

The probability of three randomly chosen points lying in on the same segment is $\frac{5}{120}$, or _____.

## Check

A *lattice* is a point at the intersection of two or more grid lines in a coordinate plane.

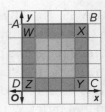

If two lattice points are chosen randomly in rectangle *ABCD*, including its sides, the probability that they are in rectangle *WXYZ*, including its sides, is _____.

If four lattice points are chosen randomly in rectangle *ABCD*, including its sides, the probability that they are *W*, *X*, *Y*, and *Z* is

_____.

## Pause and Reflect

Did you struggle with anything in this lesson? If so, how did you deal with it?

**Go Online** You can complete an Extra Example online.

---

### 🗨 Talk About It!

If you pull letter tiles from a bag containing all 26 letters, and you want to pull 3 vowels and 6 consonants, should you calculate the probability using a *permutation* or a *combination*? Explain.

Name _____ Period _____ Date _____

# Practice

**Go Online** You can complete your homework online.

## Example 1

1. **CHEER** The cheer squad is made up of 12 girls. A captain and a co-captain are selected at random. What is the probability that Chantel and Clover are chosen as leaders?

2. **BOOKS** You have a textbook for each of the following subjects: Spanish, English, Chemistry, Geometry, History, and Psychology. If you choose 4 of these at random to arrange on a shelf, what is the probability that the Geometry textbook will be first from the left and the Chemistry textbook will be second from the left?

3. **RAFFLE** Alfonso and Colton each bought one raffle ticket at the state fair. If 50 tickets were randomly sold, what is the probability that Alfonso got ticket 14 and Colton got ticket 23?

4. **SEATING** Nia and Ciro are going to a concert with their high school's key club. If they choose a seat on the row below at random, what is the probability that Ciro will be in seat C11 and Nia will be in C12?

## Examples 2 and 3

5. **PHONE NUMBERS** What is the probability that a 7-digit telephone number generated using the digits 2, 3, 2, 5, 2, 7, and 3 is the number 222-3357?

6. **DINING OUT** A group of 4 girls and 4 boys is randomly seated at a round table. What is the probability that the arrangement is boy-girl-boy-girl?

7. **IDENTIFICATION** A store randomly assigns their employees work identification numbers to track productivity. Each number consists of 5 digits ranging from 1-9. If the digits cannot repeat, find the probability that a randomly generated number is 25938.

8. **GROUPS** Two people are chosen randomly from a group of ten. What is the probability that Jimmy was selected first and George second?

9. **MODELING** The table shows the finalists for class president. The order in which they will give their speeches will be chosen randomly.

   a. What is the probability that Denny, Kelli, and Chaminade are the first 3 speakers, in any order?

   b. What is the probability that Denny is first, Kelli is second, and Chaminade is third?

| Class President Finalists |
| --- |
| Alan Shepherd |
| Chaminade Hudson |
| Denny Murano |
| Kelli Baker |
| Tanika Johnson |
| Jerome Murdock |
| Marlene Lindeman |

Example 4

10. **TROPHIES** Taryn has 15 soccer trophies but she only has room to display 9 of them. If she chooses them at random, what is the probability that each of the trophies from the school invitational from the 1st through 9th grades will be chosen?

11. **ICE CREAM** Kali has a choice of 20 flavors for her triple scoop cone. If she chooses the flavors at random, what is the probability that the 3 flavors she chooses will be vanilla, chocolate, and strawberry?

12. **PETS** Iza has a dog walking business serving 9 dogs. If she chooses 4 of the dogs at random to take an extra trip to the dog park, what is the probability that Fifi, Gordy, Spike and Fluffy are chosen?

13. **CRITIQUE** A restaurant critic has 10 new restaurants to try. If he tries half of them this week, what is the probability that he will choose The Fish Shack, Carly's Place, Chez Henri, Casa de Jorge, and Grillarious?

14. **CHARITY** Emily is giving away part of her international doll collection to charity. She has 20 dolls, each from a different country. If she selects 10 of them at random, what is the probability she chooses the ones from Ecuador, Paraguay, Chile, France, Spain, Sweden, Switzerland, Germany, Greece, and Italy?

15. **ROLLER COASTER** An amusement park has 12 roller coasters. Four are on the west side of the park, 4 are on the east side, and 4 are centrally located. The park's manager randomly chooses 4 roller coasters per month to analyze the estimated wait time for the ride. What is the probability that all 4 roller coasters on the west side are chosen in March?

## Mixed Exercises

16. **CLUBS** The Service Club is choosing members at random to attend one of four conferences in Los Angeles, Atlanta, Chicago, and New York. There are 20 members in the club. What is the probability that Jacy, Sherry, Miguel, and Jerome are chosen for these trips?

17. **FORMAL DINING** You are handed 5 pieces of silverware for the formal setting shown. If you guess their placement at random, what is the probability that the knife and spoon are placed correctly?

18. **CARDS** What is the probability in a line of these 5 cards that the ace would be first from the left and the king would be second from the left?

**19. GEOMETRY** Points *A*, *B*, *C*, *D*, and *E* are coplanar, but no 3 are collinear.

   **a.** What is the total number of lines that can be determined by these points?

   **b.** What is the probability that $\overleftrightarrow{AB}$ would be chosen at random from all of the possible lines formed?

**20. LETTERS** Jaclyn bought some decorative letters for a scrapbook project. If she selected a permutation of the letters shown, what is the probability that they would form the word "photography"?

**21. ODD JOBS** Matthew put fliers advertising his lawn service on the doors of 20 families' houses in his neighborhood. If 6 families called him, what is the probability that they were the Thompsons, the Rodriguezes, the Jacksons, the Williamses, the Kryceks, and the Carpenters?

**22. GAME SHOW** The people on the list at the right will be considered to participate in a game show. What is the probability that Wyatt, Gabe, and Isaac will be chosen as the first three contestants?

| DAY 1 STANDINGS |
| --- |
| MCAFEE, DAVID |
| FORD, GABE |
| STANDISH, TRISTAN |
| NOCHOLS, WYATT |
| PURCELL, JACK |
| ANDERSON, BILL |
| WRIGHT, ISAAC |
| FILBERT, MITCH |

**23. REASONING** As a sales strategy, the owner of a hair salon advertises that on the first day of each month, the first 6 customers will receive one of the coupons shown at the right for a discount off their total bill. Each coupon is given at random to a different customer and may be used only once.

   **a.** What is the probability that the first customer on May 1 gets the 10% discount and the second customer gets the 25% discount? Explain using favorable and possible outcomes.

   **b.** How many different groups of two coupons can the first two customers on August 1 receive regardless of order?

| | |
| --- | --- |
| 5% OFF | 10% OFF |
| 15% OFF | 20% OFF |
| 25% OFF | 50% OFF |

**24. USE TOOLS** As part of a school beautification project, 12 alumni each donated a tree to be planted on the school grounds. The types of trees are shown in the table. There will be a sign next to each tree with the donor's name.

   **a.** If the trees are planted in a row at random, what is the probability that they will be in alphabetical order by donor name? Explain.

   **b.** If 4 trees are randomly selected and planted near the school entrance, what is the probability that they will all be dogwood trees? Explain.

| Donated Trees | |
| --- | --- |
| Type | Number of Trees |
| Cherry | 5 |
| Dogwood | 4 |
| Crabapple | 2 |
| Redbud | 1 |

**25. RANDOM NUMBERS** A random number generator is a computer program that produces random numbers. What is the probability that it will produce a number less than 1000 for a 5-digit number? (Hint: 00125 = 125)

**26. UNITED NATIONS** The UN Security Council has 5 permanent members and 10 non-permanent members. Italy is one of 192 UN member states and is not a permanent member of the Security Council. What is the probability that a representative from Italy is on the Security Council?

**27. USE TOOLS** The mall has a merry-go-round with 12 horses on the outside ring. If 12 people randomly choose those horses, what is the probability they are seated in alphabetical order?

**28. PARKING STICKERS** Parking stickers contain randomly generated numbers with 5-digits ranging from 1 to 9. No digits are repeated. What is the probability that a randomly generated number is 54321?

**29. ARGUMENTS** Prove that $_nC_{n-r} = {}_nC_r$.

**30. TRAFFIC CAMERAS** A camera positioned above a traffic light photographs cars that fail to stop at a red light. In one unclear photograph, an officer could see that the first letter of the license plate was a Q, the second letter was an M or an N and the third letter was a B, P, or D. The first number was a 0, but the last two numbers were illegible.

    **a. STATE YOUR ASSUMPTION** What assumption are you making with the possible choices of letters and numbers?

    **b.** How many possible license plates fit this description?

**31. PERSEVERE** Fifteen boys and fifteen girls entered a drawing for four free movie tickets. What is the probability that all four tickets were won by girls?

**32. ANALYZE** Is the following statement sometimes, always, or never true? Explain.
$$_nP_r = {}_nC_r$$

**33. WRITE** Compare and contrast permutations and combinations.

**34. CREATE** Describe a situation in which the probability is given by $\dfrac{1}{_7C_3}$.

**35. PERSEVERE** A student claimed that permutations and combinations were related by $r! \cdot {}_nC_r = {}_nP_r$. Use algebra to show that this is true. Then explain why $_nC_r$ and $_nP_r$ differ by the factor $r!$

**36. FIND THE ERROR** Charlie claims that the number of ways $n$ objects can be arranged if order matters is equal to the number of permutations of $n$ objects taken $n - 1$ at a time. Do you agree with Charlie? Justify your answer.

# Probability and the Multiplication Rule

## Explore Independent and Dependent Events

🔾 **Online Activity** Use the video to complete the Explore.

> ⊛ **INQUIRY** How can one event affect the
> probability of a second event?   ×

## Learn Independent Events

A **compound event** or *composite event* consists of two or more
simple events. **Independent events** are two or more events in
which the outcome of one event does not affect the outcome of
the other events.

Suppose a coin is tossed and the spinner shown
is spun. The sample space for this experiment is
{(H, 1), (H, 2), (H, 3), (T, 1), (T, 2), (T, 3)}.

Using the sample space, the probability of the
coin landing on heads and the spinner on 2 is $P(\text{H and 2}) = \frac{1}{6}$.

Notice that the same probability can be found by multiplying the
probabilities of each simple event.

$$P(H) = \frac{1}{2} \qquad P(2) = \frac{1}{3} \qquad P(\text{H and 2}) = \frac{1}{2} \cdot \frac{1}{3} \text{ or } \frac{1}{6}$$

This example illustrates the first of two Multiplication Rules
for Probability.

| Key Concept • Probability of Two Independent Events | |
|---|---|
| **Words** | The probability that two independent events both occur is the product of the probabilities of each individual event. |
| **Symbols** | If two events $A$ and $B$ are independent, the $P(A \text{ and } B) = P(A) \cdot P(B)$. |

This rule can be extended to any number of events.

Consider choosing objects from a group of objects. If you
replace the object each time, choosing additional objects are
independent events.

**Today's Standards**
S.CP.2; S.CP.8
MP1, MP4

**Today's Vocabulary**
compound event
independent events
dependent events

**Study Tip**

**and** The word *and*
often illustrates
compound events.
For example, if you
roll a die, finding the
probability of getting
an odd number *and*
getting a number
greater than 5
indicates that the
probabilities of the
individual events
should be multiplied.

**Go Online**
An alternate method is available for this example.

## Example 1 Probability of Independent Events

**GAMING** **Ana is a member of a gaming Web site that randomly pairs users together to solve puzzles. Of the 50 other players currently online, Ana is friends with 10 of them. Suppose Ana is paired with a player for a game. Not liking the outcome, she disconnects and is paired with another player.**

a. **What is the probability that neither player that Ana is paired with is a friend of hers?**

These events are independent since the set of possible matches is reset to 50 once Ana disconnects. Let *F* represent a player who is Ana's friend and *NF* represent a player who is not Ana's friend.

Complete the equation to determine the probability of independent events.

User 1   User 2

$$P(NF \text{ and } NF) = P(\underline{\quad}) \cdot P(\underline{\quad})$$

$$= \frac{40}{50} \cdot \underline{\quad} \qquad P(NF) = \frac{40}{50}$$

$$= \frac{}{2500} \text{ or } \frac{16}{} \qquad \text{Simplify.}$$

So, the probability that neither of the two players is Ana's friend is $\frac{16}{25}$ or ____%.

b. **What assumption do you have to make in order to solve this problem?**

We assume that the same 50 players remain in the set for both selections. If the number of available players _____, or the number of available players who are friends with Ana _____, the probability will change.

**Alternate Method**
You can also use an area model to calculate the probability that neither player is a friend of Ana's.

The probability that a player is not a friend of Ana's is ____ or ____.

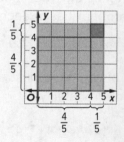

The blue region represents the probability of two sequential players not being friends with Ana. The area of the blue region is ____ of the entire shaded region.

The area of the orange region represents the probability of two sequential players being friends with Ana. The area of the orange region is ____ of the entire shaded region.

## Check

**WEATHER** Paola's weather app tells her that there is a 20% chance of rain on Tuesday and a 50% chance of rain on Wednesday. What is the probability that it will rain on both Tuesday and Wednesday? ____ or ____

**Go Online** You can complete an Extra Example online.

# Learn Dependent Events

**Dependent events** are two or more events in which the outcome of one event affects the outcome of the other events.

Suppose a marble is chosen from the bag and placed on the table. Then, another marble is chosen from the bag. The sample space for this experiment is {(R, B), (R, R), (B, R), (B, B)}.

When choosing the first marble, there are 7 possible outcomes. The probability of choosing a red marble on the first draw is $\frac{5}{7}$. Since that marble is not returned to the bag, there are only 6 possible outcomes for the second drawing. If a red marble has already been chosen, the probability of choosing a red marble on the second draw is $\frac{4}{6}$.

The second of the Multiplication Rules of Probability addresses the probability of two dependent events.

| Key Concept • Probability of Two Dependent Events | |
|---|---|
| Words | The probability that two dependent events both occur is the product of the probability that the first event occurs and the probability that the second event occurs *after* the first event has already occurred. |
| Symbols | If two events $A$ and $B$ are dependent, then $P(A \text{ and } B) = P(A) \cdot P(B|A)$. |

This rule can be extended to any number of events.

## Example 2 Independent and Dependent Events

**Determine whether the events are *independent* or *dependent*. Explain your reasoning.**

**a. One spinner is spun twice.**
The outcome of the first spin in no way changes the probability of the outcome of the second spin. Therefore, these two events are

_____.

**b. In a raffle, one ticket is drawn for the first place prize, and then another ticket is drawn for the second place prize.**
After the first place prize ticket is drawn, the ticket is removed and cannot be chosen again. This affects the probability of the second place prize winning ticket, because the sample space is reduced by one ticket. Therefore, these two events are _____.

**c. A random number generator generates two numbers.**
The number for the first generation has no bearing on the number for the second generation. Therefore, these two events are

_____.

**Go Online** You can complete an Extra Example online.

**Study Tip**
**Conditional Notation**
The notation $P(B|A)$ is read the probability that event $B$ occurs given that event $A$ has already occurred. The "|" symbol should not be interpreted as a division symbol.

**Go Online**
to watch a video to see how to use the Multiplication Rule to find the probability of two independent events.

**Think About It!**
Write an example of a series of three dependent events.

## Check

Determine whether the events are *independent* or *dependent*.

| | Independent | Dependent |
|---|---|---|
| Of the $100 that Rei has to spend, she wants to spend $59 on a blouse and $44 on some jeans. | | |
| Rei asks each of three store associates which handbag they prefer. | | |
| Rei purchases a handbag and a belt. | | |

## 🌐 Example 3 Probability zof Dependent Events

**FOOD The pizza that José and Tessa are eating is half cheese, half mushroom. Tessa spins the pizza around and randomly selects a slice of mushroom pizza. If José spins the pizza and selects a slice after that, what is the probability that both he and Tessa select a slice of mushroom pizza?**

These events are dependent because Tessa does not replace the slice she selected. Let $M$ represent a slice of mushroom pizza and $C$ represent a slice of cheese pizza.

$P(M \text{ and } M) = P(M) \cdot P(M|C)$    Probability of dependent events

$= \dfrac{5}{} \cdot \dfrac{}{9} \text{ or } \dfrac{2}{9}$    After the first slice of mushroom pizza is selected, 9 total pieces remain, and 4 of those slices have mushrooms.

So, the probability that both friends randomly select slices with mushrooms is _____ or about _____%.

## Check

**SCHOOL On a math test, 5 out of 20 students got all the questions correct. If three students are chosen at random without replacement, what is the probability that all three got all the questions correct on the test?** _____

## Pause and Reflect

Did you struggle with anything in this lesson? If so, how did you deal with it?

Record your observations here

🌐 **Go Online** You can complete an Extra Example online.

**Module 12** · Probability

**Go Online**
You can go online to check your answer with a diagram.

**Go Online** You can go online to see a common error to avoid.

Name _____ Period _____ Date _____

# Practice

**⊙ Go Online** You can complete your homework online.

**Example 1**

1. **SOCKS** George has two pairs of red socks and two pairs of white socks in a drawer. He has a drawer with 2 red T-shirts and 1 white T-shirt. If he chooses a pair of socks from the sock drawer and a T-shirt from the T-shirt drawer, what is the probability that he gets a pair of red socks and a white T-shirt?

2. **COINS** Phyllis drops a penny in a pond, and then she drops a nickel in the pond. What is the probability that both coins land with tails showing?

3. **DICE** A die is rolled and a penny is dropped. Find the probability of rolling a two and showing a tail.

4. **MARBLES** A bag contains 3 red marbles, 2 green marbles, and 4 blue marbles. A marble is drawn randomly from the bag and replaced before a second marble is chosen. Find the probability that both marbles are blue.

5. **WEATHER** The forecast predicts a 40% chance of rain on Tuesday and a 60% chance on Wednesday. If these probabilities are independent, what is the chance that it will rain on both days?

**Example 2**

**Determine whether the events are *independent* or *dependent*. Explain your reasoning.**

6. In a game, you roll an even number on a die and then spin a spinner numbered 1 through 5 and get an odd number.

7. An ace is drawn, without replacement, from a deck of 52 cards. Then, a second ace is drawn.

8. In a bag of 3 green and 4 blue marbles, a blue marble is drawn and not replaced. Then, a second blue marble is drawn.

9. You roll two dice and roll a 5 on each.

**Example 3**

10. **DRAWING** Mr. Hanes places the names of four of his students, Joe, Juanita, Hayden, and Bonita, on slips of paper. From these, he intends to randomly select two students to represent his class at the robotics convention. He draws the name of the first student, sets it aside, then draws the name of the second student. What is the probability he draws Juanita, then Joe?

11. **CARDS** A card is drawn from a standard deck of playing cards and is not replaced. Then a second card is drawn. Find the probability the first card is a jack of spades and the second card is black.

**Lesson 12-5 • Probability and the Multiplication Rule** **739**

**12. CAMPAIGNS** The table shows the number of each color of Student Council campaign buttons Clemente has to give away. If given away at random, what is the probability that the first and second buttons given away are both red?

| Button Color | Amount |
|---|---|
| blue | 20 |
| white | 15 |
| red | 25 |
| black | 10 |

**Mixed Exercises**

**13. RAFFLES** Sara and her daughter Alexis are attending a fundraising concert. In addition to the concert, there are also raffle tickets being sold for a drawing at the end of the concert. There are two prizes: a two-night trip to the Smoky Mountains, and a grand prize one-week trip to the Bahamas. Sara and Alexis each buy one ticket, of the total 60 tickets that were sold. If a ticket is not replaced once it is drawn, then what is the probability that both Sara and Alexis win?

**14. SPORTS** The World Series is a "best of seven" game championship in which up to seven games are played, and one team has to win four games to win the Series. In the 2007 World Series, the Boston Red Sox swept the Colorado Rockies, meaning that they won the series by winning the first four games. If you assume that each team has an equal chance of winning each game, what is the probability of a World Series sweep?

**15. BUSINESS** At Corrugated Packaging, Inc., a team of six employees is in charge of marketing and selling the company's products; three are women and three are men. The president of the company decides to send four team members to a national conference. He wants to make sure he chooses names fairly, so he decides to put the names of his six sales employees in a hat and draw four names to see who will go to the conference. What is the probability that the team will consist of three women and one man?

**16. REASONING** Bruce has just purchased 100 shares of stock in a company that is seeking FDA approval for a treatment that relieves the common cold. According to pharmaceutical experts, there is a 50% chance that FDA approval will be granted, and according to market insiders there is an 85% chance the value of the stock will double if FDA approval is granted.

   **a.** What is the probability that the FDA will approve the treatment and the value of Bruce's investment will double?

   **b.** What is the probability that the FDA will approve the treatment and the value of your stock will not double?

   **c.** Are these events dependent or independent? Explain.

**17. MAGIC** Iris performs a magic trick in which she holds a standard deck of cards and has each of three people randomly choose a card from the deck. Each person keeps his or her card as the next person draws. What is the probability that all three people will draw a heart?

**18. SCHEDULES** The probability that a student takes geometry and French at Satomi's school is 0.064. The probability that a student takes French is 0.45. What is the probability that a student takes geometry if the student takes French?

**19. EXTRA-CURRICULARS** At Bell High School, 43% of the students are in an after-school club and 28% play sports. What is the probability that a student is in an after-school club if he or she also plays a sport?

**Two boxes each contain computer chips, some of which are defective, as shown in the table. Use this information for Exercises 20 and 21.**

**20. REGULARITY** Derek randomly selects one chip from Box B, puts it in his pocket, and then randomly selects another chip from Box B. What is the probability that both chips are defective? Explain.

|  | Number of Chips | Defective |
|---|---|---|
| Box A | 100 | 4% |
| Box B | 150 | 2% |

**21. STRUCTURE** Sunita randomly selects one chip from Box A, and then she randomly selects another chip from Box A. The probability that both chips are defective is $\frac{1}{625}$. Did Sunita replace the first chip before selecting the second one? Explain.

**22. USE TOOLS** Each square of the area model represents 1 square on a wall at a carnival game. For each time the game is played, a random square is to be chosen. A square is to be chosen for a minor prize and then another square chosen for a major prize. What is the probability that shaded squares are chosen for both prizes? Explain.

**23. TILES** Kirsten and José are playing a game. Kirsten places tiles numbered 1 to 50 in a bag. José selects a random tile. If he selects a prime number or a number greater than 40, then he wins the game. What is the probability that José will win on his first turn?

**24. TRAVEL** A survey was conducted to determine the way in which families travel to their vacation destinations. The results indicated that $P(D) = 0.6$, $P(D \cap F) = 0.2$, and the probability that a family did not vacation is 0.1.

Fly F    Drive D

**a.** What is the probability that a family reached their vacation destination by flying?

**b.** What is the probability that a family that drives will also fly?

**25. ECONOMIC IMPACT** You are trying to decide whether you should expand a business. If you do not expand and the economy remains good, you expect $2 million in revenue. If the economy is bad, you expect $0.5 million. The cost to expand is $1 million, but the expected revenue after the expansion is $4 million in a good economy and $1 million in a bad economy. You assume that the chances of a good and a bad company are 30% and 70%, respectively. Use a probability tree to explain what you should do.

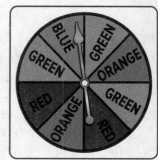

26. **REASONING** If Fred spins the spinner twice, determine the probability that he lands on sections labeled "orange" and "green." Show that $P(\text{orange}) \cdot P(\text{green}|\text{orange}) = P(\text{green}) \cdot P(\text{orange}|\text{green})$ in this situation and explain why this is true in terms of the model. Justify your answer.

27. **TRICKS** Sam is doing a trick with a standard deck of 52 playing cards where she begins each trick with a fresh deck of cards. Her friend Tracy randomly selects a card, looks at it, and puts it back in the deck. Then Sam randomly selects the same card. Are the events independent? What is the probability that they both pick the queen of spades? Explain.

28. **ARGUMENTS** Use the formula for the probability of two dependent events $P(A \text{ and } B)$ to derive the conditional probability formula for $P(B|A)$.

29. **MODELING** A double fault in tennis is when the serving player fails to land their serve "in" without stepping on or over the service line in two chances. Kelly's first serve percentage is 40%, while her second serve percentage is 70%.

   a. Draw a probability tree that shows each outcome.

   b. What is the probability that Kelly will double fault?

30. **ANALYZE** There are $n$ different objects in a bag. The probability of drawing object $A$ and then object $B$ without replacement is about 2.4%. What is the value of $n$? Explain.

31. **WRITE** A medical journal reports the chance that a person smokes given that his or her parent smokes. Explain how you could determine the likelihood that a person's smoking and their parent's smoking are independent events.

32. **CREATE** Describe a pair of independent events and a pair of dependent events. Explain your reasoning.

33. **PERSEVERE** If $P(A|B)$ is the same as $P(A)$, and $P(B|A)$ is the same as $P(B)$, what can be said about the relationship between events $A$ and $B$?

34. **FIND THE ERROR** George and Aliyah are determining the probability of randomly choosing a blue or red marble from a bag of 8 blue marbles, 6 red marbles, 8 yellow marbles, and 4 white marbles. Is either of them correct? Explain.

| George | Aliyah |
|---|---|
| $P(\text{blue or red}) = P(\text{blue}) \cdot P(\text{red})$ | $P(\text{blue or red}) = P(\text{blue}) + P(\text{red})$ |
| $= \dfrac{8}{26} \cdot \dfrac{6}{26}$ | $= \dfrac{8}{26} + \dfrac{6}{26}$ |
| $= \dfrac{48}{676}$ | $= \dfrac{14}{26}$ |
| about 7% | about 54% |

# Probability and the Addition Rule

## Explore Mutually Exclusive Events

**Online Activity** Use the guiding exercises to complete the Explore.

**⊙ INQUIRY** How can you find the probability of two or more events that may not share common outcomes?

**Today's Standards**
S.CP.7
MP3, MP4

**Today's Vocabulary**
mutually exclusive

## Learn Probability of Mutually Exclusive Events

To find the probability that one event occurs or another event occurs, you must know how the two events are related. If the two events cannot happen at the same time, they are said to be **mutually exclusive**. That is, the two events have no outcomes in common.

One way of finding the probability of two mutually exclusive events occurring is to examine their sample space.

When a die is rolled, what is the probability of getting a 3 or a 4? From the Venn diagram, you can see that there are two outcomes that satisfy this condition, 3 and 4. So, $P(3 \text{ or } 4) = \frac{2}{6}$ or $\frac{1}{3}$.

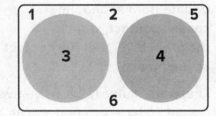

Notice that this same probability can be found by adding the probabilities of each simple event.

$$P(3) = \frac{1}{6} \qquad P(4) = \frac{1}{6} \qquad P(3 \text{ or } 4) = \frac{1}{6} + \frac{1}{6} = \frac{2}{6} \text{ or } \frac{1}{3}$$

This example illustrates the first of two Addition Rules for Probability.

| Key Concepts • Probability of Mutually Exclusive Events | |
|---|---|
| **Words** | If two events $A$ and $B$ are mutually exclusive, then the probability that $A$ or $B$ occurs is the sum of the probabilities of each individual event. |
| **Symbols** | If two events $A$ and $B$ are mutually exclusive, then $P(A \text{ or } B) = P(A) + P(B)$. |

This rule can be extended to any number of events.

## Example 1 Identify Mutually Exclusive Events

**A card is drawn from a standard deck of 52 cards. Determine whether the events are *mutually exclusive* or not *mutually exclusive*. Explain your reasoning.**

*(continued on the next page)*

**Study Tip**

**or** The word *or* is a key word indicating that at least one of the events occurs. $P(A \text{ or } B)$ is read as *the probability that A occurs or that B occurs.*

**a. drawing a 3 or a 2**

There are no common outcomes – a card cannot be both a 2 and a 3. These events _____ mutually exclusive.

**b. drawing a 7 or a red card**

The 7 of diamonds is an outcome that both events have in common. These events _____ mutually exclusive.

**c. drawing a queen or a spade**

Since the queen of spades represents both events, they _____ mutually exclusive.

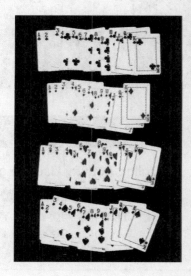

### 🌐 Example 2 Probability of Mutually Exclusive Events

**SOCIAL MEDIA Daniel organizes all of his social media contacts into three groups. If the program sends Daniel an update from a randomly chosen contact, what is the probability that the contact is either a close friend or acquaintance?**

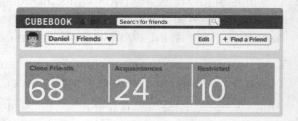

These are mutually exclusive events, since the contacts selected cannot be a close friend and an acquaintance.

Let event $F$ represent selecting a close friend. Let event A represent selecting an acquaintance. There are a total of $10 + 68 + 24$ or 102 contacts.

Because the events are mutually exclusive, you know that

$P(F \text{ or } A) = \underline{\quad} + \underline{\quad}$.

$$P(F \text{ or } A) = P(F) + P(A) \qquad \text{Probability of mutually exclusive events}$$

$$= \frac{68}{102} + \frac{24}{102} \qquad P(F) = \frac{68}{102} \text{ and } P(A) = \frac{24}{102}$$

$$= \frac{92}{102} \text{ or } \frac{46}{51} \qquad \text{Add.}$$

So the probability that the update is from a close friend or acquaintance is $\frac{46}{51}$ or about _____%.

🍩 **Think About It!**

What assumption did you make in order to solve this problem?

🧭 **Go Online** You can complete an Extra Example online.

# Check

**BIODIVERSITY** The International Union for Conservation of Nature collects and analyzes biological data. Of the more than 79,800 species on their Red List of Threatened Species, seabirds are of particular interest because they are indicators of broader marine health issues. The circle graph shows the

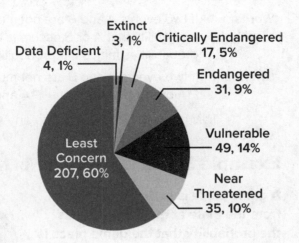

Extinct
3, 1%

Data Deficient
4, 1%

Critically Endangered
17, 5%

Endangered
31, 9%

Least Concern
207, 60%

Vulnerable
49, 14%

Near Threatened
35, 10%

proportion of seabird species in each Red List category. What is the probability that a randomly selected species of seabird is on the critically endangered list or the endangered list?

---

# Learn Probability of Events that are Not Mutually Exclusive

If two events can happen at the same time, they are not mutually exclusive.

When a die is rolled, what is the probability of getting a number greater than 2 or an even number? From the Venn diagram, you can see that there are 5 numbers on a die that are either greater than 2 or are an even number: 2, 3, 4, 5, and 6. So, $P$(greater than 2 or even) $= \frac{5}{6}$.

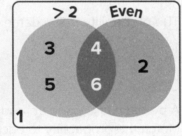

> 2    Even

3    4    2

5    6

1

Because it is possible to roll a number that is greater than 2 and an even number, these events are not mutually exclusive. Consider the probability of each individual event.

$P$(greater than 2) $= \frac{4}{6}$        $P$(even) $= \frac{3}{6}$

Adding these probabilities results in a number greater than 1 because two of the outcomes, 4 and 6, are in the intersection of the sample spaces—they are both greater than 2 and even. To account for the intersection, subtract the probability of the common outcomes.

$P$(greater than 2 or even) $= P$(greater than 2) $+ P$(even) $- P$(greater than 2 and even) $\frac{4}{6} + \frac{3}{6} - \frac{2}{6}$ or $\frac{5}{6}$

This leads to the second of the Addition Rules for Probability.

*(continued on the next page)*

*(continued on the next page)*

📡 **Go Online** You can complete an Extra Example online.

| Key Concepts • Probability of Events That Are Not Mutually Exclusive | |
| --- | --- |
| **Words** | If two events $A$ and $B$ are not mutually exclusive, then the probability that $A$ or $B$ occurs is the sum of their individual probabilities minus the probability that both $A$ and $B$ occur. |
| **Symbols** | If two events $A$ and $B$ are not mutually exclusive, then $P(A \text{ or } B) = P(A) + P(B) - P(A \text{ and B})$. |

This rule can be extended to any number of events.

## Example 3 Events That Are Not Mutually Exclusive

**A game piece is selected at random from the plate at the right. What is the probability that the game piece is round and orange?**

Since some of the game pieces are both round and orange, these events are not mutually exclusive. Use the rule for two events that are not mutually exclusive. The total number of game pieces from which to choose is 10.

$P(\text{round and orange}) = P(\text{round}) + P(\text{orange}) - P(\text{round and orange})$

$$= \frac{5}{} + \frac{}{10} - \frac{2}{10}$$

$$= \frac{}{10}$$

The probability that a game piece will be round or orange is $\frac{7}{10}$ or \_\_\_\_\_%.

## Check

A polygon is chosen at random. Match the probability to each set of events.

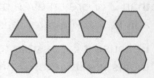

| | | |
| --- | --- | --- |
| choosing a figure that has more than 4 lines of symmetry or more than 7 sides | ⟷ | \_\_\_\_\_ |
| choosing a figure that has more than 15 diagonals or a total interior angle measure greater than 900° | ⟷ | \_\_\_\_\_ |
| choosing a figure that has more than 2 pairs of parallel sides or at least 1 diagonal | ⟷ | \_\_\_\_\_ |

| | |
| --- | --- |
| 37.5% | 25% |
| 62.5% | 87.5% |
| 50% | 75% |

**Go Online** You can complete an Extra Example online.

Name _____ Period _____ Date _____

# Practice

🔊 **Go Online** You can complete your homework online.

### Example 1

**Determine whether the events are *mutually exclusive* or *not mutually exclusive*. Explain your reasoning.**

1. **SOUVENIRS** At the ballpark souvenir shop, there are 15 posters of the first baseman, 20 of the pitcher, 14 of the center fielder, and 12 of the shortstop. A fan chooses a poster at random and gets a poster of the center fielder or the shortstop.

2. **VENDORS** A street vendor is selling T-shirts outside of a concert arena. The colors and sizes of the T-shirts he has are shown in the table. The vendor randomly selects one of his shirts to put on display, and the T-shirt he selects is blue or large.

| | Red | Blue | White |
|---|---|---|---|
| Small | 1 | 2 | 2 |
| Medium | 3 | 2 | 4 |
| Large | 4 | 5 | 6 |
| Extra Large | 7 | 6 | 3 |

3. **ANIMAL SHELTER** An animal shelter has 15 cats, 25 dogs, 9 rabbits, and 3 horses. An animal is randomly selected from the shelter to greet guests at an open house and it is a dog or a cat.

4. **GAME** A die is rolled while a game is being played. The result of the next roll is a 6 or an even number.

### Examples 2 and 3

**Determine whether the events are *mutually exclusive*. Then find the probability. Round to the nearest hundredth.**

5. **AWARDS** The student of the month gets to choose his or her award from 9 gift certificates to area restaurants, 8 T-shirts, 6 water bottles, or 5 gift cards to the mall. What is the probability that the student of the month chooses a T-shirt or a water bottle?

6. **STUDENT COUNCIL** What is the probability that a person on a student council committee is a junior or on the service committee?

| | Soph. | Junior | Junior |
|---|---|---|---|
| Service | 4 | 5 | 6 |
| Advertising | 3 | 2 | 2 |
| Dance | 4 | 8 | 6 |
| Administrative | 1 | 1 | 4 |

7. **SPINNER** Audrey spins a 3 or an odd number on a spinner with 8 equally-sized sections that are numbered 1–8.

8. **GRAB BAG** From a grab bag, a customer gets to randomly pick one of twenty envelopes, 10 of which contain store discount coupons, 8 of which contain gift cards, and 2 of which contain cash. What is the probability a customer selects a gift card or cash?

9. **MAGIC SET** Madeline just purchased a new magic set. One of the tricks involves asking a volunteer to select one card from a standard deck of playing cards. What is the probability the volunteer randomly selects an ace or the queen of hearts?

10. **CLASS** In a math class, 12 out of 15 girls are 14 years old and 14 out of 17 boys are 14 years old. What is the probability of selecting a girl or a 14-year old from this class?

## Mixed Exercises

11. **TRAFFIC** If the chance of making a green light at a certain intersection is 35%, what is the probability of arriving when the light is yellow or red?

12. **STUDENTS** In a group of graduate students, 4 out of the 5 females are international students, and 2 out of the 3 men are international students. What is the probability of selecting a graduate student from this group that is a male or an international student?

**CARDS** **Suppose you pull a card from a standard 52-card deck. Find the probability of each event.**

13. The card is a 4.

14. The card is red.

15. The card is a face card.

16. The card is not a face card.

17. $P$(queen or heart)

18. $P$(jack or spade)

19. $P$(five or prime number)

20. $P$(ace or black)

21. **LOTTERY** A lottery drawing is to take place where one ticket is to be drawn from a set of 80 tickets numbered 1 to 80. If a lottery ticket is drawn at random, what is the probability that the number drawn is a multiple of 4 or a factor of 12?

22. **ACTIVITIES** The extracurricular activities in which the seniors at Valley View High School participate are shown in the Venn diagram.

    a. How many students are in the senior class?

    b. How many students participate in athletics?

    c. If a student is randomly chosen, what is the probability that the student participates in athletics or drama?

    d. If a student is randomly chosen, what is the probability that the student participates in only drama and band?

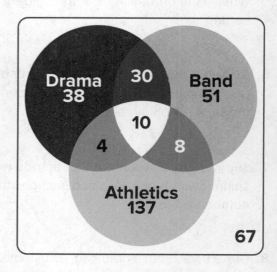

23. **BOWLING** Cindy's bowling records indicate that for any frame, the probability that she will bowl a strike is 30%, a spare 45%, and neither 25%. What is the probability that she will bowl either a spare or a strike for any given frame?

24. **SPORTS CARDS** The Dario owns 145 baseball cards, 102 football cards, and 48 basketball cards. What is the probability that he randomly selects a baseball or a football card?

25. **SCHOLARSHIPS** A review committee received 3000 essays for a $5000 college scholarship. 2865 essays were the required length, 2577 of the applicants had the minimum required grade-point average, and 2486 had the required length and minimum grade-point average. What is the probability that an essay selected at random will have the required length or the required gradepoint average?

26. **KITTENS** Ruby's cat had 8 kittens. The litter included 2 orange females, 3 mixed-color females, 1 orange male, and 2 mixed-color males. Ruby wants to keep one kitten. What is the probability that she randomly chooses a kitten that is female or orange?

27. **SPORT** The table includes programs offered at a sports complex and the number of participants aged 14–16. What is the probability that a player is 14 or plays basketball?

| Mason Sports Complex | | | |
|---|---|---|---|
| Age | Soccer | Volleyball | Basketball |
| 14 | 28 | 36 | 42 |
| 15 | 30 | 26 | 33 |
| 16 | 35 | 41 | 29 |

28. **REASONING** An exchange student is moving back to Italy, and her homeroom class wants to get her a going away present. The teacher takes a survey of the class of 32 students and finds that 10 people chose a card, 12 chose a T-shirt, 6 chose a video, and 4 chose a bracelet. If the teacher randomly selects the present, what is the probability that the exchange student will get a card or a bracelet?

29. **USE A SOURCE** Go online to research a lottery game in your state. Describe one such game and determine if the events that win are mutually exclusive.

30. **MODELING** Dennis and Kelly are designing a board game. They decide that the game will use a pair of dice and the players will have to find the sum of the numbers rolled. Dennis and Kelly created the table shown to help determine probabilities. Each player will roll the pair of dice twice during that player's turn.

| | | | | | |
|---|---|---|---|---|---|
| 1, 1 | 1, 2 | 1, 3 | 1, 4 | 1, 5 | 1, 6 |
| 2, 1 | 2, 2 | 2, 3 | 2, 4 | 2, 5 | 2,6 |
| 3, 1 | 3, 2 | 3, 3 | 3, 4 | 3, 5 | 3, 6 |
| 4, 1 | 4, 2 | 4, 3 | 4, 4 | 4, 5 | 4, 6 |
| 5, 1 | 5, 2 | 5, 3 | 5, 4 | 5, 5 | 5, 6 |
| 6, 1 | 6, 2 | 6, 3 | 6, 4 | 6, 5 | 6, 6 |

   a. What is the probability of rolling a pair, or two numbers that have a sum of seven?

   b. **PRECISION** What is the probability of rolling two numbers whose sum is an even number or not rolling a 2? Round to the nearest thousandth.

31. **CHESS** A chess board has 64 squares, 32 white and 32 black, and is played with 16 black and 16 white pieces. If all the pieces are placed randomly on the board, what is the probability of two white knights being on black squares or a black bishop being on a black square?

32. An analyst drew a Venn diagram to show that 176 of 1200 voters surveyed were under age 26 and that 28 of those were also voters from Maine. If a voter from the survey is randomly selected, what is the probability that the voter is under age 26 or from Maine?

**33. PARKS** The table shows Parks and Recreation Department classes and the number of participants ages 7–9. What is the probability that a participant chosen at random is in Drama or is an 8-year-old?

| Age | Swimming | Drama | Art |
|---|---|---|---|
| 7 | 40 | 35 | 25 |
| 8 | 30 | 21 | 14 |
| 9 | 20 | 44 | 11 |

**34. FLOWER GARDEN** Erin is planning her summer garden. The table shows the number of bulbs she has according to type and color of flower. If Erin randomly selects one of the bulbs, what is the probability that she selects a bulb for a yellow flower or a dahlia?

| Flower | Orange | Yellow | White |
|---|---|---|---|
| Dahlia | 5 | 4 | 3 |
| Lily | 3 | 1 | 2 |
| Gladiolus | 2 | 5 | 6 |
| Iris | 0 | 1 | 4 |

**35. PERSEVERE** You roll 3 dice. What is the probability that the outcome of at least two of the dice will be less than or equal to 4? Explain your reasoning.

**36. FIND THE ERROR** Tetsuya and Mason want to determine the probability that a red marble will be chosen out of a bag of 4 red, 7 blue, 5 green, and 2 purple marbles. Is either of them correct? Explain your reasoning.

| Tetsuya | Mason |
|---|---|
| $P(R) = \frac{4}{17}$ | $P(R) = 1 - \frac{4}{18}$ |

**ANALYZE** Determine whether the following are mutually exclusive. Explain.

**37.** choosing a quadrilateral that is a square and a quadrilateral that is a rectangle

**38.** choosing a triangle that is equilateral and a triangle that is equiangular

**39.** choosing a complex number and choosing a natural number

**40. WRITE** Explain why the sum of the probabilities of two mutually exclusive events is not always 1.

**41.** Each year, one student attending the annual high school picnic is randomly selected to win a free copy of the yearbook. Numbers of students attending the picnic are given in the table.

**a. REASONING** Clarissa wants to know the probability that a junior or senior will win. Are these events mutually exclusive? Find the probability and explain your reasoning.

**b. CREATE** Draw a Venn diagram to represent the situation. How did you decide how many circles to draw and how the circles are related?

| Number of Students Attending the Annual School picnic | |
|---|---|
| Class | Number of Students |
| Freshman | 112 |
| Sophomore | 78 |
| Junior | 124 |
| Senior | 226 |
| Total | 540 |

# Conditional Probability

## Explore Conditional Probability

🧭 **Online Activity** Use the guiding exercises to complete the Explore.

> @ **INQUIRY** How can you find the probability of an event given that another event has already occurred?

## Learn Conditional Probabilities

The **conditional probability** of an event $B$ is the probability that the event will occur given that an event $A$ has already occurred. In addition to finding the probability of two or more dependent events, conditional probability can be used when additional information is known about an event.

Suppose two dice are rolled and it is known that one of the die shows a 5. What is the probability that the sum of the numbers rolled is 7? Since one event, rolling a 5, has already occurred, the sample space for the other event is reduced from 36 to 11 outcomes. This example leads to the following formula.

### Key Concept • Conditional Probability

The conditional probability of $B$ given $A$ is $P(B|A) = \dfrac{P(A \text{ and } B)}{P(A)}$, where $P(A) \neq 0$.

The Venn diagram shows the sample space for both events. The probability of rolling a sum of 7 on two dice given that one die shows a 5, is represented by the probability of the intersection of the two events divided by the probability of the given event.

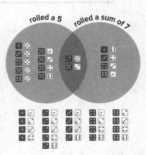

In this case, if event $A$ is rolling a 5 and event $B$ is rolling a sum of 7,

$$P(B|A) = \frac{P(A \text{ and } B)}{P(A)} = \frac{\frac{2}{36}}{\frac{11}{36}} = \frac{2}{11}.$$

Because conditional probability reduced the sample space, the Venn diagram from the example can be simplified as shown, with the intersection of the two events representing those outcomes in $A$ and $B$.

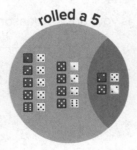

**Study Tip**

**Conditional Probability**
$P(5|odd)$ is read *the probability that the number rolled is a 5 given that the number rolled is odd.*

🧭 **Go Online**
Go online to watch a video to see how to find a conditional probability.

## 🌐 **Example 1** Conditional Probability

**GROCERY SHOPPING** There are currently 16 customers in line at the deli counter, each holding a numbered ticket from 179 to 194. Naveen will help customers holding tickets with even numbers, and Ellie will help customers holding tickets with odd numbers. If a customer is helped by Naveen, what is the probability that the customer is holding ticket 190?

**Understand**

What do you know?

_____

_____

_____

What do you need to find?

_____

**Plan**

**Step 1** Find $P(A)$, the probability that the number on the ticket is even.

**Step 2** Find $P(A \text{ and } B)$, the probability that the ticket is both 190 and even.

**Step 3** Use the formula for conditional probability.

**Solve**

**Step 1** There are 16 tickets. The sample space from event $A$ contains 8 outcomes. From least to greatest, these outcomes are:

{____, ____, ____, ____, ____, ____, ____, ____}.

So, $P(A) = \frac{8}{16}$ or $\frac{1}{2}$.

**Step 2** There are 16 tickets. The sample space for $P(A \text{ and } B)$ contains 1 outcome: {____}.

So, $P(B) = \frac{1}{16}$.

**Step 3**

$$P(B|A) = \frac{P(A \text{ and } B)}{P(A)} \qquad \text{Formula for conditional probability}$$

$$= \frac{\frac{1}{16}}{\frac{1}{2}} \qquad \text{Substitution}$$

$$= \frac{1}{8} \qquad \text{Simplify.}$$

**Check**

Is your answer reasonable? Explain.

_____

_____

_____

🖱️ **Go Online**
to practice what you've learned in Lessons 12-2 and 12-5 through 12-7.

🖱️ **Go Online** You can complete an Extra Example online.

# Practice

🧭 **Go Online** You can complete your homework online.

**Example 1**

**1.** A blue marble is selected at random from a bag of 3 red and 9 blue marbles and not replaced. What is the probability that a second marble selected will be blue?

**2.** A die is rolled. If the number rolled is less than 5, what is the probability that it is the number 2?

**3.** A quadrilateral has a perimeter of 16 and all side lengths are even integers. What is the probability that the quadrilateral is a square?

**4.** A spinner numbered 1 through 12 is spun. Find the probability that the number spun is an 11 given that the number spun was an odd number.

**5.** CLUBS The Spanish Club is having a Cinco de Mayo fiesta. The 10 students randomly draw cards numbered with consecutive integers from 1 to 10. Students who draw odd numbers will bring main dishes. Students who draw even numbers will bring desserts. If Cynthia is bringing a dessert, what is the probability that she drew the number 10?

**6.** CARDS A card is randomly drawn from a standard deck of 52 cards. What is the probability that the card is a king of diamonds, given that the card drawn is a king?

**7.** GAME In a game, a spinner with the 7 colors of the rainbow is spun. Find the probability that the color spun is blue, given the color is one of the three primary colors (red, yellow, or blue).

**8.** DRAWINGS Fifteen cards numbered 1–15 are placed in a hat. What is the probability that the card has a multiple of 3 on it, given that the card picked is an odd number?

**Mixed Exercises**

**9.** If two dice are rolled, what is the probability that the sum of the faces is 8, given that the first die rolled is even?

**10.** PICNIC A school picnic offers students hamburgers, hot dogs, chips, and a drink.

**a.** At the picnic, 60% of the students order a hamburger and 48% of the students order a hamburger and chips. What is the conditional probability that a student who orders a hamburger also orders chips?

**b.** If 50% of the students ordered chips, are the events of ordering a hamburger and ordering chips independent? Explain.

**c.** If 80% of the students who ordered a hot dog also ordered a drink and 35% of all the students ordered a hot dog, find the probability that a student at the picnic orders a hot dog and drink. Explain.

11. **CARNIVAL** Twenty balls numbered 1–20 are placed in a box.

   a. If a ball is randomly chosen from the box and has an odd number on it, find the conditional probability that the ball has a prime number on it.

   b. If a ball is randomly chosen from the box and has a multiple of 5 on it, find the conditional probability that the ball has a multiple of 3 on it.

   c. Two balls are randomly chosen from the box without replacing the first one. If the first ball chosen has an even number on it, find the probability that the second ball has an odd number on it.

12. **HEALTH** There are three types of flu viruses that cause seasonal influenza: A, B, and C. Jane conducted a survey at her school and found that the probability of a student contracting influenza last year at her school was 5%. She also found the probability of a student contracting influenza A at her school last year was 1%. What is the probability that if a student contracts influenza, it will be type A?

13. **MODELING** The Venn diagram represents the results of a random survey about where students study for final exams. Let L represent the library and H represent at home.

   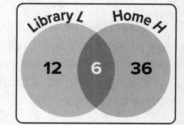

   a. A total of 60 students responded to the survey. Determine the number of students who replied that they study neither at the library nor at home.

   b. **REASONING** What is the probability that if a student selected the library, he or she selected the library and at home? Explain.

   c. **ARGUMENTS** A student says that selecting the library and selecting at home are independent events. Do you agree? Explain.

14. **WRITE** Let event A be owning a house and event B be owning a car. Do you expect the two events to be independent or dependent? How do you think P(A|B) compares to P(B|A)? Explain your reasoning.

15. **PERSEVERE** Of all the students at North High School, 25% are enrolled in Algebra and 20% are enrolled in Algebra and Health.

   a. If a student is enrolled in Algebra, find the probability that the student is enrolled in Health as well.

   b. If 50% of the students are enrolled in Health, are being enrolled in Algebra and being enrolled in Health independent events? Explain.

   c. Of all the students, 20% are enrolled in Accounting and 5% are enrolled in Accounting and Spanish. If being enrolled in Accounting and being enrolled in Spanish are independent events, what percent of students are enrolled in Spanish? Explain.

16. **ANALYZE** In a standard deck of playing cards, the face-value cards are the cards numbered 2–10. Two cards are to be randomly drawn without replacing the first card. Find the probability of drawing two face-value cards and the conditional probability that exactly one of those cards is a 4. Explain.

# Two-Way Frequency Tables

## Explore Two-Way Frequency Tables

**Today's Standards**
S.CP.4; S.CP.6
MP2, MP4

**Today's Vocabulary**
two-way frequency table
marginal frequencies
joint frequencies
relative frequency

🔾 **Online Activity** Use the tables to complete the Explore.

> ⓠ **INQUIRY** How can you use data in a two-way frequency table to determine whether two events are independent?

## Learn Independent Events in Frequency Tables

A **two-way frequency table**, or contingency table, is used to show the frequencies of data from a survey or experiment classified according to two variables, with the rows indicating one variable and the columns indicating the other.

The frequencies reported in the *Totals* row and *Totals* column are called **marginal frequencies**, with the bottom rightmost cell reporting the total number of observations. Marginal frequencies allow you to analyze with respect to one variable.

The frequencies reported in the interior of the table are called **joint frequencies**. These show the frequencies of all possible combinations of the categories for the first variable with the categories for the second variable.

A **relative frequency** is the ratio of the number of observations in a category to the total number of observations.

When survey results are classified according to variables, you may want to decide whether these variables are independent of each other. Variable $A$ is considered independent of variable $B$ if

$P(A \text{ and } B) = P(A) \cdot P(B).$

## 🌐 Example 1 Frequency and Relative Frequency Tables

BREAKFAST **Francesca asks a random sample of 140 upperclassmen at her high school whether they prefer eating breakfast at home or at school. She finds that 55 juniors and 23 seniors prefer eating breakfast at home before school, while 12 juniors and 50 seniors prefer eating breakfast at school.**

a. **Organize the responses in a two-way frequency table.**

Identify the variables. The students surveyed can be classified according to *class* and *preference*. Because the survey included only

*(continued on the next page)*

upperclassmen, the variable *class* has two categories: senior or _____. The variable *preference* also has two categories: *prefers eating breakfast at home* and *prefers eating breakfast at school.*

Create a two-way frequency table. Let the rows of the table represent *class* and the columns represent *preference.* Then fill in the cells of the table with the information given.

|  | Breakfast at Home | Breakfast at School | Totals |
|---|---|---|---|
| Senior | 23 |  | 73 |
| Junior |  | 12 | 67 |
| Totals | 78 | 62 |  |

Add a Totals row and a Totals column to your table and fill in these cells with the correct sums.

**b. Construct a relative frequency table.**

To complete a relative frequency table for these data, start by dividing the frequency reported in each cell by the total number of respondents, 140. Then, write each fraction as a percent rounded to the nearest tenth.

|  | Breakfast at Home | Breakfast at School | Totals |
|---|---|---|---|
| Senior | $\frac{23}{140} = 6.4\%$ | $\frac{50}{140} = 35.7\%$ | $\frac{73}{140} = 52.1\%$ |
| Junior | $\frac{55}{140} = $ % | $\frac{12}{140} = $ % | $\frac{67}{140} = $ % |
| Totals | $\frac{78}{140} = $ % | $\frac{62}{140} = $ % | $\frac{140}{140} = $ % |

**Think About It!**

What is the probability that a surveyed student is a junior who prefers eating breakfast at home?

## Check

**CONCERTS** At the end of a symphony performance, each audience member is asked to text the answer to the following two questions.

- Text "PUBLIC" if you took public transportation to this event or "PRIVATE" if you drove.

- Text "BEETHOVEN" if you want to hear his Symphony No. 5 or "CHOPIN" if you want to hear his Nocture No. 2 for the encore.

The results of the survey in the table are displayed for the audience.

|  | Beethoven | Chopin | Totals |
|---|---|---|---|
| Public Transportation | 185 | 158 | 343 |
| Private Transportation | 520 | 615 | 1135 |
| Totals | 705 | 773 | 1478 |

Construct a relative frequency table. Round each percent to the nearest tenth.

|  | Beethoven | Chopin | Totals |
|---|---|---|---|
| Public Transportation | % | % | % |
| Private Transportation | % | % | % |
| Totals | % | % | % |

**Go Online** You can complete an Extra Example online.

## 🌐 Example 2 Independence and Relative Frequency

**MUSIC Anaud polls 240 of his friends on social media about what grade they are in and whether they prefer electronic dance music (EDM) or hip-hop. He posts the results of the survey in the table.**

|  | EDM | Hip-Hop | Totals |
|---|---|---|---|
| College Freshman | 100 | 107 | 207 |
| High School Senior | 16 | 17 | 33 |
| Totals | 116 | 124 | 240 |

**Use the table to determine whether a respondent's musical preference is independent of his or her grade level.**

In a two-way frequency table, you can test for the independence of two variables by comparing the joint relative frequencies with the products of the corresponding marginal relative frequencies.

Divide each reported frequency by 240 to convert the frequency table to a relative frequency table. Enter each fraction as a percent rounded to the nearest tenth. Complete the table.

|  | EDM | Hip-Hop | Totals |
|---|---|---|---|
| College Freshman | $\frac{100}{240} =$ ___% | $\frac{107}{240} =$ ___% | $\frac{207}{240} =$ ___% |
| High School Senior | $\frac{16}{240} =$ ___% | $\frac{17}{240} =$ ___% | $\frac{33}{240} =$ ___% |
| Totals | $\frac{116}{240} =$ ___% | $\frac{124}{240} =$ ___% | $\frac{240}{240} = 100\%$ |

Calculate the expected joint relative frequencies if the two variables were independent. Then compare them to the actual relative frequencies.

For example, if 86.3% of respondents were college freshmen and 48.3% of respondents prefer EDM, then one would expect that 86.3% · 48.3% or about 41.7% of respondents are college freshmen who prefer EDM. The table below shows the expected joint relative frequencies.

|  | EDM | Hip-Hop | Totals |
|---|---|---|---|
| College Freshman | 41.7% | 44.6% | 86.3% |
| High School Senior | 6.7% | 7.1% | 13.8% |
| Totals | 48.3% | 51.7% | 100% |

Comparing the two tables, the expected and actual joint relative frequencies are the same. Therefore, the musical preferences for these respondents are independent of grade level.

🔎 **Go Online** You can complete an Extra Example online.

## Check

**SCHOOL** Immediately after a physics test, the entire class sits together at lunch and discusses how long each of them studied and how many questions they guessed on. The table shows the responses from the classmates.

True or False: For these classmates, guessing on more than 5 problems on the physics test is independent of studying 4 hours or less. _____

---

### 🌐 **Example 3** Conditional Probability with Two-Way Frequency Tables

You can use joint and marginal relative frequencies to approximate conditional probabilities.

**MEMES Abu posts a question to an online forum about the originality of posts to the site. Of the 55 respondents who have posted viral memes, 27 photos and 15 videos were not original content, while 3 photos and 10 videos were original content.**

**a. Construct a relative frequency table of the data. Round each percent to the nearest tenth.**

|  | Not Original Content | Original Content | Totals |
|---|---|---|---|
| Video | 27.3% | 18.2% | % |
| Photo | % | % | % |
| Totals | % | % | 100% |

**b.** Find the probability that a viral meme on the forum is not original content given that is it a photo.

The probability that a meme is not original content given that it is a photo is the conditional probability P(not original content|photo).

$$P(\text{not original content|photo}) = \frac{P(\text{not original content and photo})}{P(\text{photo})}$$

$$\approx \frac{0.491}{0.546} \text{ or } \_\_\_\_\%$$

|  | Guessed on < 5 Problems | Guessed on > 5 Problems | Totals |
|---|---|---|---|
| Studied 4 Hours or Less | 9 | 3 | 12 |
| Studied More Than 4 Hours | 12 | 4 | 16 |
| Totals | 21 | 7 | 28 |

So the probability that a viral meme on the forum is not original content given that it is a photo is 89.9%.

🌀 **Go Online** You can complete an Extra Example online.

💭 **Think About It!**

Why do we divide by 0.546 when finding the conditional probability in the example?

# Practice

**Go Online** You can complete your homework online.

**Example 1**

1. VEHICLES One hundred people were surveyed about the vehicle they drive. The survey shows that 15 males and 40 females drive SUVs, while 35 males and 10 females drive trucks.

   a. Make a two-way frequency table of the data.

   b. Find the number of males surveyed. Is this a *marginal* or *joint* frequency?

   c. Find the number of females that drive a truck. Is this a *marginal* or *joint* frequency?

   d. Use your two-way frequency table to construct a relative frequency table of the data.

2. FRUIT One hundred students were survey about whether they like strawberries. The surveyed shows that 25 boys and 35 girls like strawberries, while 25 boys and 15 girls do not like strawberries.

   a. Make a two-way frequency table of the data.

   b. Find the number of girls surveyed? Is this a *marginal* or *joint* frequency?

   c. Find the number of boys that do not like strawberries. Is this a *marginal* or *joint* frequency?

   d. Use your two-way frequency table to construct a relative frequency table of the data.

**Example 2**

3. REASONING A random sample of 2000 college freshmen is asked the following questions at the end of the academic calendar year.

   • Are you attending college in or out of state?

   • Throughout the school year, did your number of visits home exceed 4, or were they 4 or fewer?

   The survey found that of the 1260 college freshmen who attended an in-state college, 928 visited home more than 4 times and 332 visited home four or less times. Of the 740 college freshmen who attended an out-of-state college, 118 visited home more than four times and 622 visited home four or less times.

   a. Organize the responses in a two-way frequency table.

   b. Use your two-way frequency table to construct a relative frequency table. Enter each fraction as a percent rounded to the nearest tenth, if necessary.

   c. Suppose you let event *A* represent whether the students attend an in-state or an out-of-state college, and event *B* represent whether the students visit home more than four times or four or fewer times. Use the table to determine whether the number of visits home is independent or dependent on whether the student is attending college at an in-state or out-of-state institution. Justify your response.

**4. COURSES** Salina surveyed a random sample of students to determine how many take Spanish or French. The data are shown in the two-way frequency table.

|  | Spanish | French | Totals |
| --- | --- | --- | --- |
| Girls | 75 | 45 | 120 |
| Boys | 60 | 20 | 80 |
| Totals | 135 | 65 | 200 |

**a.** Construct a relative frequency table of the data. Enter each fraction as a percent rounded to the nearest tenth, if necessary.

**b.** Use the relative frequency table to determine whether gender is independent of taking Spanish or French. Explain.

**Example 3**

**5. MODELING** Fifty students were surveyed about after school activities. Four play an instrument and play sports. Sixteen play an instrument and do not play sports. Twenty-four do not play an instrument and play sports. Six do not play an instrument and do not play sports.

**a.** Construct a relative frequency table of the data.

**b.** Find the probability that a surveyed student does not play sports given that they play an instrument.

**c.** Find the probability that a surveyed student plays an instrument given that they play sports.

**6. TICKETS** A movie theater is keeping track of the last 800 tickets it sold to two different movies. Of the 578 adult tickets sold, 136 of them were for the animated film and 442 were for the documentary film. Of the 222 student tickets sold, 181 of them were for the animated film and 41 were for the documentary film.

**a.** Complete the two-way frequency table shown.

|  | Adult |  | Totals |
| --- | --- | --- | --- |
|  | 181 |  |  |
| Documentary |  |  |  |
|  | 578 |  | 800 |

**b.** Construct a relative frequency table of your completed two-way frequency table. Round each percent to the nearest tenth, as necessary.

**c.** Find the probability that a ticket sold is an adult ticket given that it is a documentary ticket. Show your work by writing the formula that you used to perform the calculation.

## Mixed Exercises

**7.** **REGULARITY** Nico surveyed 100 students to determine if they prefer a dog or cat as a pet. Of the 55 ninth graders surveyed, 20 preferred a cat. Of the 65 people that preferred a dog, 30 were tenth graders. Make a two-way frequency table of the data.

**8.** **EXAMS** The data about homework completeness and exam score of the students in Mr. VanWinkle's class are shown in the two-way frequency table. How many students completed their homework and passed the exam? Tell whether you are using a marginal frequency or joint frequency.

|  | Completed Homework | Did Not Complete Homework | Totals |
|---|---|---|---|
| Passed Exam | 18 | 2 | 20 |
| Did Not Pass Exam | 4 | 2 | 6 |
| Totals | 22 | 4 | 26 |

**9.** **VOTING** A recent poll was taken to determine whether registered voters would vote "yes" or "no" for Issue 1. Of the 250 people that were polled, 130 were females and 120 were males. Of the males that were surveyed, 80 said they would vote "yes" and 40 said they would vote "no" for Issue 1. Of the females that were surveyed, 95 said they would vote "yes" and 35 said they would vote "no" for Issue 1. Make a relative frequency table of the data.

**10.** **STRUCTURE** Rachel surveyed 160 people to determine if they preferred Drama or Comedy movies. The relative frequency table shows the data collected from the survey. Determine whether gender is independent of movie type preference. Explain.

|  | Drama | Comedy | Totals |
|---|---|---|---|
| Males | 12.5% | 25% | 37.5% |
| Females | 46.9% | 15.6% | 62.5% |
| Totals | 59.4% | 40.6% | 100% |

**11.** **TECHNOLOGY** For a business report on technology use, Darnell asks a random sample of 72 shoppers whether they own a smart phone and whether they own a tablet computer. His survey shows that out of 51 shoppers who own smart phones, 9 of them also own a tablet, while out of 21 shoppers who do not own smart phones, 15 of them do not own tablets either.

  **a.** Organize Darnell's data into the two-way frequency table.

  **b.** Convert the table from part **a** to a two-way frequency table. Round to the nearest tenth of a percent. What is the probability that the shopper has a smart phone and a tablet computer? Explain.

  **c.** Find the conditional probability that a shopper has a tablet computer, given that he or she has a smart phone. Explain.

**12.** **ARGUMENTS** Amanda asked a random sample of seniors at her high school whether they own a car and whether they have a job. The results of the survey are shown in the two-way relative frequency table. Amanda says that the conditional probability that a student has a job given that they have a car is 46.7%. Do you agree?

| Car | Has a Job | Does Not Have a Job | Totals |
|---|---|---|---|
| Has a Car | 21.9% | 12.5% | 34.4% |
| Does Not Have a Car | 25% | 40.6% | 65.6% |
| Totals | 46.9% | 53.1% | 100% |

**13.** **PERSEVERE** Suppose an exit poll held outside a voting area on the day of an election produced these results.

| Age and Gender | Votes for Candidate A | Votes for Candidate B |
|---|---|---|
| 18–30 Male | 19 | 32 |
| 18–30 Female | 31 | 18 |
| 31–45 Male | 51 | 12 |
| 31–45 Female | 43 | 20 |
| 46–60 Male | 42 | 35 |
| 46–60 Female | 20 | 42 |
| 60+ Male | 45 | 21 |
| 60+ Female | 27 | 18 |

a. Which events are mutually exclusive?

b. Find the probability that a male between the ages of 46 and 60 would vote for Candidate A.

c. Find the probability that a female would vote for Candidate A.

d. Find the probability that someone who voted for Candidate B was a female and age 18–30.

e. According to the data, on which demographic(s) does Candidate A need to focus campaign efforts?

f. According to the data, on which demographic(s) does Candidate B need to focus campaign efforts?

**14.** A market research firm asks a random sample of 240 adults and students at a movie theater whether they would rather see a new summer blockbuster in 2-D or 3-D. The survey shows that 64 adults and 108 students prefer 3-D, while 42 adults and 26 students prefer 2-D.

a. **CREATE** Organize the responses into a two-way frequency table.

b. **USE TOOLS (ESTIMATION)** Convert the table from **part a** into a two-way relative frequency table. Round to the nearest tenth of a percent. Out of every 10 people surveyed, about how many would prefer to see the movie in 3-D? Explain.

c. Find the probability that a person surveyed prefers seeing the movie in 3-D, given that he or she is an adult. Write the formula that you used to perform the calculation.

d. **FIND THE ERROR** An analyst at the firm claims that the probability that a person surveyed is a child given that he or she does not prefer to see the movie in 3-D is 10.8%. Do you agree? Justify your answer.

e. **ANALYZE** Is a preference for 2-D or 3-D movies independent of age? Explain your reasoning.

## Essential Question

How can you use measurements to find probabilities?

## Module Summary

### Lessons 12-1 through 12-3

Probability of Simple Events

- The number of possible outcomes in a sample space can be found by multiplying the number of possible outcomes from each stage or event.
- For the probability of the intersection of two or more events, find the ratio of the number of outcomes in both events to the total number of possible outcomes.
- When two events $A$ and $B$ occur, the union of $A$ and $B$ is the set of all outcomes in the sample space of event $A$ combined with all outcomes in the sample space of event $B$.
- If region $A$ contains a region $B$ and a point $E$ in region $A$ is chosen at random, then the probability that point $E$ is in region $B$ is
$\frac{\text{area of region } B}{\text{area of region } A}$.

### Lesson 12-4

Permutations and Combinations

- The number of distinguishable permutations of $n$ objects in which one object is repeated $r_1$ times, another is repeated $r_2$ times, and so on, is
$\frac{n!}{r_1! r_2! \cdot \ldots \cdot r_k!}$
- The number of combinations of $n$ distinct objects taken $r$ at a time is denoted by $_nC_r$ and is given by $_nC_r = \frac{n!}{(n-r)! r!}$.

### Lessons 12-5 through 12-7

Probability of Compound Events

- If two events $A$ and $B$ are independent, the $P(A \text{ and } B) = P(A) \cdot P(B)$.
- If two events $A$ and $B$ are dependent, then $P(A \text{ and } B) = P(A) \cdot P(B \mid A)$.
- If two events $A$ or $B$ are mutually exclusive, then $P(A \text{ or } B) = P(A) + P(B)$.
- If two events $A$ or $B$ are not mutually exclusive, then $P(A \text{ or } B) = P(A) + P(B) - P(A \text{ and } B)$.
- The conditional probability of $B$ given $A$ is
$P(B \mid A) = \frac{P(A \text{ and } B)}{P(A)}$, where $P(A) \neq 0$.

### Lesson 12-8

Frequency Tables

- The frequencies in the Totals row and Totals column are marginal frequencies.
- The frequencies in the interior of the table are joint frequencies.
- A relative frequency is the ratio of the number of observations in a category to the total number of observations.

### Study Organizer

Foldables

Use your Foldable to review this module. Working with a partner can be helpful. Ask for clarification of concepts as needed.

# Test Practice

**1. MULTIPLE CHOICE** Two dice are tossed. Which is the sample space of the event that the sum of the outcomes is 5? (Lesson 12-1)

ⓐ {(5, 5)}

ⓑ {(1, 4), (2, 3)}

ⓒ {(1, 4), (2, 3), (3, 2), (4, 1)}

ⓓ {(1, 5), (2, 5), (3, 5), (4, 5), (5, 5)}

**2. OPEN RESPONSE** A restaurant has a special deal where you can build your own meal from certain selections in the menu. The number of selections available in each category is shown in the table.

| Item | Number of Choices |
|------|-------------------|
| Drink | 12 |
| Appetizer | 7 |
| Main Entrée | 8 |
| Side Dishes | 14 |
| Dessert | 9 |

If a person selects one of each item, how many different meals can be ordered? (Lesson 12-1)

**3. MULTIPLE CHOICE** An integer between 1 and 12 is generated using a random number generator. Let $A$ be the event of generating a multiple of 4, and let $B$ be the event of generating a factor of 12. Which of the following represents $A \cap B$? (Lesson 12-2)

ⓐ {4, 12}

ⓑ {4, 8, 12}

ⓒ {1, 2, 3, 4, 6, 12}

ⓓ {1, 2, 3, 4, 6, 8, 12}

**4. OPEN RESPONSE** Miko's birthday is in May. Let $W$ be the event that her birthday lands on a weekend. Let $P$ be the event that her birthday is a prime number.

What is $W \cap P$? (Lesson 12-2)

| May | | | | | | |
|-----|-----|------|-----|-------|-----|-----|
| Sun | Mon | Tues | Wed | Thurs | Fri | Sat |
| | 1 | 2 | 3 | 4 | 5 | 6 |
| 7 | 8 | 9 | 10 | 11 | 12 | 13 |
| 14 | 15 | 16 | 17 | 18 | 19 | 20 |
| 21 | 22 | 23 | 24 | 25 | 26 | 27 |
| 28 | 29 | 30 | 31 | | | |

**5. GRIDDED RESPONSE** Point $J$ will be placed randomly on $\overline{AB}$.

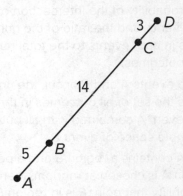

What is the probability that point $J$ is on $\overline{AC}$ to the nearest percent? (Lesson 12-3)

**6. MULTIPLE CHOICE** Stacy is at a carnival trying to win a prize. She must toss a bean bag in the hole to win.

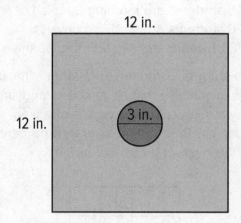

12 in.

12 in.

3 in.

What is the probability that when tossed randomly, the bean bag lands in the hole? Assume that the beanbag lands on the board. (Lesson 12-3)

- Ⓐ 4.9%
- Ⓑ 7.065%
- Ⓒ 19.625%
- Ⓓ 28.26%

**7. MULTIPLE CHOICE** If three points are randomly chosen from on the vertices of hexagon *ABCDEF*, what is the probability that they all lie on the same line segment? (Lesson 12-4)

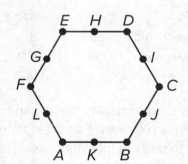

- Ⓐ $\frac{1}{1320}$
- Ⓑ $\frac{1}{220}$
- Ⓒ $\frac{1}{216}$
- Ⓓ $\frac{3}{110}$

**8. OPEN RESPONSE** The numbers 0–39 are used to create a locker combination. Show how to determine the probability that the combination is 20 − 21 − 22. (Lesson 12-4)

**9. TABLE ITEM** Identify each situation as *independent* or *dependent* events. (Lesson 12-5)

| Situation | Independent | Dependent |
|---|---|---|
| Two dice are tossed. | | |
| Two marbles are selected from a bag. | | |
| Two students are chosen as the captains of a team. | | |
| A coin is tossed and a card is chosen. | | |
| Three books are selected from the library. | | |
| One student from each class is chosen to collect papers. | | |

**10. OPEN RESPONSE** The table shows the books of several different genres available to read on Yasmin's bookshelf.

| Genre | Number of Books |
|---|---|
| Action | 8 |
| Mystery | 5 |
| Romance | 2 |
| Science fiction | 12 |
| Horror | 3 |

Yasmin selects two different books to read this month. What is the probability, as a fraction, that Yasmin selects two mysteries? (Lesson 12-5)

**11.** **MULTI-SELECT** Suppose a die is tossed once. Which of these events are mutually exclusive? Select all that apply. (Lesson 12-6)

Ⓐ Landing on a 4 or a 5

Ⓑ Landing on a 2 or an even

Ⓒ Landing on a 2 or a prime

Ⓓ Landing on 4 or an odd

Ⓔ Landing on an odd or a prime

**12.** **MULTIPLE CHOICE** A group of college students was surveyed about their browser use. The results are shown on the circle graph.

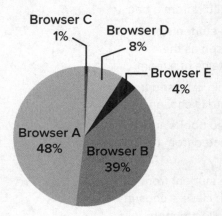

Browser C 1%
Browser D 8%
Browser E 4%
Browser A 48%
Browser B 39%

What is the probability that a student selected randomly will use either Browser A or Browser D? (Lesson 12-6)

Ⓐ $\frac{47}{100}$

Ⓑ $\frac{12}{25}$

Ⓒ $\frac{14}{25}$

Ⓓ $\frac{87}{100}$

**13.** **OPEN RESPONSE** Two dice have been tossed. What is the probability of tossing a sum of 8 given that at least one die landed on an odd number? (Lesson 12-7)

**14.** **GRIDDED RESPONSE** In a recent survey, 1650 students were asked what they are studying. Of the 1650 students, 948 students are learning Spanish. 426 students are studying physics, 378 of whom are also learning Spanish.

If a student is chosen at random, what is the probability that he or she is studying physics, given the student is studying Spanish? Enter the answer as a percent to the nearest tenth. (Lesson 12-7)

**15.** **MULTIPLE CHOICE** The table shows the results of a survey asking whether the respondent preferred to use the Internet over a phone or laptop.

| Age | Phone | Laptop | Total |
| --- | --- | --- | --- |
| 12–29 years old | 85 | 21 | 106 |
| 30+ years old | 124 | 87 | 211 |
| Total | 209 | 108 | 317 |

What is the probability that a participant is 30+ years old given that they prefer a laptop to use the Internet? (Lesson 12-8)

Ⓐ 27.4%

Ⓑ 41.2%

Ⓒ 59.3%

Ⓓ 80.6%

# Selected Answers

## Module 7

### Module 7 Opener

**1.** 150 **3.** 1

### Lesson 7-1

**1.** 2520° **3.** $m\angle Q = 121°$, $m\angle R = 58°$, $m\angle S = 123°$, $m\angle T = 58°$ **5.** $m\angle A = 90°$, $m\angle B = 90°$, $m\angle C = 128°$, $m\angle D = 74°$, $m\angle E = 158°$ **7.** 135° **9.** 128.6° **11.** 10 **13.** 18 **15.** 6 **17.** 37 **19.** 44 **21.** 72° **23.** 60° **25.** 40° **27.** 51.4°, 128.6° **29.** 25.7°, 154.3° **31.** 30 **33.** 186°, 137°, 40°, 54°, 123° **35.** 360° **37.** Consider the sum of the measures of the exterior angles $N$ for an $n$-gon. $N$ = sum of measures of linear pairs − sum of measures of interior angles

$$= 180n - 180(n - 2)$$
$$= 180n - 180n + 360$$
$$= 360$$

So, the sum of the exterior angle measures is 360 for any convex polygon. **39.** 15
**41.** 360° **43.** Liam: by the Exterior Angle Sum Theorem, the sum of the measures of any convex polygon is 360.

**45.** 8; Sample answer:

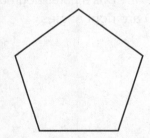

Interior angles sum = $(5 - 2) \cdot 180$ or 540. Twice this sum is 2(540) or 1080. A polygon with this interior angles sum is the solution to $(n - 2) \cdot 180 = 1080$, $n = 8$

**47.** Always; by the Exterior Angles Sum Theorem, $m\angle QPR = 60$ and $m\angle QRP = 60$. Since the sum of the interior angle measures of a triangle is 180, the measure of $\angle PQR = 180 - m\angle QPR - m\angle QRP = 180 - 60 - 60 = 60$. So, $\triangle PQR$ is an equilateral triangle.

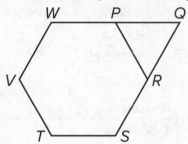

### Lesson 7-2

**1.** 52° **3.** 5 **5.** 3 mi **7.** Proof:

| Statements | Reasons |
|---|---|
| 1. *WXTV* and *ZYVT* are parallelograms | 1. Given |
| 2. $\overline{WX} \cong \overline{VT}$, $\overline{VT} \cong \overline{YZ}$ | 2. Opposite sides of a parallelogram are congruent. |
| 3. $\overline{WX} \cong \overline{YZ}$ | 3. Transitive Property |

**9.** Given: ▱*ABCD*
Prove: $\angle A$ and $\angle B$ are supplementary. $\angle B$ and $\angle C$ are supplementary. $\angle C$ and $\angle D$ are supplementary. $\angle D$ and $\angle A$ are supplementary.
Proof: It is given ▱*ABCD*, so $\overline{AB} \parallel \overline{CD}$ and $\overline{BC} \parallel \overline{DA}$ by the definition of a parallelogram. If two parallel lines are cut by a transversal, then consecutive interior angles are supplementary. So, $\angle A$ and $\angle B$, $\angle B$ and $\angle C$, $\angle C$ and $\angle D$, and $\angle D$ and $\angle A$ are pairs of supplementary angles.
**11.** $a = 7$, $b = 11$ **13.** $x = 5$, $y = 17$
**15.** $x = 58$, $y = 63.5$ **17.** (3, 2)
**19.** (3.5, 2) **21.** (0, 1)

**23.** Proof:

| Statements | Reasons |
| --- | --- |
| **1.** $\square PQRS$ | **1.** Given |
| **2.** Draw an auxiliary segment $\overline{PR}$ | **2.** Diagonal of $PQRS$. |
| **3.** $\overline{PQ} \parallel \overline{SR}$, $\overline{PS} \parallel \overline{QR}$ | **3.** Opposite sides of a parallelogram are parallel. |
| **4.** $\angle QPR \cong \angle SRP$, $\angle SPR \cong \angle QRP$ | **4.** Alternate Interior Angles Theorem |
| **5.** $\overline{PR} \cong \overline{PR}$ | **5.** Reflexive Property |
| **6.** $\triangle QPR \cong \triangle SRP$ | **6.** ASA |
| **7.** $\overline{PQ} \cong \overline{SR}$, $\overline{QR} \cong \overline{SP}$ | **7.** CPCTC |

**25.** Proof:

| Statements | Reasons |
| --- | --- |
| **1.** $\square GKLM$ | **1.** Given |
| **2.** $\overline{GK} \parallel \overline{ML}$, $\overline{GM} \parallel \overline{KL}$ | **2.** Opposite sides of a parallelogram are parallel. |
| **3.** $\angle G$ and $\angle K$ are supplementary $\angle K$ and $\angle L$ are supplementary $\angle L$ and $\angle M$ are supplementary $\angle M$ and $\angle G$ are supplementary | **3.** Consecutive interior angles are supplementary. |

**27.** 3  **29.** 131°  **31.** 29°  **33a.** $JP = \sqrt{13}$, $LP = \sqrt{13}$, $MP \sqrt{34}=$ , $KP = \sqrt{34}$; since $JP = LP$ and $MP = KP$, the diagonals bisect each other. **33b.** No; $JP + LP \neq MP + KP$. **33c.** No; the slope of $\overline{JK} = 0$ and the slope of $\overline{JM} = 2$. The slopes are not negative reciprocals of each other. **35.** 7  **37.** Sample answer: In a parallelogram, the opp. sides and $\angle$s are $\cong$. Two consecutive $\angle$s *in a* $\square$ *are* supplementary. If one angle of a $\square$ is right, then all the angles are right. The diagonals of a parallelogram bisect each other.
**39.** $m\angle 1 = 116$, $m\angle 10 = 115$; sample answer: $m\angle 8 = 64$ because alternate interior angles are congruent. $\angle 1$ is supplementary to $\angle 8$ because consecutive angles in a parallelogram are supplementary, so $m\angle 1$ is 116. $\angle 10$ is supplementary to the 65 degree angle because consecutive angles in a parallelogram are supplementary, so $m\angle 10$ is $180 - 65$ or 115.

# Lesson 7-3

**1.** Yes; a pair of opposite sides are parallel and congruent.  **3.** No; none of the tests for parallelograms is fulfilled.  **5.** Yes; the diagonals bisect each other.  **7.** $x = 24, y = 19$
**9.** $x = 45, y = 20$  **11.** $x = -6, y = 13$
**13.** Method 1: Use the Slope Formula, $m = \dfrac{y_2 - y_1}{x2 - x_1}$.
slope of $\overline{AD} = \dfrac{3 - 0}{-2 - (-3)} = \dfrac{3}{1} = 3$  slope of $\overline{BC} = \dfrac{2 - (-1)}{3 - 2} \dfrac{3}{1} = 3$  slope of $\overline{AB} = \dfrac{2 - 3}{3 - (-2)} = \dfrac{1}{5}$  slope of $\overline{CD} = \dfrac{-1 - 0}{2 - (-3)} = -\dfrac{1}{5}$
Since opposite sides have the same slope, $\overline{AB} \parallel \overline{CD}$ and $\overline{AD} \parallel \overline{BC}$. Therefore, $ABCD$ is a parallelogram by definition.

Method 2: Use the Distance Formula, $d = \sqrt{(x2 - x_1)^2 + (y_2 - y_1)^2}$.
$AB = \sqrt{(-2 - 3)^2 + (3 - 2)^2} = \sqrt{25 + 1}$ or $\sqrt{26}$
$CD = \sqrt{(-2 - (-3))^2 + (-1 - 0)^2} = \sqrt{25 + 1}$ or $\sqrt{26}$
$AD = \sqrt{(-2 - (-3))^2 + (3 - 0)^2} = \sqrt{1 + 9}$ or $\sqrt{10}$
$BC = \sqrt{(3 - 2)^2 + (2 - (-1))^2} = \sqrt{1 + 9}$ or $\sqrt{10}$
Since both pairs of opposite sides have the same length, $\overline{AB} \cong \overline{CD}$ and $\overline{AD} \cong \overline{BC}$. Therefore, $ABCD$ is a parallelogram by Theorem 6.9.

**15.** Yes; $SR = ZT$ and the slopes of $\overline{SR}$ and $\overline{ZT}$ are equal, so one pair of opposite sides is parallel and congruent.  **17.** No; slope $\overline{XY} = -\dfrac{3}{5}$ and slope of $\overline{WZ} = -\dfrac{1}{3}$, so opposite sides are not parallel.  **19.** Given: $ABCD$ is a parallelogram. Prove: $\angle B$, $\angle C$, and $\angle D$ are right angles.

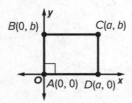

Proof:
slope of $\overline{AD} = \dfrac{0 - 0}{a - 0} = 0$
slope of $\overline{BC} = \dfrac{b - b}{a - 0} = 0$
slope of $\overline{AB}$ is undefined
slope of $\overline{CD}$ is undefined
Therefore, $\overline{AD} \perp \overline{CB}$, $\overline{RC} \perp \overline{AD}$, and $\overline{AB} \parallel \overline{BC}$. So, $\angle B$, $\angle C$, and $\angle D$ are right angles.

**21.** $Y(a - b, c)$, $X(a, 0)$   **23.** $(4, -1)$, $(0, 3)$, or $(-4, -5)$   **25.** $-4$   **27.** 28 cm   **29.** Yes; the lengths of the opposite sides are congruent. When the coordinate plane is placed over the map, the street lamps align perfectly with the corners of the grid.   **31.** No; Madison and Angela have to be the same distance and Nikia and Shelby have to be the same distance but Nikia and Shelby's distance from the center does not have to be equal to Madison and Angela's distance.

**33.** You cannot prove a general statement with an example. Counterexample:

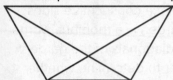

**35.** Sample answer: The theorems are converses of each other. The hypothesis of Theorem 6.3 is "a figure is a parallelogram", and the hypothesis of 6.9 is "both pairs of opposite sides of a quadrilateral are congruent". The conclusion of Theorem 6.3 is "opposite sides are congruent", and the conclusion of 6.9 is "the quadrilateral is a parallelogram".   **37.** $(-3, 1)$ and $(-2, -2)$

## Lesson 7-4

**1.** 2 ft   **3.** 50°   **5.** 52°   **7.** 38°   **9.** $x = 7$
**11.** 77°   **13.** 180°   **15.** Proof:

| Statements | Reasons |
|---|---|
| **1.** $ABCD$ is a rectangle. | **1.** Given |
| **2.** $ABCD$ is a parallelogram. | **2.** Definition of rectangle. |
| **3.** $\overline{AD} \cong \overline{BC}$ | **3.** Opposite sides of a parallelogram are congruent. |
| **4.** $\overline{DC} \cong \overline{CD}$ | **4.** Reflexive Property |
| **5.** $\overline{AC} \cong \overline{BD}$ | **5.** Diagonals of a rectangle are congruent. |
| **6.** $\triangle ADC \cong \triangle BCD$ | **6.** SSS |

**17.** Yes; sample answer: Opposite sides are parallel and consecutive sides are perpendicular.   **19.** No; sample answer:

Diagonals are not congruent.   **21.** Yes; sample answer: Both pairs of opposite sides are congruent and diagonals are congruent.
**23.** 50°   **25.** 40°   **27.** 50°

**29.**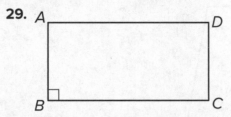

$ABCD$ is a parallelogram and $\angle B$ is a right angle. Since $ABCD$ is a parallelogram and has one right angle, then it has four right angles. So, by the definition of a rectangle, $ABCD$ is a rectangle.   **31.** No; if you only know that opposite sides are congruent and parallel, the most you can conclude it that the plot is a parallelogram.

**33.** Sample answer: Since $\overline{RP} \perp \overline{PQ}$ and $\overline{SQ} \perp \overline{PQ}$, $m\angle P = m\angle Q = 90$. Lines that are perpendicular to the same line are parallel, so $\overline{RP} \parallel \overline{SQ}$. The same compass setting was used to locate points $R$ and $S$, so $\overline{RP} \cong \overline{SQ}$ If one pair of opposite sides of a quadrilateral is both parallel and congruent, then the quadrilateral is a parallelogram. A parallelogram with right angles is a rectangle. Thus, $PRSQ$ is a rectangle.

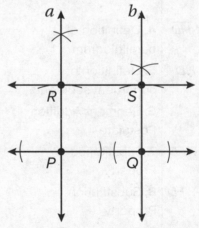

**35.** 5   **37.** They should be equal.   **39.** $(-6, 3)$
**41.** $x = 6$, $y = -10$   **43.** Sample answer: All rectangles are parallelograms because, by definition, both pairs of opposite sides are parallel. Parallelograms with right angles are rectangles, so some parallelograms are rectangles, but others with non-right angles are not.

**45.** No; Sample answer: a rectangle must be a parallelogram in addition to having congruent diagonals. An isosceles trapezoid is a counterexample.

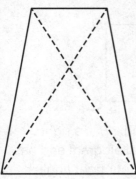

## Lesson 7-5

**1.** 60°  **3.** 24  **5.** 32°  **7.** 10  **9.** 21

**11.** Proof:

| Statements | Reasons |
|---|---|
| **1.** *ACDH* and *BCDF* are parallelograms, $\overline{BF} \cong \overline{AB}$ | **1.** Given |
| **2.** $\overline{BF} \cong \overline{CD}$; $\overline{CD} \cong \overline{AH}$ | **2.** Definition of parallelogram |
| **3.** $\overline{BF} \cong \overline{AH}$ | **3.** Transitive Property |
| **4.** $\overline{BC} \cong \overline{FD}$; $\overline{AC} \cong \overline{HD}$ | **4.** Definition of parallelogram |
| **5.** $BC = FD$, $AC = HD$ | **5.** Definition of congruent segments |
| **6.** $AC = AB + BC$, $HD = HF + FD$ | **6.** Segment Addition Postulate |
| **7.** $AB + BC = HF + FD$ | **7.** Substitution Property |
| **8.** $AB + FC = HF + FD$ | **8.** Substitution Property |
| **9.** $AB = HF$ | **9.** Subtraction Property of Equality |
| **10.** $\overline{AB} \cong \overline{HF}$ | **10.** Definition of congruent segments |
| **11.** $\overline{AH} \cong \overline{BF}$, $\overline{AB} \cong \overline{HF}$ | **11.** Substitution Property |
| **12.** *ABFH* is a rhombus | **12.** Definition of rhombus |

**13.** Proof:

| Statements | Reasons |
|---|---|
| **1.** $\overline{WZ} \parallel \overline{XY}$, $\overline{WX} \parallel \overline{ZY}$, $\overline{WX} \cong \overline{ZY}$ | **1.** Given |
| **2.** *WXYZ* is a parallelogram. | **2.** Both pairs of opposite sides are parallel. |
| **3.** *WXYZ* is a rhombus. | **3.** If one pair of consecutive sides of a parallelogram are congruent, then the parallelogram is a rhombus. |

**15.** Sample answer: Since consecutive sides are congruent, the garden is a rhombus. Jorge needs to know if the diagonals of the garden are congruent in order to determine whether it is a square.

**17.** Rectangle, rhombus, square; the four sides are congruent and consecutive sides are perpendicular.  **19.** Rhombus; all sides are congruent and the diagonals are perpendicular, but not congruent.

**21.** Rhombus, rectangle, square; all sides are congruent and the diagonals are perpendicular and congruent.  **23.** 55  **25.** 31  **27.** 18  **29.** 90°  **31.** 6  **33.** $3\sqrt{2}$  **35.** 45°

**37.** The figure consists of 5 congruent rhombi.

**39.** square  **41.** Proof:

| Statements | Reasons |
|---|---|
| **1.** *RSTU* is a parallelogram, $\overline{RS} \cong \overline{ST}$ | **1.** Given |
| **2.** $\overline{RS} \cong \overline{UT}$, $\overline{RU} \cong \overline{ST}$ | **2.** Definition of a parallelogram |
| **3.** $\overline{UT} \cong \overline{RS} \cong \overline{ST} \cong \overline{RU}$ | **3.** Substitution Property |
| **4.** *RSTU* is a rhombus | **4.** Definition of a rhombus |

**43.**

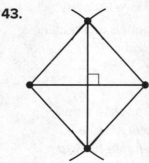

Sample answer: The diagonals bisect each other, so the quadrilateral is a parallelogram. Since the diagonals of the parallelogram are perpendicular to each other, the parallelogram is a rhombus.   **45.** right triangles

**47.** Parallelogram: Opposite sides of a parallelogram are parallel and congruent. Opposite angles of a parallelogram are congruent. The diagonals of a parallelogram bisect each other and each diagonal separates a parallelogram into two congruent triangles.

Rectangle: A rectangle has all the properties of a parallelogram. A rectangle has four right angles. The diagonals of a rectangle are congruent.

Rhombus: A rhombus has all the properties of a parallelogram. All sides of a rhombus are congruent. The diagonals of a rhombus are perpendicular and bisect the angles of the rhombus.

Square: A square has all of the properties of a parallelogram. A square has all the properties of a rectangle. A square has all of the properties of a rhombus.   **49.** True; sample answer: A rectangle is a quadrilateral with four right angles and a square is both a rectangle and a rhombus, so a square is always a rectangle.

Converse: If a quadrilateral is a rectangle then it is a square. False; sample answer: A rectangle is a quadrilateral with four right angles. It is not necessarily a rhombus, so it is not necessarily a square.

Inverse: If a quadrilateral is not a square, then it is not a rectangle. False; sample answer: A quadrilateral that has four right angles and two pairs of congruent sides is not a square, but it is a rectangle.

Contrapositive: If a quadrilateral is not a rectangle, then it is not a square. True; sample answer: If a quadrilateral is not a rectangle, it is also not a square by definition.

## Lesson 7-6

**1a.** 9   **1b.** 74°   **1c.** 110   **3.** 112°   **5.** 28

**7.** 13   **9.** 20.6   **11.** 101°   **13a.** $\sqrt{65}$   **13b.** 42.2

**15.** 15   **17.** 100°   **19.** 17.5 ft   **21a.** rectangle

**21b.** 420 ft   **21c.** 10,400 ft$^2$   **23.** Never; exactly two pairs of adjacent sides are congruent.

**25.** Sometimes; if the rectangle has 4 congruent sides, then it is a square. Otherwise, it is not a square.

**27.** 2   **29.** 20   **31.** 5

**33a.** Point $D$ should be moved to modify the design; (20, 6).

**33b.** Sample answer: The slope of $\overline{AB}$ is $\dfrac{34-20}{20-4} = \dfrac{7}{8}$, and the slope of $\overline{DC}$ is $\dfrac{20-6}{36-20} = \dfrac{7}{8}$ Using the Distance Formula, $\overline{AB} = \sqrt{(20-4)^2 + (34-20)^2} = 2\sqrt{113}$ and $DC = \sqrt{(36-20)^2 + (20-6)^2} = 2\sqrt{113}$ . By Theorem 7.12, $ABCD$ is a parallelogram.

**35.** Belinda; $m\angle D = m\angle B$. $m\angle A + m\angle B + m\angle C + m\angle D = 360°$ or $m\angle A + 100 + 45 + 100 = 360°$. $m\angle A = 115°$   **37.** A quadrilateral must have exactly one pair of sides parallel to be a trapezoid. If the legs are congruent, then the trapezoid is an isosceles trapezoid. If a quadrilateral has exactly two pairs of consecutive congruent sides with the opposite sides not congruent, the quadrilateral is a kite. A trapezoid and a kite both have four sides. In a trapezoid and isosceles trapezoid, both have exactly one pair of parallel sides.

## Module 7 Review

**1.** 120   **3.** 144   **5.** $m\angle BIF = 126°$, $m\angle JBC = 63°$, and $m\angle BJD = 117°$   **7.** (6, 4)
**9.** $m\angle WYX = 63°$ and $m\angle WVZ = 54°$   **11.** B
**13.** D   **15.** 94

## Module 8

## Module 8 Opener

1. $\frac{1}{4}$   3. $\frac{3}{5}$   5. $\frac{4}{5}$   7. 40 in.

## Lesson 8-1

1. enlargement; 3   3. reduction; $\frac{2}{3}$

5.

7a.

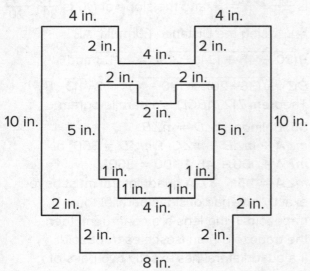

7b. 52 units

9.

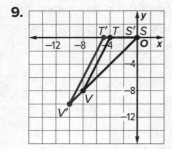

11.

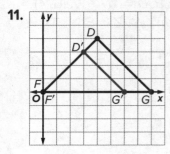

13.

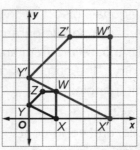

15. 1.5   17. enlargement; $\frac{3}{2}$

19. enlargement; $\frac{7}{3}$

21.

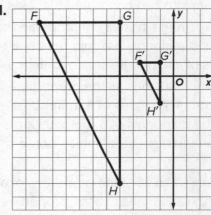

23.

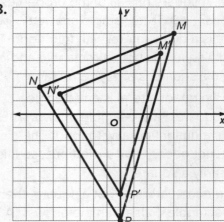

25. $\frac{2}{3}$   27. enlargement; 2

29. enlargement; $\frac{4}{3}$

31.

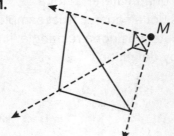

33a. The slope is $\frac{b}{a}$; Sample answer: The coordinates of $P'$ are $(ka, kb)$. The slope of $\overleftrightarrow{PP'}$ is $\frac{kb - b}{ka - a} = \frac{b(k-1)}{a(k-1)} = \frac{b}{a}$.

**33b.** If $P$ lies on the $y$-axis, then its image $P'$ will also lie on the $y$-axis. In this case, $\overleftrightarrow{PP'}$ will be vertical and the slope will be undefined.

**35.** $y = 4x - 3$

**37a.** remain invariant under the dilation.

**37b.** Always; Sample answer; Since the rotation is centered at $B$, point $B$ will always remain invariant under the rotation.

**37c.** Sometimes; Sample answer; If one of the vertices in on the $x$-axis, then that point will remain invariant under reflection. If two vertices are on the $x$-axis, then the two vertices located on the $x$-axis will remain invariant under refection.

**37d.** Never; When a figure is translated, all points move an equal distance. Therefore, no points can remain invariant under translation.

**37e.** Sometimes; Sample answer; If one of the vertices of the triangle is located at the origin, then that vertex would remain invariant under the dilation. If none of the points of $\triangle XYZ$ are located at the origin, then no points will remain invariant under the dilation.

**39.** Sample answer: Translations, reflections, and rotations produce congruent figures because the sides and angles of the preimage are congruent to the corresponding sides and angles of the image.

**41a.** Always; the line $y = cx$ passes through the origin and a dilation leaves lines through the center of dilation unchanged.

**41b.** Never; $A'B' = k(AB)$, and $k > 1$, so $A'B' \neq AB$. This means $\overline{AB}$ cannot be congruent to its image.

**41c.** Never; If the scale factor is between 0 and 1, then it is a rational number between 0 and 1, which means the perimeter of parallelogram $A'B'C'D'$ will always be less than the perimeter of $ABCD$.

## Module 8-2

**1.** $\angle A \cong \angle W$, $\angle B \cong \angle X$, $\angle C \cong \angle Y$, $\angle D \cong \angle Z$, $\frac{AB}{WX} = \frac{BC}{XY} = \frac{CD}{YZ} = \frac{AD}{WZ}$

**3.** $\angle F \cong \angle J$, $\angle G \cong \angle K$, $\angle H \cong \angle L$, $\frac{FG}{JK} = \frac{GH}{KL} = \frac{FH}{JL}$

**5.** No; $\angle W \cong \angle L$

**7.** No; $\frac{MN}{GH} \neq \frac{NP}{HJ}$   **9.** 5   **11.** about 158.2 inches

**13.** 64.5 cm   **15.** $\angle A \cong \angle F$, $\angle B \cong \angle G$, $\angle C \cong \angle H$, $\angle D \cong \angle J$, $\frac{AB}{FG} = \frac{BC}{GH} = \frac{CD}{HJ} = \frac{AD}{FJ}$

**17.** Yes; $\angle A \cong \angle D$, $\angle C \cong \angle F$, $\angle B \cong \angle E$, $\frac{AB}{DE} = \frac{BC}{EF} = \frac{CA}{FD}$; $\frac{3}{2}$   **19.** 3   **21.** 5   **23.** 18.9

**25.** $x = 63$, $y = 32$   **27.** No; none are similar.

**29.** $\frac{3}{2}$   **31.** Yes; the ratio of the longer dimensions of the rinks is $\frac{20}{17}$ and the ratio of the smaller dimensions of the rinks is $\frac{20}{17}$.

**33.** 4

**35.** Sample Answer:

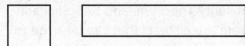

**37.** Sample answer: The figures could be described as congruent if they are the same size and shape, similar if their corresponding angles are congruent and their corresponding sides are proportional, and equal if they are the same exact figure.

**39.** No; he set up an incorrect proportion. The proportion should be $\frac{x}{12} = \frac{17}{8}$. Solving the proportion shows that the correct value of $x$ is 25.5.

## Module 8-3

**1.** Yes; $\triangle FGH \sim \triangle JHK$ by AA Similarity

**3.** No; The triangles would be similar by AA Similarity if $\overline{AB} \parallel \overline{DF}$.   **5.** 13.5 in.   **7.** 50 ft.

**9.** yes; AA   **11.** $\triangle ABC \sim \triangle DBE$; 16

**13.** $\triangle DEF \sim \triangle GEH$; 9   **15.** Sample answer: $m\angle ADB = 108$, so $m\angle DBA = 36$ (base of isosceles $\triangle ABD$). Thus, $m\angle ABC = 72$. Similarly, $m\angle DCB = 36$ and $m\angle ACB = 72$. So, $m\angle BEC = 72$ and $m\angle BAC = 36$. Therefore, $\triangle ABC$ and $\triangle BCE$ are similar.

**17.** The triangles are similar. It is given that $KM \perp JL$ and $JK \perp KL$. $\angle JKL \cong \angle JMK$ since they are both right angles. By the Reflexive Property, we know $\angle J \cong \angle J$. Therefore, by the AA Similarity Postulate, we can conclude $\triangle JKL \sim \triangle JMK$.

## Lesson 8-4

**1.** Yes; $\triangle RST \sim \triangle WSX$ (or $\triangle XSW$) by SAS Similarity  **3.** Yes; $\triangle STU \sim \triangle JPM$ by SAS Similarity  **5.** $\triangle HIJ \sim \triangle KLJ$; 5  **7.** $\triangle RST \sim \triangle UVW$; 11.25  **9.** 5 ft  **11.** 202.2 in.

**13.** No; SAS does not apply.

**15.** Sample answer: $m\angle TSU = m\angle QSR$ because they are vertical angles. $\frac{ST}{SQ}$ is proportional to $\frac{SU}{SR}$. Therefore, $\triangle STU$ and $\triangle SQR$ are similar.  **17.** 24 ft

**19.** Sample answer: The AA Similarity Postulate, SS Similarity Theorem, and SAS Similarity Theorem are all tests that can be used to determine whether two triangles are similar. The AA Similarity Postulate is used when two pairs of congruent angles of two triangles are given. The SSS Similarity Theorem is used when the corresponding side lengths of two triangles are given. The SAS Similarity Theorem is used when two proportional side lengths and the included angle of two triangles are given.  **21.** 6

## Lesson 8-5

**1.** 6  **3.** 12  **5.** yes; $\frac{PN}{NM} = \frac{QR}{RM} = \frac{1}{2}$
**7.** no; $\frac{PN}{NM} \neq \frac{QR}{RM}$  **9.** 50  **11.** 1.35  **13.** 1.12 km  **15.** 34.5 ft  **17.** 18  **19.** 5.6
**21.** yes; $\frac{AD}{DB} = \frac{AE}{EC} = \frac{5}{4}$  **23.** yes; $\frac{AE}{EC} = \frac{AD}{DB} = \frac{2}{1}$
**25.** 18  **27.** 20  **29.** $x = 2$, $y = 4$

**31a.** Since $\frac{AD}{DB} = \frac{AE}{EC}$, an equivalent proportion is $\frac{DB}{AD} = \frac{EC}{AE}$. Now add 1 to each side of the proportion as follows: $\frac{DB}{AD} + \frac{AD}{AD} = \frac{EC}{AE} + \frac{AE}{AE}$. Therefore, $\frac{DB + AD}{AD} = \frac{EC + AE}{AE}$. By the Segment Addition Postulate, this is equivalent to $\frac{AB}{AD} = \frac{AC}{AE}$.  **31b.** Since $\frac{AB}{AD} = \frac{AC}{EC}$ and $\angle A \cong \angle A$ by the Reflexive Property of Congruence, $\triangle ADE \sim \triangle ABC$ by the SAS Similarity Theorem. Therefore, $\angle ADE \cong \angle ABC$ since they are corresponding angles; so $\overline{DE} \parallel \overline{BC}$, since if corresponding angles are congruent, then the lines are parallel.

**33.**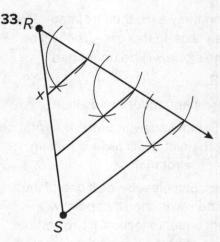

**35.** 39 cm; Since $J$, $K$, and $L$ are midpoints of their respective sides, $JK = \frac{1}{2}QR$, $KL = \frac{1}{2}PQ$, and $JL = \frac{1}{2}PR$ by the Triangle Midsegment Theorem. So $JK + KL + JL = \frac{1}{2}(QR + PQ + PR) = \frac{1}{2}(78) = 39$.  **37a.** $ST = 8$  $UV = 4$  $WX = 2$

**37b.** Based on the pattern, the length of the midsegment of $\triangle WXR = 1$.

**39.** Always; Sample answer: $FH$ is a midsegment. Let $BC = x$, then $FH = \frac{1}{2}x$. FHCB is a trapezoid, so $DE = \frac{1}{2}(BC + FH) = \frac{1}{2}\left(x + \frac{1}{2}x\right) = \frac{1}{2}x + \frac{1}{4}x = \frac{3}{4}x$. Therefore, $DE = \frac{3}{4}BC$.  **41.** By Corollary 8.1, $\frac{a}{b} = \frac{c}{d}$.

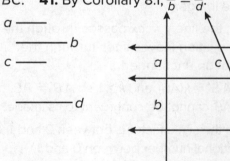

**43.** Mark the midpoint of each side of $\triangle PQR$. Connect these points to form $\triangle XYZ$. Since each segment of $\triangle XYZ$ is a midsegment of $\triangle PQR$, its length will be half the corresponding side, and therefore the perimeter will be half the perimeter of $\triangle PQR$.

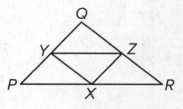

## Module 8-6

**1.** 16.5  **3.** 11  **5.** 8.4  **7.** 10 ft  **9.** $17\frac{1}{3}$

**11.** 5.1 cm  **13.** 15  **15.** 2:1

**17.** Sample answer: Origami is the art of paper folding. Using the sample drawing, $\overline{SR}$ is the bisector of $\angle R$, so $\frac{PS}{QS} = \frac{PR}{QR} = \frac{2.5}{2} = 1.25$. Therefore, $\frac{PS}{QS} > 1$, which means $PS > QS$.

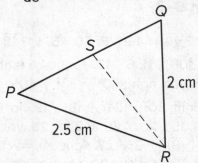

**19.** No; the triangle may be just isosceles.

**21.** $\frac{AB}{BC} = \frac{XW}{YZ}$, but $\triangle ABC$ is not similar to $\triangle XYZ$.

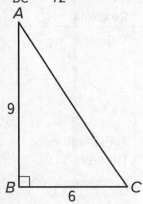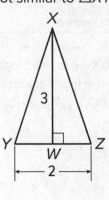

## Module 8 Review

**1.** D  **3.** $A'$(4, 3.2), $B'$(8, 6.4), and $C'$(16, 0)

**5.** $\angle F$  **7.** 21  **9.** Sample answer: Yes, $\triangle ABC$ is similar to $\triangle DEF$. The measure of angle $C$ is $180 - 44 - 56 = 80$ degrees, so angle $C$ is congruent to angle $F$. Thus the triangles are similar by the Angle-Angle Similarity Postulate.

**11.** Sample answer: Corresponding angles are not congruent.  **13.** A, B  **15.** A

## Module 9

### Module 9 Opener

**1.** $x = 18$ **3.** 10 **5.** 14 **7.** 17

### Lesson 9-1

**1.** $\sqrt{24}$ or $2\sqrt{6} \approx 4.9$ **3.** 10 **5.** $\sqrt{51} \approx 7.1$
**7.** $\triangle ACB \sim \triangle CDB \sim \triangle ADC$
**9.** $\triangle EGF \sim \triangle GHF \sim \triangle EHG$
**11.** $x = \sqrt{184}$ or $2\sqrt{46} \approx 13.6$; $y = \sqrt{248}$ or $2\sqrt{62} \approx 15.7$; $z = \sqrt{713} \approx 26.7$
**13.** $x = 4.5$; $y = \sqrt{13} \approx 3.6$; $z = 6.5$
**15.** 7.2 ft **17.** 11
**19.** Given: $\triangle ADC$ is a right triangle. $\overline{DB}$ is an altitude of $\triangle ADC$.
Prove: $\dfrac{AB}{DB} = \dfrac{DB}{CB}$

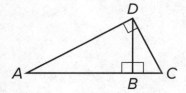

Proof:
It is given that $\triangle ADC$ is a right triangle and $\overline{DB}$ is an altitude of $\triangle ADC$. $\angle ADC$ is a right angle by the definition of a right triangle. Therefore, $\triangle ADB \sim \triangle DCB$, because if the altitude is drawn from the vertex of the right angle to the hypotenuse of a right triangle, then the two triangles formed are similar to the given triangle and to each other. So $= \dfrac{AB}{DB} = \dfrac{DB}{CB}$ by definition of similar triangles.
**21.** 2.88 mi **23.** $x = 5.2$, $y = 6.8$, $z = 11$
**25.** Sample answer: 9 and 4, 8 and 8; In order for two whole numbers to result in a whole-number geometric mean, their product must be a perfect square.
**27a.** Never; sample answer: The geometric mean of two consecutive integers is $\sqrt{x(x + 1)}$, and the average of
two consecutive integers is $\dfrac{x + (x + 1)}{2}$. If you set the two expressions equal to each other, the equation has no solution.

**27b.** Always; sample answer: Since $\sqrt{ab}$ is equal to $\sqrt{a} \cdot \sqrt{b}$, the geometric mean for two perfect squares will always be the product of two positive integers, which is a positive integer.
**27c.** Sometimes; sample answer: When the product of two integers is a perfect square, the geometric mean will be a positive integer.

### Lesson 9-2

**1.** $\sqrt{18}$ or $3\sqrt{2} \approx 4.2$ **3.** 60 **5.** $\sqrt{1345} \approx 36.7$
**7.** 15 **9.** 100 **11.** 6 **13.** about 4.6 feet high **15.** yes, right; $50^2 = 30^2 + 40^2$
**17.** yes, right; $30^2 = 24^2 + 18^2$ **19.** no; $6 + 12 = 18$ **21.** 15 **23.** $4\sqrt{6} \approx 9.8$ **25.** yes, right; $(\sqrt{12})^2 = (\sqrt{8})^2 + 2^2$ **27.** right; $XY = \sqrt{8}$, $YZ = \sqrt{2}$, $XZ = \sqrt{10}$ **29.** obtuse; $XY = 5$, $YZ = 2$, $XZ = \sqrt{41}$; $(\sqrt{41})^2 > 5^2 + 2^2$
**31.** Proof:

| Statements | Reasons |
|---|---|
| **1.** In $\triangle ABC$, $c^2 < a^2 + b^2$ where $c$ is the length of the longest side. In $\triangle PQR$, $\angle R$ is a right angle. | **1.** Given |
| **2.** $a^2 + b^2 = c^2$ | **2.** Pythagorean Theorem |
| **3.** $c^2 < x^2$ | **3.** Substitution Property |
| **4.** $c < x$ | **4.** A property of square roots. |
| **5.** $m\angle R = 90°$ | **5.** Definition of right angle |
| **6.** $m\angle C < m\angle R$ | **6.** Converse of the Hinge Theorem |
| **7.** $m\angle C < 90°$ | **7.** Substitution Property |
| **8.** $\angle C$ is an acute angle | **8.** Definition of acute angle |
| **9.** $\triangle ABC$ is an acute triangle. | **9.** Definition of acute triangle |

**33.** 30 ft **35.** 2120 **37.** P = 36 units; A = 60 units$^2$ **39.** 15 **41a.** Because the height of the tree is 20 m, $JL = 20 - x$. By the Pythagorean Theorem, $16^2 + x^2 = (20 - x)^2$.

**41b.** $16^2 + x^2 = (20 - x)^2$, so $256 + x^2 = 400 - 40x + x^2$. Subtracting $x^2$ from both sides gives $256 = 400 - 40x$. Therefore, $-144 = -40x$. Dividing both sides by $-40$ gives $3.6 = x$. The stump of the tree is 3.6 meters tall.   **43.** 10   **45.** $\frac{1}{2}$

**47.** 12.04 or about 12 miles

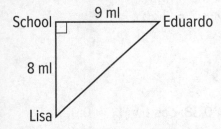

**49a.** Sample answer: $a^2 + b^2 = (m^2 - n^2)^2 + (2mn)^2 = m^4 - 2m^2n^2 + n^4 + 4m^2n^2 = m^4 + 2m^2n^2 + n^4$ and $c^2 = (m^2 + n^2)^2 = m^4 + 2m^2n^2 + n^4$. This means that $a^2 + b^2 = c^2$, so that $a$, $b$, and $c$ do form the sides of a right triangle by the converse of the Pythagorean Theorem.

**49b.**

| m | n | a | b | c |
|---|---|----|----|----|
| 2 | 1 | 3 | 4 | 5 |
| 3 | 1 | 8 | 6 | 10 |
| 3 | 2 | 5 | 12 | 13 |
| 4 | 1 | 15 | 8 | 17 |
| 4 | 2 | 12 | 16 | 20 |
| 4 | 3 | 7 | 24 | 25 |
| 5 | 1 | 24 | 10 | 26 |

**49c.** Sample answer: Take $m = 24$ and $n = 7$ to get $a = 527$, $b = 336$, and $c = 625$.

**51.**

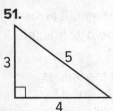

Right; sample answer: If you double or halve the side lengths, all three sides of the new triangles are proportional to the sides of the original triangle. Using the Side-Side-Side Similarity Theorem, you know that both of the new triangles are similar to the original triangle, so they are both right.

**53.** Sample answer: Incommensurable magnitudes are magnitudes of the same kind that do not have a common unit of measure. Irrational numbers were invented to describe geometric relationships, such as ratios of incommensurable magnitudes that cannot be described using rational numbers. For example, to express the measures of the sides of a square with an area of 2 square units, the irrational number $\sqrt{2}$ is needed.

## Lesson 9-3

**1.**

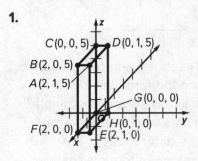

**3.**

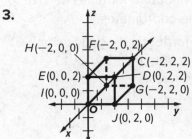

**5.**

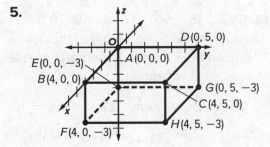

**7.** $\sqrt{29}$   **9.** $\sqrt{46}$   **11.** 20.2 mi   **13.** $(1, -1, -2)$

**15.** $\left(4, \frac{7}{2}, \frac{17}{2}\right)$   **17.** $\left(5, 4, \frac{19}{2}\right)$

**19.** $PQ = \sqrt{36}$ or 6; $(-3, -1, 1)$

**21.** $FG = \sqrt{10}$; $\left(\frac{3}{10}, \frac{3}{2}, \frac{2}{5}\right)$

**23.** $BC = \sqrt{39}$; $-\left(\frac{\sqrt{3}}{2}, 3, 3\sqrt{2}\right)$

**25.**

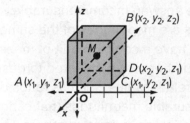

In $\triangle ACD$, $DC = (x_2 - x_1)$ and $AC = (y_2 - y_1)$. By the Pythagorean Theorem $(AD)^2 = (DC)^2 + (AC)^2$. Thus, $(AD)^2 = (x_2 - x_1)^2 + (y_2 - y_1)^2$. In $\triangle ADB$, $BD = (z_2 - z_1)$. By the Pythagorean Theorem, $(AB)^2 = (AD)^2 + (BD)^2$. Thus, $(AB)^2 = [(x_2 - x_1)^2 + (y_2 - y_1)^2] + (z_2 - z_1)^2$. Therefore, $AB = \sqrt{(x_2 - x_1)^2 + (y_2 - y_1)^2 + (z_2 - z_1)^2}$.

The formulas for the coordinate plane involve two coordinates and the formulas for three-dimensional space involve three coordinates. Both distance formulas involve square root of the squares of the differences of the coordinates. Both midpoint formulas involve the averages of the coordinates.

**27.** Telon; Camilla forgot the negative sign when subtracting $y_1$ from $y_2$. **29.** $2\sqrt{19}$

## Lesson 9-4

**1.** $7\sqrt{2}$ **3.** $6\sqrt{2}$ **5.** 10 **7.** $18\sqrt{2}$ **9.** $25\sqrt{2}$ **11.** 2 **13.** $5\sqrt{2}$ **15.** $2\sqrt{2}$ **17.** 16 **19.** 20 ft **21.** No; sample answer: The height of the frame is only 10 cm, and the height of the certificate is about 10.4 cm, so it will not fit.

**23.** $\frac{9\sqrt{2}}{2}$ **25.** $6\sqrt{2}$ or 8.5 cm **27.** $x = 8$; $y = 16$

**29.** $x = \frac{15\sqrt{3}}{2}$; $y = \frac{15}{2}$ **31.** $x = 24\sqrt{3}$; $y = 48$

**33.** $12\sqrt{3}$ or 20.8 ft **35.** $6\sqrt{2}$ yd or about 8.49 yd **37a.** $x$ feet **37b.** $\sqrt{3}$ ft **37c.** 20.5 ft **39.** $x = \frac{13\sqrt{2}}{2}$; $y = 45$ **41.** $x = 3$; $y = 1$ **43.** $x = 6\sqrt{3}$; $y = 3$ **45.** (6, 9)

**47.** $RS \approx 5.464$ ft; $ST \approx 9.464$ ft; $RT \approx 10.928$ ft **49.** 59.8 **51.** Carmen; Sample answer: Since the three angles of the larger triangle are congruent, it is an equilateral triangle and the right triangles formed by the altitude are 30°-60°-90° triangles. The hypotenuse is 6, so the shorter leg is 3 and the longer leg $x$ is $3\sqrt{3}$.

**53.** Sample Answer:

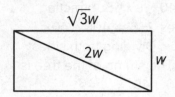

Let $l$ represent the length.
$l^2 + w^2 = (2w)^2$; $l^2 = 3w^2$; $l = w\sqrt{3}$

## Lesson 9-5

**1.** $\sin L = \frac{5}{13} \approx 0.38$; $\cos L = \frac{12}{13} \approx 0.92$; $\tan L = \frac{5}{12} \approx 0.42$; $\sin M = \frac{12}{13} \approx 0.92$; $\cos M = \frac{5}{13} \approx 0.38$; $\tan M = \frac{12}{5} \approx 2.4$ **3.** $\sin R = \frac{8}{17} \approx 0.47$; $\cos R = \frac{15}{17} \approx 0.88$; $\tan R = \frac{8}{15} \approx 0.53$; $\sin S = \frac{15}{17} \approx 0.88$; $\cos S = \frac{8}{17} \approx 0.47$; $\tan S = \frac{15}{8} \approx 1.88$ **5.** $\frac{1}{2}$; 0.5 **7.** $\frac{1}{2}$; 0.5 **9.** $\frac{\sqrt{3}}{3}$; 0.58 **11.** 22.55 **13.** 24.15 **15.** 15.5 ft **17.** 33.6° **19.** 35.5° **21.** 67.0° **23.** $WX = 15.1$; $XZ = 9.8$; $m\angle W = 33$ **25.** $RT = 3.7$; $ST = 5.9$; $m\angle R = 58$ **27.** $NQ = 25.5$; $MQ = 18.0$; $m\angle N = 45$ **29.** 56.3° **31.** 18° **33.** 28.53 cm; 23.46 cm² **35a.** $\cos 26°$ or $\sin 64°$ **35b.** $\cos 64°$ or $\sin 26°$ **35c.** $\tan 64°$ **37.** Sample answer: The sine of an angle equals the cosine of its complement, so. $\sin \beta = \cos \alpha = 0.7660$. Similarly, $\cos \beta = \sin \alpha = 0.6428$. **39.** $x = 18.8$; $y = 25.9$

**41.** $x = 9.2$; $y = 11.7$

**43.** Yes, they are both correct. Sample answer: Because $27 + 63 = 90$, the sine of 27° is the same ratio as the cosine of 63°.

**45.** Sample answer: Yes; since the values of sine and cosine are both calculated by dividing one of the legs of a right triangle by the hypotenuse, and the hypotenuse is always the longest side of a right triangle, then values will always be less than 1. You will always be dividing the smaller number by the larger number.

**47.** Sample answer: To find the measure of an acute angle of a right triangle, you can find the ratio of the leg opposite the angle to the hypotenuse and use a calculator to find the inverse sine of the ratio; you can find the ratio of the leg adjacent to the angle to the hypotenuse and use a calculator to find the inverse cosine of the ratio; or you can find the ratio of the leg opposite the angle to the leg adjacent to the angle and use a calculator to find the inverse tangent of the ratio.

## Lesson 9-6

**1.** 674.5 ft   **3.** about 21 ft   **5.** about 57.7°
**7.** 12°   **9.** 14.8°   **11.** 10.4 feet tall   **13.** 19 ft
**15.** 62.3 units$^2$   **17.** 71.5 units$^2$   **19.** 10.0 ft$^2$
**21.** 32.9 cm$^2$   **23.** 106.4 ft$^2$   **25.** 41,028 ft
**27.** about 35°   **29.** 365 m   **31.** 4.8 cm$^2$
**33.** 9.0 ft$^2$   **35.** 22.4 ft$^2$   **37.** Sample answer: What is the relationship between the angle of elevation and angle of depression?
**39.** 7.8

## Lesson 9-7

**1.** $x \approx 102.1$   **3.** $x \approx 22.9$   **5.** $x \approx 4.1$
**7.** $x \approx 22.8$   **9.** $x \approx 15.1$   **11.** $x \approx 2.0$
**13.** 6.2 mi   **15.** two solutions; $B \approx 64°$, $C \approx 66°$, $c \approx 40.4$; $B \approx 116°$, $C \approx 14°$, $c \approx 11.0$   **17.** one solution; $B \approx 34°$, $C \approx 21°$, $c \approx 9.6$   **19.** one solution; $B = 90°$, $C = 60°$, $c \approx 3.5$   **21.** one solution; $B \approx 34°$, $C \approx 108°$, $c \approx 15.4$   **23.** one solution; $B \approx 35°$, $C \approx 12°$, $c \approx 2.6$   **25.** one solution; $B \approx 31°$, $C \approx 40°$, $c \approx 16.4$   **27a.** Definition of sine
**27b.** Multiplication Property   **27c.** Substitution
**27d.** Division Property   **29.** 24.3   **31a.** 402 m
**31b.** 676 m   **33a.** 24°   **33b.** 19.6 m
**35.** Sample answer: In 1464 Johann Muller first presented the Law of Sines in his De Triangulus Omnimodis trigonometry book.
**37.** 2; $b \sin A = 16 \sin 55°$, or about 13.1 Since $\angle A$ is acute, 14 < 16, and 14 > 13.1, the measures define 2 triangles.   **39.** 2; $b \sin A = 25 \sin 39°$, or about 15.7 Since $\angle A$ is acute, 22 < 25, and 25 > 15.7, the measures define 2 triangles.

**41.** 1; Since $\angle A$ is acute and $a = b = 10$, the measures define 1 triangle.
**43.** 0; $b \sin A = 17 \sin 52°$ or about 13.4. Since $\angle A$ is acute and 13 < 13.4, the measures define 0 triangles.   **45.** 2; $b \sin A = 15 \sin 33°$ or about 8.2. Since $\angle A$ is acute, 10 < 15, and 10 > 8.2, the measures define 2 triangles.
**47.** 25.4 m

**51.** Solution 1: $m\angle B \approx 70°$, $m\angle C \approx 68°$, $c \approx 20.9$; Solution 2: $m\angle B \approx 110°$, $m\angle C \approx 28°$, $c \approx 10.4$   **53a.** Sample answer: $a = 22$, $b = 25$, $m\angle A = 70°$   **53b.** Sample answer: $a = 25$, $b = 22$, $m\angle A = 95°$   **53c.** Sample answer: $a = 22$, $b = 30$, $m\angle A = 43°$

## Lesson 9-8

**1.** $x \approx 29.9$   **3.** $x \approx 74$   **5.** $x \approx 20$   **7.** 69°
**9.** $A \approx 41°$, $C \approx 54°$, $b \approx 6.1$   **11.** $c \approx 12.8$, $A \approx 67°$, $B \approx 33°$   **13.** $N = 42°$, $n \approx 35.8$, $m \approx 24.3$   **15.** 275.2   **17.** sines; $B = 142°$, $a \approx 21.0$, $b \approx 67.8$   **19.** sines; $B \approx 33°$, $C \approx 110°$, $c \approx 31.2$   **21a.** 11.3 yards   **21b.** 40.9°
**23.** 2.5 M   **25.** Sample answer: When solving a right triangle, you can use the Pythagorean Theorem to find missing side lengths and trigonometric ratios to find missing side lengths or angle measures. To solve any triangle, you can use the Law of Sines or the Law of Cosines, depending on what measures you are given.   **27.** 5.6

## Module 9 Review

**1.** A   **3.** 52   **5.** C(9, −7, 0)   **7.** 16.5   **9.** 8.7 feet
**11.** $\frac{\sqrt{2}}{2}$   **13.** 45 ft   **15.** D, E   **17.** C

# Module 10

## Module 10 Opener

**1.** $x = 5$   **3.** 9.2   **5.** 3

## Lesson 10-1

**1.** $\odot O$   **3.** $\overline{AB}$ and $\overline{CD}$   **5.** $\overline{PA}$, $\overline{PB}$, and $\overline{PC}$
**7.** $\overline{AB}$   **9.** 9 mm   **11.** Yes; all diameters of the same circle are congruent.   **13.** 9.5 m
**15.** 13 in.; 81.68 in.   **17.** 12.73 in.; 6.37 in.
**19.** 4.97 m; 2.49 m   **21.** 25.31 yd; 12.65 yd
**23.** 0.5   **25.** 6   **27.** 4.25 in.; 26.70 in.
**29.** 200 m; 100 m   **31.** 11.14$x$ cm; 5.57$x$ cm
**33.** congruent   **35.** neither   **37.** $9\sqrt{2}\pi$ in.
**39.** 11$\pi$ yd   **41.** 2$\pi$ cm   **45.** chord
**47.** 449.6 ft   **49.** 24 units   **51.** Sample answer: First apply the translation $(x, y) \rightarrow (x - 3, y - 4)$ to $\odot D$. The dilations should have a scale factor of 3 and be centered at $(-1, -1)$. The translation and dilation are similarity transformations, so $\odot D$ is similar to $\odot E$.

## Lesson 10-2

**1.** 80   **3.** 138   **5.** minor arc; 50°   **7.** major arc; 210°   **9.** semicircle; 180°   **11.** 108°
**13.** 180°   **15.** 500°   **17.** 130°   **19.** 320°
**21.** 2.09 yd   **23.** 12.57 ft   **25.** 13.09 in.
**27.** $\frac{\pi}{4}$ radians   **29.** $\frac{\pi}{2}$ radians   **31.** $\frac{5\pi}{4}$ radians
**33.** 270°   **35.** 150°   **37.** 15°   **39.** 2.79 in.
**41.** 5.24 in.   **43.** 98.4°   **45a.** 62.83 in.
**45b.** 7.85 in.   **45c.** 57.3°
**47.** Proof:

| Statements | Reasons |
|---|---|
| **1.** $\angle BAC \cong \angle DAE$ | **1.** Given |
| **2.** $m\angle BAC \cong m\angle DAE$ | **2.** Definition of congruent angles |
| **3.** $m\widehat{BAC} = m\widehat{BC}$ and $m\widehat{DAE} = m\widehat{DE}$ | **3.** Definition of arc measure |
| **4.** $m\widehat{BC} = m\widehat{DE}$ | **4.** Substitution Property |
| **5.** $\widehat{BC} \cong \widehat{DE}$ | **5.** Definition of congruent arcs |

**49.** 150°   **51.** 52   **53.** 128
**55a.** 10$\pi$ or about 31.4 ft   **55b.** Sample answer: She can find the circumference of the entire circle and then use the arc measure to find the appropriate fraction of the circle. If the radius of the circle is $r$ and the measure of the arc is $x$, then the arc length is $\frac{x}{360} \cdot 2\pi r$.
**57.** Never; obtuse angles intersect arcs that measure between 90° and 180°.
**59.** Selena; the circles are not congruent because they do not have congruent radii. So, the arcs are not congruent.
**61.** Sample answer

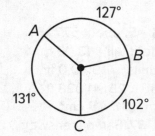

**63.** $m\widehat{LM} = 150$, $m\widehat{MN} = 90$, $m\widehat{NL} = 120$

## Lesson 10-3

**1.** 148   **3.** 82   **5.** 7   **7.** 2   **9.** 13   **11.** 5
**13.** 45   **15.** 4.47   **17.** 5.5   **19.** 4.8 cm
**21.** Proof:

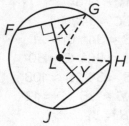

| Statements | Reasons |
|---|---|
| **1.** $\overline{LG} \cong \overline{LH}$ | **1.** All radii of a circle are congruent. |
| **2.** $\overline{LX} \perp \overline{FG}$, $\overline{LY} \perp \overline{JH}$, $\overline{LX} \perp \overline{LY}$ | **2.** Given |
| **3.** $\angle LXG$ and $\angle LYH$ are right angles. | **3.** Definition of perpendicular lines |
| **4.** $\triangle XGL \cong \triangle YHL$ | **4.** HL |
| **5.** $\overline{XG} \cong \overline{YH}$ | **5.** CPCTC |
| **6.** $XG = YH$ | **6.** Definition of congruent segments |

**7.** $2(XG) = 2(YH)$ | **7.** Multiplication Property of Equality

**8.** $\overline{LX}$ bisects $\overline{FG}$; $\overline{LY}$ bisects $\overline{JH}$ | **8.** $\overline{LX}$ and $\overline{LY}$ are contained in radii. A radius perpendicular to a chord bisects the chord.

**9.** $FG = 2(XG)$; $JH = 2(YH)$ | **9.** Definition of segment bisector

**10.** $FG = JH$ | **10.** Substitution

**11.** $\overline{FG} \cong \overline{JH}$ | **11.** Definition of congruent segments

**23.** Sample answer: Neil can draw a line perpendicular to the line he just drew through the mark he made. If the midpoint of the first line is also the midpoint of the second line, it is a diameter.

**25.** No; sample answer: In a circle with a radius of 12, an arc with a measure of 60 determines a chord of length 12. (The triangle related to a central angle of 60 is equilateral.) If the measure of the arc is tripled to 180, then the chord determined by the arc is a diameter and has a length of 2(12) or 24, which is not three times as long as the original chord.

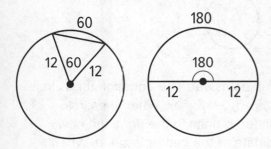

## Lesson 10-4

**1.** 72° **3.** 226° **5.** 81° **7.** 17° **9.** 43°

**11.** 58° **13.** 95° **15.** Proof: it is given that $m\angle T = \frac{1}{2}m\angle S$. This means that $m\angle S = 2m\angle T$. Since $m\angle S = \frac{1}{2}m\widehat{TUR}$ and $m\angle T = \frac{1}{2}m\widehat{URS}$, the equation becomes $\frac{1}{2}m\widehat{TUR} = 2\left(\frac{1}{2}m\widehat{URS}\right)$. Multiplying each side of the equation by 2 results in $m\widehat{TUR} = 2m\widehat{URS}$.

**17.** 39 **19.** 16 **21.** 58° **23.** 69° **25.** 105°
**27.** 72° **29.** 1.5 km **31.** 82° **33.** 158°
**35.** 65° **37.** 110° **39.** 98° **41.** Since *PQRS* is inscribed in a circle, $\angle Q$ is supplementary to $\angle S$ because they are opposite angles. Since $\angle Q \cong \angle S$, $\angle Q$ and $\angle S$ must be right angles. So, $\angle Q$ intercepts $\widehat{PSR}$, $m\angle Q = 90$ and $m\widehat{PSR} = 180$ by the Inscribed Angle Theorem. This means that $\widehat{PSR}$ is a semicircle, so $\overline{PR}$ must be a diameter. **43a.** $m\angle ABD = 90°$, $m\angle BAD = 41°$, and $m\angle BDA = 49°$
**43b.** 10.6 m

**43c.** Sample answer: If the radius is 7.1 m, then the diameter is 14.2 m. The diameter must be the longest chord in any circle, but 14.2 < 16, so the diameter would be short than the chord $\overline{AB}$. Therefore, the answer is not reasonable.

**43d.** Sample answer: The circumference of the circle is $(2\pi)(10.6) = 66.6$ m. $AD = 2(10.6) = 21.2$ m. $AB = 16$ m. $BD = \sqrt{21.2^2 - 16^2} = 13.9$m; the total is $66.6 + 21.2 + 16 + 13.9 = 117.7$ m.
**45.** By Theorem 10.7 $m\angle ADC = \frac{1}{2}m\widehat{ADC}$ and $m\angle ADC = x$. So $m\widehat{ADC} = 2x$.
**47.** $m\angle SRU = 80°$; $x = 12$ **49.** Always; rectangles have right angles at each vertex, therefore each pair of opposite angles will be supplementary and inscribed in a circle.

**51.** Sometimes; a rhombus can be inscribed in a circle as long as it is a square. Since the opposite angles of rhombi that are not squares are no supplementary, they cannot be inscribed in a circle.

**53.** $\frac{\pi}{2}$ **57.** Never: $\widehat{PR}$ is a minor arc, so $m\widehat{PR} < 180$. By the Inscribed Angle Theorem, $m\angle PQR < 90$. **59.** Always; each angle of the triangle measures 60°, so each intercepted arc measures 120°.

## Lesson 10-5

**1.** 4 **3.** 3 **5.** yes; $9^2 + 40^2 = 41^2$ **7.** yes; $14^2 + 48^2 = 50^2$ **9.** 25 **11.** 20 **13.** 15 **15.** 4
**17.** 3 **19.** 48° **21.** 80 **23.** 24 **25.** 4; 64
**27.** They are equal. **29a.** 5, 2, and 10
**29b.** 68 **31.** 8; 52 cm

**33.** Proof:

| Statements | Reasons |
|---|---|
| **1.** Quadrilateral *ABCD* is circumscribed about ⊙*P* | **1.** Given |
| **2.** Sides $\overline{AB}$, $\overline{BC}$, $\overline{CD}$, and $\overline{DA}$ are tangent to ⊙*P* at points *H*, *G*, *F*, and *E*, respectively | **2.** Definition of circumscribed |
| **3.** $\overline{EA} \cong \overline{AH}$; $\overline{HB} \cong \overline{BG}$; $\overline{GC} \cong \overline{CF}$; $\overline{FD} \cong \overline{DE}$ | **3.** Two segments tangent to a circle from the same exterior point are congruent. |
| **4.** $AB = AH + HB$; $BC = BG + GC$; $CD = CF + FD$; $DA = DE + EA$ | **4.** Segment Addition |
| **5.** $AB + CD = AH + HB + CF + FD$; $DA + BC = DE + EA + BG + GC$ | **5.** Substitution Property |
| **6.** $AB + CD = AH + BG + GC + FD$; $DA + BC = FD + AH + BG + GC$ | **6.** Substitution Property |
| **7.** $AB + CD = FD + AH + BG + GC$ | **7.** Commutative Property of Addition |
| **8.** $AB + CD = DA + BC$ | **8.** Substitution Property |

**35.** 8.06   **37.** spoke 10   **39.** 1916 mi;

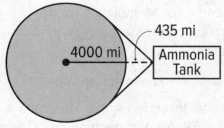

**41.** a square   **43.** By Theorem 10.12, if two segments from the same exterior point are tangent to a circle, then they are congruent. So, $\overline{XY} \cong \overline{XZ}$ and $\overline{XZ} \cong \overline{XW}$. Thus, $\overline{XY} \cong \overline{XZ} \cong \overline{XW}$.   **45.** Figure B

# Lesson 10-6

**1.** 42.5   **3.** 62°   **5.** 53°   **7.** 84   **9.** 148°
**11.** 99°   **13.** 40°   **15.** 48°   **17.** 264°   **19.** 128°

**21.** ≈ 60°   **23.** 9   **25.** 19   **27.** Find the difference of the two intercepted arcs and divide by 2.   **29.** 15

# Lesson 10-7

**1.** $x^2 + y^2 = 64$   **3.** $(x - 3)^2 + (y - 2)^2 = 4$
**5.** $(x - 3)^2 + (y + 4)^2 = 16$   **7.** $(x + 4)^2 + (y + 1)^2 = 20$   **9.** (0, 0); $r = 4$   **11.** (0, 0); $r = 2$
**13.** $(x - 2)^2 + \left(y - \dfrac{7}{2}\right)^2 = 169$   **15.** (0, 3), (−2.4, 1.8)   **17.** (−3, −4), (11, 10)   **19.** $(x + 13)^2 + (y - 6)^2 = 121$   **21.** Disagree; completing the square shows that the equation can be written as $x^2 + 4x + 4 + y^2 - 10y + 25 = k + 29$ or $(x + 2)^2 + (y - 5)^2 = k + 29$. The radius of the circle is $\sqrt{k + 29}$ and this expression results in a positive radius only when $k + 29 > 0$. So the equation represents a circle only if $k > -29$.
**23.** $(x - 4)^2 + (y - 3)^2 = 4$
**25.** 18 mi   **27.** $(x - 3)^2 + (y + 2)^2 = 4$

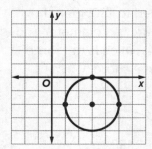

**29.** Sample answer: The equation for a circle is $(x - h)^2 + (y - k)^2 = r^2$. When the circle is translated *a* units to the right, the new *x*-coordinate of the center is $x + a$. When the circle is translated *b* units down, the new *y*-coordinate of the center is $y - b$. The new equation for the circle is $(x - (h + a))^2 + (y - (k - b))^2 = r^2$, or $(x - h - a)^2 + (y - k + b)^2 = r^2$.   **31.** Sample answer: (2, 1) falls exactly two units below the center point of (2, 3). This means that the radius of the circle is 2. If we count two units in any direction, the result will be a point on the circle. Up two units is (2, 5). To the left two units is (0, 3). To the right two units is (4, 3). These three points all have integer coordinates.

# Lesson 10-8

**1.** $y = \frac{1}{12}x^2$   **3.** $y = \frac{1}{20}x^2$   **5.** $x = \frac{1}{36}y^2$

**7.** $y^2 = 8x$   **9.** $(-1, 1), (2, 4)$   **11.** $(-2, -12), (0, 0)$

**13.** $y = -2x^2$

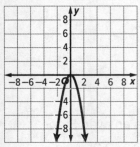

**15.** $x = \frac{1}{10}y^2$   **17.** $x = \frac{1}{3.6}y^2$

**19.** Since the directrix is vertical, the parabola must open left or right. The focus is $\left(-\frac{3}{4}, 0\right)$, which is to the left of the directrix, so the parabola opens to the left.

**21.** No; Sample answer: a parabola can have either a maximum or minimum value, but not both.

# Module 10 Review

**1.** D   **3.** B   **5.** A   **7.** D

**9.**

| Circle | Tangent? | |
|:---:|:---:|:---:|
| | yes | no |
| A. | X | |
| B. | | X |
| C. | | X |
| D. | X | |

**11.** A, E   **13.** D   **15.** $y = \frac{1}{20}x^2$

# Module 11

## Module 11 Opener

**1.** $x = 5$   **3.** 9.2   **5.** 3

## Lesson 11-1

**1.** 108 m²   **3.** 19.1 ft²   **5.** 273 mm²   **7.** 264 m²
**9.** 52.5 cm²   **11.** 480 m²   **13.** 10.6 cm,
31.7 cm   **15.** 4 m   **17.** 10 ft   **19.** 519.6 cm²
**21.** 120 m²   **23.** 33.8 in²   **25.** 136 in²
**27.** 50 in²   **29.** 11.6 m²   **31.** 15.5 in., 31.0 in.
**33.** 6 m; 24 m   **35.** 98 in²   **37.** 389.7 ft²
**39.** 24 units²   **41.** Sample answer: Using the
area formula for the given figure, you can plug
in the known area and all the other known
dimensions into the formula. Use algebraic
properties to solve for the missing variable
(dimension).   **43.** Pieces 3 and 4 are the largest;
area of top of cake = 30 in²   **45.** 8 in²
**47.** Always; Sample answer: If the areas are
equal, the perimeter of the nonrectangular
parallelogram will ways be greater because the
length of the side that is not perpendicular to the
height forms a right triangle with the height. The
height is a leg of the triangle and the side of the
parallelogram is the hypotenuse of the triangle.
Since the hypotenuse is always the longest side
of a right triangle, the non–perpendicular side
of the parallelogram is always greater than the
height. The bases of the quadrilaterals must be
the same because the areas and the heights
are the same. Since the bases are the same and
the height of the rectangles is also the length
of a side, the perimeter of the parallelogram will
always be greater.   **49.** 7.2   **51.** Sometimes;
Sample answer: If the areas are equal, it means
that the products of the diagonals are equal.
The only time that the perimeters will be equal is
when the diagonals are also equal, or when the
two rhombi are congruent.

## Lesson 11-2

**1.** center: point $Z$, radius: $\overline{ZY}$, apothem: $\overline{ZQ}$,
central angle: $<YZR$, 45°   **3.** 27.7 m²   **5.** 124.7 ft²

**7.** ≈ 181.0 ft²   **9.** 198 cm²   **11.** 49.5 cm²
**13.** 19.3 in.²   **15.** 1086.4 in²   **17.** 1089.8 in²
**19.** 112.5 m²   **21.** 128.1 ft²
**23.** Sample Answer:

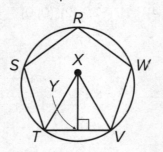

**25.** 52.0 in²   **27a.** Sample answer: Floor
area = $(7.5)(2.2) + (3.5)(4.2 - 2.2) + \frac{1}{2}$
$(7.5 - 3.5 - 2.0)(4.2 - 2.2) = 25.5$ m²;
$\frac{x}{1.0\ L} = \frac{25.5\ m^2}{4.5\ m^2}$, so $x = \frac{25.5\ m^2}{4.5\ m^2}$ (1.0 L) ≈ 5.7 L.
**27b.** Sample answer: Perimeter = 7.5 + 2.2
+ 2.0 + 2.8 + 3.5 + 4.2 = 22.2 m. Area to
be painted (ignoring door and windows) =
$(22.2)(2.6) = 57.72$ m². $\frac{x}{1.0}$ L = $\frac{57.72\ m^2}{7.5\ m^2}$, so
$x = \frac{57.72\ m^2}{7.5\ m^2} ≈ 7.7$ L.   **27c.** Sample answer:
Extra varnish might be needed; the estimate
might be high because of the door and
windows.
**29a.** Sample Answer:

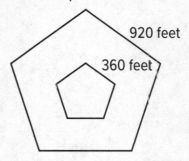

920 feet

360 feet

**29b.** ≈1,233,238.2 ft²   **31.** 754.4 in²
**33.** 591,137.7 ft²   **35.** 6.9 cm²   **37.** 3.87 cm²
**39.** Chloe; Sample answer: The measure of
each angle of a regular hexagon is 120°, so
the segments from the center of each vertex
form 60 angles. The triangles formed by the
segments from the center to each vertex are
equilateral, so each side of the hexagon is 11 in.
The perimeter of the hexagon is 66 in. Using
technology, the length of the apothem is about
9.5 in. Putting the values into the formula for
the area of a regular polygon and simplifying,
the area is about 313.5 in².

**41.** Sample Answer:

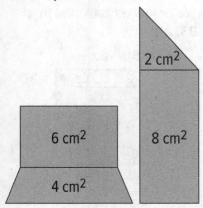

2 cm²

6 cm²

8 cm²

4 cm²

**43.** Sample answer: You can decompose the figure into shapes of which you know the area formulas. Then, you can add all of the areas to find the total area of the figure.

## Lesson 11-3

**1.** 153.9 m²   **3.** 346.4 m²   **5.** 176.7 in²
**7.** 10.9 mm   **9.** 6 in.   **11.** 15.0 in.   **13.** 1.8 m²
**15.** 331.4m²   **17.** 2520 cm²   **19.** 1625 mm²
**21.** 882 m²   **23.** 380.1 cm²   **25.** 1385.4 m²
**27.** 422.2 cm²   **29.** 21.1 in²   **31.** 8.1 ft
**33.** 14.0 in.   **35a.** 176.7 ft²   **35b.** 173.6 ft²
**37.** 88.4 cm²   **39.** The area of the circle is
36π. The area of the hexagon is $A = \frac{1}{2}aP$.
Since a regular hexagon can be divided into six equilateral triangles, use the 30°−60°−90° triangle ratios to find $a$ and the length of one side of the hexagon. So, $a = 3\sqrt{3}$ in., each side is 6 in., and $P = 36$ in. By substitution, the area of the hexagon is 93.5 in². The area of the shaded region is $36\pi - 93.5 \approx 19.6$ in².
**41.** 139.1 units²   **43.** 706.8 ft²   **45a.** red:
102.6 ft², purple: 94.2 ft², green: 84.8 ft²
**45b.** The largest sector was neither the one with the largest radius nor the one with the largest central angle. We can conclude that *both* factors affect the area of the sector.
**45c.** purple: 2.5 stars per square foot; green: 1.8 stars per square foot   **45d.** 2 × 1.8 = 3.6; 3.6(area of the red sector) ≈ 369 stars
**47.** 449.0 cm²   **49.** Sample answer: You can find the shaded area of the circle by subtraction x from 360 and using the resulting measure in the formula for the area of a sector. You could also find the shaded area by finding the area of the entire circle, finding the area

of the un-shaded sector using the formula for the area of a sector, and subtracting the area of the un−shaded sector from the area of the entire circle. The method in which you find the ratio of the area of a sector to the area of the whole circle is more efficient. It requires less steps, is faster, and there is a lower probability for error.   **51.** Sample answer: If the radius of the circle doubles, the area will not double. If the radius of the circle doubles, the area will be four times greater. Since the radius is squared, if you multiply the radius by 2, you multiply the area by 2², or 4. If the arc length of a sector is doubled, the area of the sector is doubled. Since the arc length is not raised to a power, if the arc length is doubled, the area would also be twice as large.

## Lesson 11-4

**1.** $L = 528$ yd²; $S = 768$ yd²   **3.** $L = 138$ cm²;
$S = 378$ cm²   **5.** $L = 324$ cm²; $S = 394.2$ cm²
**7.** $L \approx 377.0$ in²; $S \approx 603.2$ in²   **9.** $L \approx 226.2$ yd²;
$S \approx 282.7$ yd²   **11.** $L \approx 603.2$ in²;
$S \approx 1005.3$ in²   **13a.** $12x^2$   **13b.** 768 ft²
**15a.** $3x^2$   **15b.** 243 ft²   **17.** $S \approx 615.8$ in²
**19.** $S \approx 289.5$ mm²   **21.** 85,442.8 ft²
**23.** $S \approx 301.6$ cm²   **25.** $S \approx 972.0$ in²
**27.** 54 units²   **29.** $L = 1032$ cm²; $S = 1932$ cm²
(18 × 25 base); $L = 1332$ cm²; $S = 1932$ cm²
(25 × 12 base); $L = 1500$ cm²; $S = 1932$ cm²
(18 × 12 base)   **31.** $L \approx 996.0$ in²;
$S \approx 1686.0$ in²   **33.** $L \approx 34.7$ cm²; $S \approx 43.8$ cm²
**35a.** $S = 12.6$ cm²   **35b.** $S = 128$ ft²
**37.** 825 cm²   **39.** 2111.15 in²   **41a.** Sample answer: I drew a figure showing all of the surfaces of the sofa, using simple geometric figures with the dimensions labeled. To find the total surface area, I can find and add the areas of the simple figures.   **41b.** Sample answer:
Area = (66)(8 + 20 + 26 + 32) + 2[(30)(8)]
$+ 2[\frac{1}{2}(30 - 20)(32 - 8)] = 6396$ in²; I did not include the base in my calculation.   **43.** To find the surface area of any solid figure, find the area of the base (or bases) and add the area of the lateral faces of the figure. The lateral bases of a rectangular prism are rectangles. Since the bases of a cylinder are circles, the lateral face of a cylinder is a rectangle.

**45.** $\frac{\sqrt{3}}{2}\ell^2 + 3\ell h$; the area of an equilateral triangle of side $\ell$ is $\frac{\sqrt{3}}{4}\ell^2$ and the perimeter of the triangle is $3\ell$. So, the total surface area is $\frac{\sqrt{3}}{2}\ell^2 + 3\ell h$. **47.** Always; If the heights are radii are the same, the surface area of the cylinder will be greater since it has two circular bases and additional lateral area. **49.** Sample answer: a square pyramid with a base edge of 5 units and a slant height of 7.5 units

## Lesson 11-5

**1.** The square pyramid has 2 planes of symmetry. One plane of symmetry is perpendicular to the base of the pyramid and passes through the vertex. Another plane of symmetry is perpendicular to the base of the pyramid, perpendicular to the first plane of symmetry, and passes through the vertex.
**3.** The triangular prism has 1 plane of symmetry. The line of symmetry is parallel to the base of the prism and passes through the center of the prism. **5.** triangle **7.** triangle **9.** cylinder with a cylinder removed from the middle
**11.** cone **13.** yes **15.** The cylinder has infinite planes of symmetry. One plane of symmetry is parallel to the base of the cylinder and passes through the center of the cylinder. The other planes of symmetry are perpendicular to the base of the cylinder and passes through the center of the cylinder. **17.** The regular pentagonal prism has 6 planes of symmetry. One plane of symmetry is parallel to the base of the prism and passes through the center of the prism. The regular pentagonal prism has five planes of symmetry that are perpendicular to the base of the prism. Each plane of symmetry passes through a vertex and the midpoint of the opposite edge. **19.** rectangle
**21.** circle **23.**

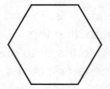

**25.** sphere **27.** yes **29.** yes **31.** Rotate a segment above or below a horizontal axis about the axis. **33.** Allen

**35a.** yes       **35b.** yes

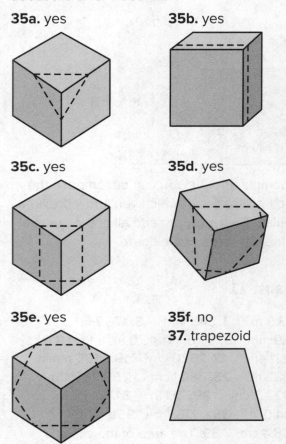

**35c.** yes       **35d.** yes

**35e.** yes       **35f.** no
                   **37.** trapezoid

**39a.** sphere; any cut **39b.** cylinder; cut parallel to base **39c.** cone; cut parallel to base
**41.** Equilateral triangular pyramid; Sample answer: Since the figure has axis symmetry of order 3, the base has to be an equilateral triangle. Because it does not have plan symmetry, you know that it is a pyramid instead of a prism. Therefore, the figure must be an equilateral triangular pyramid. **43.** The cross section is a triangle. There are six different ways to slice the pyramid so that two equal parts are formed because the figure has six planes of symmetry. In each case, the cross section is an isosceles triangle. Only the side lengths of the triangles change.

## Lesson 11-6

**1.** 2304 cm³ **3.** 90 m³ **5.** 2928.0 cm³
**7.** 1224 cm³ **9a.** $5.25x^3 + 3x^2$ **9b.** 54 in³
**11.** 66.7 ft³ **13.** 55.1 in³ **15a.** $0.75x^3 \sqrt{3}$
**15b.** 162.4 in³ **17.** 25 ft³ **19.** 301.1 in³
**21.** ≈ 14,508 in³ **23.** 74.7 cm³ **25.** 608.4 cm³

**27.** 156 cm³   **29a.** 900,000 ft³
**29b.** 883,573 ft³   **31a.** 1024 ft³   **31b.** 432 ft³
**31c.** 592 ft³   **31d.** about 26,640 pounds
**33a.** He can find the volume of the contents of each box by multiplying the length times the width times 2 less than the height of the box and then multiply by the mass per cubic centimeter of the contents.   **33b.** baking soda: volume = 8 · 4 · 10 = 320 cm³, weight = 320 · 2.2 = 704 grams; cereal: volume = 30 · 6 · 33 = 5940 cm³, weight = 5940 · 0.12 = 712.8 grams

**35a.** $V = \frac{1}{3} \cdot 2^2 \cdot 3 = 4$; the volume is 4 in³, and the price per cubic inch is $0.50.
**35b.** Sample answer: Doubling the length and the width multiplies the volume by 4. She would need to multiply the price by 4 and charge $8.00 for the larger box.   **35c.** Sample answer: base has sides of 3 inches, and the height is 2.5 inches; $V = \frac{1}{3} \cdot 3^2 \cdot 2.5 = 7.5$; the volume is 7.5 in³.   **37.** Francisco; Valerie incorrectly used $4\sqrt{3}$ as the length of one side of the triangular base. Francisco used a different approach, but his solution is correct.
**39.** Sample Answer:

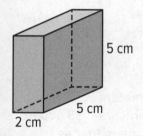

5 cm
5 cm
2 cm

**41.** The volume of the pyramid is $\frac{16 \cdot 12}{3}$ or 64 cubic units and the volume of the prism is 4 · 4 · 4 or 64 cubic units.

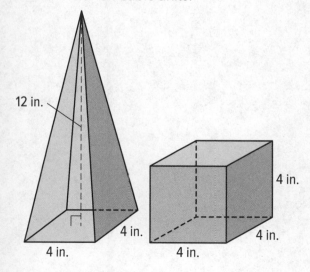

12 in.
4 in.
4 in.
4 in.
4 in.
4 in.

**43.** Because the volume of a pyramid is one-third the volume of the same prism, 3 of the pentagonal pyramids would fit in a prism of the same height.

## Lesson 11-7

**1.** 16,257.7 mm³   **3.** 1696.5 mm³   **5.** 141.4 in³
**7a.** $\pi(x + 2)^2(18)$ in³   **7b.** 2035.8 in³
**9.** 938.3 ft³   **11.** 871.3 in³   **13.** 452.4 cm³
**15.** 5.03 in³   **17.** 268.1 cm³   **19.** 1.8 yd³
**21.** 446,091.2 ft³   **23a.** $\frac{4}{3}\pi(x + 2)^3$ in³
**23b.** 33.51 in³   **25.** 1102.7 m³   **27.** 95.426 in³
**29.** $2\pi r^3 + \frac{2\pi r^3}{3} + \frac{\pi r^3}{3}$ or $3\pi r^3$ cubic units
**31.** 339.3 cm³   **33.** 2827.4 in³   **35.** 10,687.7 in³   **37.** 1210.6 mm³   **39.** 60.63 ft³   **41a.** about 382 in³; the cube root of 729 is 9, so the side length of the cube of wood in 9 inches. The largest sphere that could be carved has a radius of 4.5 inches. So, $V = \frac{4}{3}\pi(4.5)^3$ or about 382 in³.   **41b.** Yes. The sphere's radius is exactly one half of the side of the Cube. If the cube has a side $s$, then the volume of the sphere is $\frac{4}{3}\pi\left(\frac{s}{2}\right)^3 = \frac{4}{3}\pi\frac{s^3}{8} = \frac{s^3\pi}{6} = s^3\frac{\pi}{6}$. Therefore, the volume of a sphere that shares a diameter with the side of a cube is always is $\frac{\pi}{6}$ times the volume of the cube.   **43.** Sample answer: I would find the volume of the original cylinder. Then I would substitute that volume and the 8-inch height in the volume formula and solve for $r$; 3.4 inches   **45.** True; If two cylinders have the same height and the same lateral area, the circular bases must have the same area. Therefore, $\pi r^2 h$ is the same for each cylinder.   **47.** Sometimes; The statement is true if the base area of the cone is 3 times as great as the base area of the prism. For example, if the base of the prism has an area of 10 square units, then its volume is $10h$ cubic units. So, the cone must have a base area of 30 square units so that its volume is $\frac{1}{3}(30)h$ or $10h$ cubic units.   **49.** 1704 cm³; The volume of a cylinder is three times as much as the volume of a cone with the same radius and height.   **51.** 587.7 in³

**53.** Sample Answer:

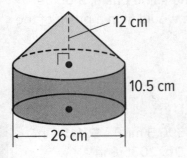

## Lesson 11-8

**1.** 320 m² **3.** 1843.2 m² **5.** $\frac{1}{3}$; 7 m **7.** $\sqrt{\frac{71}{16}}$ ;
4.5 in. **9.** 27.8 in² **11.** 5 in² **13.** small: 108 cm²;
large: 256 cm² **15.** small: 98.2 cm²; large:
785.4 **17.** 6144 ft³ of grain **19.** 27:125
**21.** 27:64 **23.** 3:2 **25.** 313 pieces of cake
**27a.** 2 in. **27b.** No, the larger sculpture has an
apothem of about 8.5 inches. This means that
the octagonal shape of the larger sculpture is
about 17 inches across. This is greater than the
diameter of the box. **29.** 6.75 in. **31.** 1.5:1
**33.** As the diameter increases, the volume
increases by the cube of the ratio of the
diameters. For example, from size Small to
Medium, the ratio of the diameters is $\frac{4.5}{3}$ or
1.5. The volume of the Medium is 4.5π(1.5)³ or
15.1875π.

| Size | Diameter (cm) | Volume (cm³) |
| --- | --- | --- |
| Small | 3 | 4.5π |
| Medium | 4.5 | 15.1875π |
| Large | 6.75 | 51.2578125π |

**35.** Neither; Sample answer: In order to find
the area of the enlarged circle, you can multiply
the radius by the scale factor and substitute it
into the area formula, or you can multiply the
area formula by the scale factor squared.
The formula for the area of the enlargement
is $A = \pi(kr)^2$ or $A = k^2r^2$. **37.** $P_{enlarged} = Q\sqrt{R}$
**39.** Sample answer: Since the ratio of the areas
should be 4:1, the ratio of the lengths of the
sides will be $\sqrt{4}:\sqrt{1}$ or 2:1. Thus, a
0.5-inch by 1-inch rectangle and a 1-inch by

2-inch rectangle are similar, and the ratio of
their areas is 4:1.

$A = 0.5$ in²     1 in.   $A = 2$ in²
0.5 in.       1 in.       2 in.

**41.** 8:135: The volume of Cylinder C is 8 times
the volume of Cylinder A, and the volume of
Cylinder D is 27 times the volume of Cylinder
B. If the original ratio of volumes was 1x:5x, the
new ratio is 8x:135x. So, the ratio of volumes is
8:135. **43.** 14 cm

## Lesson 11-9

**1.** ≈ 14,290.0 persons/mi² **3.** ≈ 13,525.6
persons/mi² **5.** Los Angeles **7.** 46.9 mi²
**9.** 20 lb/ft³ **11.** 41.67 persons/mi²
**13.** 19.75 persons/km² **15.** No; Sample
answer: Because the load will weigh 42,411.6
pounds, which exceeds the limit of 34,000
pounds. **17.** Answers will vary. **19.** Block B
has a greater density with 0.79 g/cm³ than
that of block A with a density 0.66 g/cm³.
**21.** pentagonal prism

## Module 11 Review

**1.** 480 **3.** 30 square centimeters
**5.** 942 square feet **7.** Lateral area: 33 square
inches Surface area: 55 square inches
**9.** A **11.** 480 **13.** A **15.** C **17.** B **19.** 1006

# Module 12

## Module 12 Opener

**1.** $\frac{5}{6}$ or 83%   **3.** $\frac{1}{6}$ or 17%   **5.** $\frac{1}{5}$ or 20%

**7.** $\frac{11}{20}$ or 55%

## Lesson 12-1

**1.** $S = \{H, T\}$   **3a.** $S = \{1, 2, 3, 4, 5, 6, 7, 8, 9, 10, 11, 12\}$   **3b.** $S$(even number on spinner) $= \{2, 4, 6, 8, 10, 12\}$

**5.** BB, BW, WB, WW

| Outcomes | Blue Pants | White pants |
|---|---|---|
| Blue Shirts | B, B | B, W |
| White Shirts | W, B | W, W |

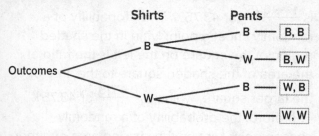

**7.** AB, AC, MB, MC

| Outcomes | Bottle | Cup |
|---|---|---|
| Apple juice | A, B | A, C |
| Milk | M, B | M, C |

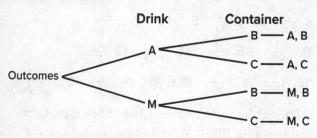

**9.** finite, $S = \{$yellow, blue, green, red$\}$

**11.** infinite; continuous

**13.** 28,800   **15.** 360   **17.** 55,296

**19.** 1296   **21a.** 4   **21b.** 12   **21c.** 16

**23.**

| | red | blue |
|---|---|---|
| glitter | R, G | B, G |
| beads | R, B | B, B |
| fasteners | R, F | B, F |

**25.** 2592

**27.** Sample answer: 6 different ways

$2(x + 4) + 4(x + 6) + 2(3);$

$2(x + 4) + 2(x + 6) + 2(x + 6) + 2(3);$

$2(x) + 2(4) + 4(x + 6) + 2(3);$

$2(x) + 2(4) + 2(x + 6) + 2(x + 6) + 2(3);$

$2(x) + 2(2) + 2(2) + 4(x + 6) + 2(3);$

$2(x) + 2(2) + 2(2) + 2(x + 6) + 2(x + 6) + 2(3)$

**29.** 12   **31a.** 5   **31b.** 18

**33.** $P = n^k$; Sample answer: The total number of possible outcomes is the product of the number of outcomes for each of the stages 1 through $k$. Since there are $k$ stages, you are multiplying $n$ by itself $k$ times which is $n^k$.

**35.** Sample answer: In an experiment, you choose between a blue box and a red box. You then remove a ball from the box that you chose without looking into the box. The blue box contains a red ball, a purple ball, and a green ball. The red box contains a yellow ball and an orange ball.

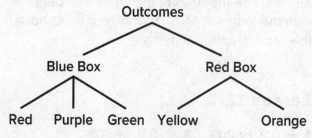

**37.** Sample answer: Never; the sample space is the set of all possible outcomes. An outcome cannot fall outside the sample space. A failure occurs when the outcome is in the sample space, but is not a favorable outcome.

## Lesson 12-2

**1.** $A \cap B = 6$   **3a.** 4, 10   **3b.** 4, 8, 12   **3c.** 4

**5.** $\frac{5}{26}$ or about 0.19   **7.** $\frac{8}{52}$ or $\frac{2}{13}$ or about 0.15

**9.** $\frac{3}{12}$ or $\frac{1}{4}$ or 0.25   **11.** $\frac{3}{12}$ or $\frac{1}{3}$ or about 0.33

**13a.** A, A, E, O, U   **13b.** J, J   **13c.** A, A, E, O, U, J, J   **15a.** 5, 10, 15, 20   **15b.** 1, 2, 3, 4, 5, 6, 7, 8, 9, 10, 11   **15c.** 1, 2, 3, 4, 5, 6, 7, 8, 9, 10, 11, 15, 20

**17.** $\frac{1}{5}$ or 0.2   **19.** $\frac{1}{2}$ or 0.5   **21.** $\frac{190}{200}$ or 0.95

**23.** $\frac{10}{100}$ or $\frac{1}{10}$ or 0.10  **25a.** $\frac{19}{120}$ or about 0.16

**25b.** $\frac{80}{120}$ or $\frac{2}{3}$ or about 0.67  **27a.** {Amy, Alex}

**27b.** $P$(one musical) $= 1 - P$(both) $= 1 - \frac{2}{9} = \frac{7}{9}$ or about 0.78  **29.** 0.3

**31.**

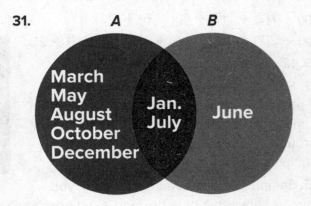

A          B

March
May
August        Jan.
October       July     June
December

**33.** Sometimes: Sample answer: While it is usually the case that the union of two sets will consist of more elements than their intersection, there are exceptions. For example, let $A$ be the days of the week that end in *day*, and let $B$ be the days of the week that begin with the letters *F, M, S, T,* or *W*. Both sets have the same number of items.

# Lesson 12-3

**1.** $\frac{1}{5}$, 0.2, or 20%  **3.** $\frac{1}{2}$, 0.5, or 50%

**5.** $\frac{5}{12} = 0.41\overline{6} \approx 42\%$  **7.** $\frac{1}{3} = 0.\overline{3}$ or about 33%

**9.** $\frac{49}{100}$, 0.49, or 49%  **11.** approximately 0.054, or about 5%

**13.** $\approx 0.054$ or about 5.4%

**15.** $\frac{\pi}{36} \approx 0.09$, or about 9%

**17.** $\frac{1}{9} = 0.\overline{1}$, or about 11%

**19.** $\frac{2}{9} = 0.\overline{2}$, or about 22%

**21.** $\frac{7}{12} = 0.583\overline{3}$, or about 58%

**23.** Sample Answer: rolling a 1 or 2

**25a.** Both buses leave from the same bus stop and would not require any additional travel time between them.

**25b.** 21%  **25c.** 75%

**27.** $\frac{\pi}{4} \approx 0.79$ or about 79%

**29.** $\frac{1}{3}$, $0.3\overline{3}$, or about 33%

**31.** $\frac{17}{35}$, $\approx 0.49$, or about 49%

**33a.** 0.842  **33b.** 0.205  **35.** 14.3%

**37.** No; Sample answer: Athletic events should not be considered random because there are other factors involved such as pressure and ability that have an impact on the success of the event.

**39.** Sample answer: The probability of a randomly chosen point lying in the shaded region of the square on th left is found by subtracting the area of the unshaded square from the area of the larger square and finding the ration of the difference of the areas to the area of the larger square. The probability is $\frac{1^2 - 0.75^2}{1^2}$ or 43.75%. The probability of a randomly chosen point lying in the shaded region of the square on the left is the ratio of the area of the shaded square to the area of the larger square, which is $\frac{0.4375}{1}$ or 43.75%. Therefore, the probability of a randomly chosen point lying in the shaded area of either square is the same.

# Lesson 12-4

**1.** $\frac{1}{66}$  **3.** $\frac{1}{2450}$  **5.** $\frac{1}{420}$  **7.** $\frac{1}{15,120}$

**9a.** $\frac{1}{35}$  **9b.** $\frac{1}{210}$  **11.** $\frac{1}{1140}$  **13.** $\frac{1}{252}$

**15.** $\frac{1}{495}$  **17.** $\frac{1}{20}$  **19a.** 10  **19b.** $\frac{1}{10}$  **21.** $\frac{1}{38,760}$

**23a.** $\frac{1}{30}$; Sample answer: There are 6 possible coupons for the first customer, 5 possible for the second customer, and so forth, so the total number of possible outcomes is 6! = 720. If the first customer gets the 10% coupon and the second customer gets the 25% coupon, then there are 4! = 24 ways the remaining four customers can get coupons so the total number of favorable outcomes is 24. The probability is $\frac{\text{number of favorable outcomes}}{\text{number of possible outcomes}}$ $= \frac{24}{720} = \frac{1}{30}$.

**23b.** $_6C_2 = 15$   **25.** $\dfrac{1}{100}$   **27.** $\dfrac{1}{39{,}916{,}800}$

**29.**
$$C(n, n - r) \overset{?}{=} C(n, r)$$

$$\dfrac{n!}{[n - (n - r)!](n - r)!} \overset{?}{=} \dfrac{n!}{(n - r)!r!}$$

$$\dfrac{n!}{r!(n - r)!} \overset{?}{=} \dfrac{n!}{(n - r)!r!}$$

$$\dfrac{n!}{(n - r)!r!} = \dfrac{n!}{(n - r)!r!} \checkmark$$

**31.** $\dfrac{13}{261}$

**33.** Sample answer: Both permutations and combinations are used to find the number of possible arrangements of a group of objects. The order of the objects is important in permutations, but not in combinations.

**35.** Sample answer:

$$r! \cdot {}_nC_r = r! \cdot \dfrac{n!}{(n - r)!r!}$$

$$= \dfrac{n!r!}{(n - r)!r!}$$

$$= \dfrac{n!}{(n - r)!}$$

$$= {}_nP_r$$

$_nC_r$ and $_nP_r$ differ by the factor $r!$ because there are always $r!$ ways to order the groups that are selected. Therefore, there are $r!$ permutations of each combination.

## Lesson 12-5

**1.** $\dfrac{1}{6}$ or about 17%   **3.** $\dfrac{1}{12}$ or about 8%

**5.** $\dfrac{6}{25}$ or 24%

**7.** dependent; Sample answer: Since the first ace drawn was not replaced, the probability of drawing the second card is affected.

**9.** independent; Sample answer: These two rolls have no bearing on each other.

**11.** $\dfrac{25}{2652}$ or about 1%

**13.** $\dfrac{1}{3540}$ or about 0.03%

**15.** $\dfrac{1}{5}$ or 0.2 or 20%

**17.** $\dfrac{11}{850}$ or 0.0129 or 1.29%   **19.** 0.43

**21.** Yes; Sample answer: The probability of selecting a defective chip from Box A is $\dfrac{1}{25}$. Since $\dfrac{1}{625} = \dfrac{1}{25} \cdot \dfrac{1}{25}$, $P(A \text{ and } B) = P(A) \cdot P(B)$ and the events are independent.

**23.** $\dfrac{11}{25}$ or 44%

**25.**

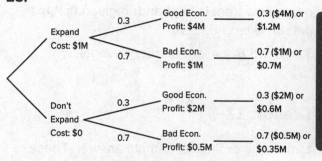

Sample answer: The expected value of choosing to expand is $1.2M + $0.7M or $1.9M, and the expected value of choosing not to expand is $0.95M. When we subtract the costs of expanding and of not expanding, we find the net expected value of expanding is $1.9M − $1M or $0.9M and the net expected value of not expanding is $0.95M − $0 or $0.95M. Since $0.9M < $0.95M, you should not expand the business.

**27.** $\dfrac{1}{2704}$; Sample answer: The card is replaced, so the events are independent; the probability of each picking the queen of spaces is $\dfrac{1}{52}$, so the probability of them both picking it is $\dfrac{1}{52} \cdot \dfrac{1}{52} = \dfrac{1}{2704}$.

**29a.** The first branch of the tree should be the first serve, while the second set of branches should be the second serve. The second serve only occurs when the first serve is a fault.

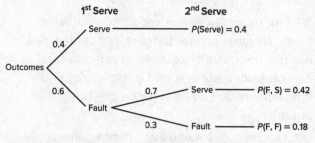

**29b.** A double fault is back-to-back faults, or $P(F, F)$, which equals 0.18 or 18%

**31.** Sample answer: In order for the events to be independent, two things must be true: 1) the chance that a person is left-handed is the same as the chance that a person is left-handed given that the person's parent is left-handed, and 2) the chance that a person's parent is left-handed is the same as the chance that a person's parent is left-handed given that the person is left-handed.

**33.** $A$ and $B$ are independent events.

# Lesson 12-6

**1.** mutually exclusive; Sample answer: These are mutually exclusive events because the posters are two different players.

**3.** Mutually exclusive; Sample answer: These are mutually exclusive events because a dog and a cat are two different animals.

**5.** mutually exclusive, 0.50   **7.** not mutually exclusive, 0.50   **9.** mutually exclusive, 0.10

**11.** 0.65   **13.** $\frac{1}{13}$ or 7.7%   **15.** $\frac{3}{13}$ or 23.1%

**17.** $\frac{4}{13}$ or about 31%   **19.** $\frac{4}{13}$ or about 31%

**21.** 0.30   **23.** $\frac{3}{4}$ or 75%   **25.** $\frac{739}{750}$; about 98.5%

**27.** 56%   **29.** Answers will vary.   **31.** 0.726

**33.** 0.6

**35.** 0.74; sample answer: There are three outcomes in which the values of two or more of the dice are less than or equal to 4 and one outcome where the values of all three of the dice are less than or equal to 4. You have to find the probability of each of the four scenarios and add them together.

**37.** Not mutually exclusive; sample answer: Since squares are rectangles, but rectangles are not necessarily squares, a quadrilateral can be both a square and a rectangle, and a quadrilateral can be a rectangle but not a square.

**39.** Not mutually exclusive; sample answer: A natural number is also a complex number.

**41a.** Sample answer: The winner being a junior or being a senior are mutually exclusive events, since the winner cannot be a junior and a senior. So, $P(junior\ or\ senior) = P(junior) + P(senior) = \frac{124}{540} + \frac{226}{540} = \frac{350}{540}$ or about 64.8%.

**41b.** There are two nonoverlapping circles because the Venn diagram represents two events that are mutually exclusive.

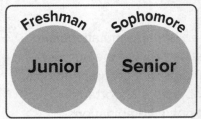

# Lesson 12-7

**1.** $\frac{8}{11}$ or about 73%   **3.** $\frac{1}{4}$ or 25%   **5.** $\frac{1}{5}$ or 20%

**7.** $\frac{1}{3}$   **9.** $\frac{1}{6}$   **11a.** $\frac{7}{10}$   **11b.** $\frac{1}{4}$   **11c.** $\frac{10}{19}$

**13a.** 6 students

**13b.** $P(H|L) = \frac{1}{3}$; $P(L) = \frac{3}{10}$ and $P(H\ and\ L) = \frac{1}{10}$.

So, $P(H|L) = \frac{P(H\ and\ L)}{P(L)} = \frac{\frac{1}{10}}{\frac{3}{10}}$ or $\frac{1}{3}$.

**13c.** No; sample answer: If $P(H|L)$ is the same as $P(H)$ or $P(L|H)$ is the same as $P(L)$, then $H$ and $L$ are independent. Since $P(H|L)$ is $\frac{1}{3}$ and $P(H)$ is $\frac{7}{10}$, $H$ and $L$ are not independent.

**15a.** 80%; $P(Algebra) = 0.25$ and $P(Algebra\ and\ Health) = 0.2$, so $P(Health|Algebra) = \frac{P(Algebra\ and\ Health)}{P(Algebra)} = \frac{0.2}{0.25} = 0.8$.

**15b.** No; $P(Health|Algebra) = 0.8$ and $P(Health) = 0.5$. Because the two probabilities are not equal, the two events are dependent.

**15c.** 25%; $P(Accounting) = 0.2$ and $P(Accounting\ and\ Spanish) = 0.05$, so $P(Spanish|Accounting) = \frac{P(Accounting\ and\ Spanish)}{P(Accounting)} = \frac{0.05}{0.2} = 0.25$. Because the events are independent, $P(Spanish|Accounting) = P(Spanish) = 0.25$.

# Lesson 12-8

**1a.**

|  | SUV | Truck | Totals |
|---|---|---|---|
| Males | 15 | 35 | 50 |
| Females | 40 | 10 | 50 |
| Totals | 55 | 45 | 100 |

**1b.** 50; marginal    **1c.** 10; joint

**1d.**

|  | SUV | Truck | Totals |
|---|---|---|---|
| Males | $\frac{15}{100} = 15\%$ | $\frac{35}{100} = 35\%$ | $\frac{50}{100} = 50\%$ |
| Females | $\frac{40}{100} = 40\%$ | $\frac{10}{100} = 10\%$ | $\frac{50}{100} = 50\%$ |
| Totals | $\frac{55}{100} = 55\%$ | $\frac{45}{100} = 45\%$ | $\frac{100}{100} = 100\%$ |

**3a.**

|  | More Than 4 | 4 or Fewer | Totals |
|---|---|---|---|
| In state | 928 | 332 | 1260 |
| Out of state | 118 | 622 | 740 |
| Totals | 1046 | 954 | 2000 |

**3b.**

|  | More Than 4 | 4 or Fewer | Totals |
|---|---|---|---|
| In State | $\frac{928}{2000} = 46.4\%$ | $\frac{332}{2000} = 16.6\%$ | $\frac{1260}{2000} = 63\%$ |
| Out of State | $\frac{118}{2000} = 5.9\%$ | $\frac{622}{2000} = 31.1\%$ | $\frac{740}{2000} = 37\%$ |
| Totals | $\frac{1046}{2000} = 52.3\%$ | $\frac{954}{2000} = 47.7\%$ | $\frac{2000}{2000} = 100\%$ |

**3c.** Sample answer: 63% of the college students surveyed are attending an in-state college.52.3% of the college students surveyed visited home more than 4 times. It would be expected that 63% · 52.3%, or about 33% would be in-state students that visited home more than 4 times. Since $P(A \text{ and } B) \neq P(A) \cdot P(B)$, or 33% ≠ 46.4%, the events are dependent.

**5a.**

|  | Play Sports | Do Not Play Sports | Totals |
|---|---|---|---|
| Play an Instrument | 8% | 32% | 40% |
| Do Not Play an Instrument | 48% | 12% | 60% |
| Totals | 56% | 44% | 100% |

**5b.** 80%    **5c.** 14.3%

**7.**

|  | Dog | Cat | Totals |
|---|---|---|---|
| 9th Grade | 35 | 20 | 55 |
| 10th Grade | 30 | 15 | 45 |
| Totals | 65 | 35 | 100 |

**9.**

|  | Yes | No | Totals |
|---|---|---|---|
| Males | 32% | 16% | 48% |
| Females | 38% | 14% | 52% |
| Totals | 70% | 30% | 100% |

**11a.**

| Device | Tablet | No Tablet | Totals |
|---|---|---|---|
| Smart Phone | 9 | 42 | 51 |
| No Smart Phone | 6 | 15 | 21 |
| Totals | 15 | 57 | 72 |

**11b.**

| Device | Tablet | No Tablet | Totals |
|---|---|---|---|
| Smart Phone | 12.5% | 58.3% | 70.8% |
| No Smart Phone | 8.3% | 20.8% | 29.2% |
| Totals | 20.8% | 79.2% | 100% |

**12.** 5%; P(has a smart phone and has a tablet) = 12.5%

**11c.** 17.7%; P(has a tablet I has a smart phone) = $\frac{P(\text{has a tablet and has a smart phone})}{P(\text{has a smart phone})} \approx \frac{0.125}{0.708}$ or 17.7%

Selected Answers

**13a.** voting for candidate A or B

**13b.** 55%  **13c.** 55%  **13d.** 9%

**13e.** Sample answer: males aged 18-30 and females aged 46-60

**13f.** Sample answer: females aged 18-30, males and females aged 31–45, and anyone over 60

## Module 12 Review

**1.** C  **3.** A  **5.** 86  **7.** A

**9.**

| Situation | Independent | Dependent |
|---|---|---|
| Two dice are tossed. | X | |
| Two marbles are selected from a bag. | | X |
| Two students are chosen as the captains of a team. | | X |
| A coin is tossed and a card is chosen. | X | |
| Three books are selected from the library. | | X |
| One student from each class is chosen to collect papers. | X | |

**11.** A, B  **13.** $\frac{2}{27}$  **15.** D

| English | Español |
|---|---|

## A

**30°-60°-90° triangle** (Lesson 9–4)   A right triangle with two acute angles that measure 30° and 60°.

**triángulo 30°-60°-90°**   Un triángulo rectángulo con dos ángulos agudos que miden 30° y 60°.

**45°-45°-90° triangle** (Lesson 9–4)   A right triangle with two acute angles that measure 45°.

**triángulo 45°-45°-90°**   Un triángulo rectángulo con dos ángulos agudos que miden 45°.

**accuracy** (Lesson 2–7)   The nearness of a measurement to the true value of the measure.

**exactitud**   La proximidad de una medida al valor verdadero de la medida.

**adjacent angles** (Lesson 2–1)   Two angles that lie in the same plane and have a common vertex and a common side but have no common interior points.

**ángulos adyacentes**   Dos ángulos que se encuentran en el mismo plano y tienen un vértice común y un lado común, pero no tienen puntos comunes en el interior.

**adjacent arcs** (Lesson 10–2)   Arcs in a circle that have exactly one point in common.

**arcos adyacentes**   Arcos en un circulo que tienen un solo punto en común.

**alternate exterior angles** (Lesson 3–7)   When two lines are cut by a transversal, nonadjacent exterior angles that lie on opposite sides of the transversal.

**ángulos alternos externos**   Cuando dos líneas son cortadas por un ángulo transversal, no adyacente exterior que se encuentran en lados opuestos de la transversal.

**alternate interior angles** (Lesson 3–7)   When two lines are cut by a transversal, nonadjacent interior angles that lie on opposite sides of the transversal.

**ángulos alternos internos**   Cuando dos líneas son cortadas por un ángulo transversal, no adyacente interior que se encuentran en lados opuestos de la transversal.

**altitude of a parallelogram** (Lesson 11–1)   A perpendicular segment between any two parallel bases.

**altitud de un paralelogramo**   Un segmento perpendicular entre dos bases paralelas.

**altitude of a prism or cylinder** (Lesson 11–4)   A segment perpendicular to the bases that joins the planes of the bases.

**altitud de un prisma o cilindro**   Un segmento perpendicular a las bases que une los planos de las bases.

**altitude of a pyramid or cone** (Lesson 11–4)   A segment perpendicular to the base that has the vertex as one endpoint and a point in the plane of the base as the other endpoint.

**altitud de una pirámide o cono**   Un segmento perpendicular a la base que tiene el vértice como un punto final y un punto en el plano de la base como el otro punto final.

**altitude of a triangle** (Lesson 6–3)   A segment from a vertex of the triangle to the line containing the opposite side and perpendicular to that side.

**altitud de triángulo**   Un segmento de un vértice del triángulo a la línea que contiene el lado opuesto y perpendicular a ese lado.

ambiguous case (Lesson 9–7)   When two different triangles could be created or described using the given information.

analytic geometry (Lesson 1–1)   The study of geometry that uses the coordinate system.

angle (Lesson 2–1)   The intersection of two noncollinear rays at a common endpoint.

angle bisector (Lesson 2–1)   A ray or segment that divides an angle into two congruent angles.

angle of depression (Lesson 9–6)   The angle formed by a horizontal line and an observer's line of sight to an object below the horizontal line.

angle of elevation (Lesson 9–6)   The angle formed by a horizontal line and an observer's line of sight to an object above the horizontal line.

angle of rotation (Lesson 2–4)   The angle through which a figure rotates.

apothem (Lesson 11–2)   A perpendicular segment between the center of a regular polygon and a side of the polygon or the length of that line segment.

approximate error (Lesson 2–7)   The positive difference between an actual measurement and an approximate or estimated measurement.

arc (Lesson 10–2)   Part of a circle that is defined by two endpoints.

arc length (Lesson 10–2)   The distance between the endpoints of an arc measured along the arc in linear units.

area (Lesson 2–3)   The number of square units needed to cover a surface.

auxiliary line (Lesson 5–1)   An extra line or segment drawn in a figure to help analyze geometric relationships.

axiom (Lesson 1–1)   A statement that is accepted as true without proof.

axiomatic system (Lesson 1–1)   A set of axioms from which theorems can be derived.

caso ambiguo   Cuando dos triángulos diferentes pueden ser creados o descritos usando la información dada.

geometría analítica   El estudio de la geometría que utiliza el sistema de coordenadas.

ángulo   La intersección de dos rayos no colineales en un extremo común.

bisectriz de un ángulo   Un rayo o segmento que divide un ángulo en dos ángulos congruentes.

ángulo de depresión   El ángulo formado por una línea horizontal y la línea de visión de un observador a un objeto por debajo de la línea horizontal.

ángulo de elevación   El ángulo formado por una línea horizontal y la línea de visión de un observador a un objeto por encima de la línea horizontal.

ángulo de rotación   El ángulo a través del cual gira una figura.

apotema   Un segmento perpendicular entre el centro de un polígono regular y un lado del polígono o la longitud de ese segmento de línea.

error aproximado   La diferencia positiva entre una medida real y una medida aproximada o estimada.

arco   Parte de un círculo que se define por dos puntos finales.

longitude de arco   La distancia entre los extremos de un arco medido a lo largo del arco en unidades lineales.

área   El número de unidades cuadradas para cubrir una superficie.

línea auxiliar   Una línea o segmento extra dibujado en una figura para ayudar a analizar las relaciones geométricas.

axioma   Una declaración que se acepta como verdadera sin prueba.

sistema axiomático   Un conjunto de axiomas de los cuales se pueden derivar teoremas.

**axis of a cone** (Lesson 11–4)   The segment with endpoints at the vertex and the center of the base.

**eje de un cono**   El segmento con puntos finales en el vértice y el centro de la base.

**axis of a cylinder** (Lesson 11–4)   The segment with endpoints that are centers of the bases.

**eje de un cilindro**   El segmento con puntos finales que son centros de las bases.

**axis symmetry** (Lesson 11–5)   If a figure can be mapped onto itself by a rotation between 0° and 360° in a line.

**eje simetría**   Si una figura puede ser asignada sobre sí misma por una rotación entre 0° y 360° en una línea.

## B

**base angles of a trapezoid** (Lesson 7–6)   The two angles formed by the bases and legs of a trapezoid.

**ángulos de base de un trapecio**   Los dos ángulos formados por las bases y patas de un trapecio.

**base angles of an isosceles triangle** (Lesson 5–6)   The two angles formed by the base and the congruent sides of an isosceles triangle.

**ángulo de la base de un triángulo isosceles**   Los dos ángulos formados por la base y los lados congruentes de un triángulo isosceles.

**base edge** (Lesson 11–4)   The intersection of a lateral face and a base in a solid figure.

**arista de la base**   La intersección de una cara lateral y una base en una figura sólida.

**base of a parallelogram** (Lesson 11–1)   Any side of a parallelogram.

**base de un paralelogramo**   Cualquier lado de un paralelogramo.

**bases of a polyhedron** (Lesson 2–5)   The two parallel congruent faces of a polyhedron.

**base de poliedro**   Las dos caras congruentes paralelas de un poliedro.

**bases of a trapezoid** (Lesson 7–6)   The parallel sides in a trapezoid.

**bases de un trapecio**   Los lados paralelos en un trapecio.

**betweenness of points** (Lesson 1–3)   Point $C$ is between $A$ and $B$ if and only if $A$, $B$, and $C$ are collinear and $AC + CB = AB$.

**intermediación de puntos**   El punto $C$ está entre $A$ y $B$ si y sólo si $A$, $B$, y $C$ son colineales y $AC + CB = AB$.

**biconditional statement** (Lesson 3–2)   The conjunction of a conditional and its converse.

**declaración bicondicional**   La conjunción de un condicional y su inverso.

**bisect** (Lesson 1–7)   To separate a line segment into two congruent segments.

**bisecar**   Separe un segmento de línea en dos segmentos congruentes.

## C

**center of a circle** (Lesson 10–1)   The point from which all points on a circle are the same distance.

**centro de un círculo**   El punto desde el cual todos los puntos de un círculo están a la misma distancia.

**center of dilation** (Lesson 8–1)   The center point from which dilations are performed.

**centro de dilatación**   Punto fijo en torno al cual se realizan las homotecias.

**center of a regular polygon** (Lesson 11–2)   The center of the circle circumscribed about a regular polygon.

**centro de un polígono regular**   El centro del círculo circunscrito alrededor de un polígono regular.

center of rotation (Lesson 2–4) The fixed point about which a figure rotates.

centro de rotación El punto fijo sobre el que gira una figura.

center of symmetry (Lesson 4–6) A point in which a figure can be rotated onto itself.

centro de la simetría Un punto en el que una figura se puede girar sobre sí misma.

central angle of a circle (Lesson 10–2) An angle with a vertex at the center of a circle and sides that are radii.

ángulo central de un círculo Un ángulo con un vértice en el centro de un círculo y los lados que son radios.

central angle of a regular polygon (Lesson 11–2) An angle with its vertex at the center of a regular polygon and sides that pass through consecutive vertices of the polygon.

ángulo central de un polígono regular Un ángulo con su vértice en el centro de un polígono regular y lados que pasan a través de vértices consecutivos del polígono.

centroid (Lesson 6–3) The point of concurrency of the medians of a triangle.

baricentro El punto de intersección de las medianas de un triángulo.

chord of a circle or sphere (Lessons 10–1, 11–4) A segment with endpoints on the circle or sphere.

cuerda de un círculo o esfera Un segmento con extremos en el círculo o esfera.

circle (Lesson 10–1) The set of all points in a plane that are the same distance from a given point called the center.

círculo El conjunto de todos los puntos en un plano que están a la misma distancia de un punto dado llamado centro.

circumcenter (Lesson 6–1) The point of concurrency of the perpendicular bisectors of a triangle.

circuncentro El punto de concurrencia de las bisectrices perpendiculares de un triángulo.

circumference (Lesson 2–3) The distance around a circle.

circunferencia La distancia alrededor de un círculo.

circumscribed angle (Lesson 10–5) An angle with sides that are tangent to a circle.

ángulo circunscrito Un ángulo con lados que son tangentes a un círculo.

circumscribed polygon (Lesson 10–5) A polygon with vertices outside the circle and sides that are tangent to the circle.

poligono circunscrito Un polígono con vértices fuera del círculo y lados que son tangentes al círculo.

collinear (Lesson 1–2) Lying on the same line.

colineal Acostado en la misma línea.

combination (Lesson 12–4) A selection of objects in which order is not important.

combinación Una selección de objetos en los que el orden no es importante.

common tangent (Lesson 10–5) A line or segment that is tangent to two circles in the same plane.

tangente común Una línea o segmento que es tangente a dos círculos en el mismo plano.

complement of $A$ (Lesson 12–2) All of the outcomes in the sample space that are not included as outcomes of event $A$.

complemento de $A$ Todos los resultados en el espacio muestral que no se incluyen como resultados del evento $A$.

complementary angles (Lesson 2–2) Two angles with measures that have a sum of 90°.

ángulo complementarios Dos ángulos con medidas que tienen una suma de 90°.

**component form** (Lesson 2–4) A vector written as $\langle x, y \rangle$, which describes the vector in terms of its horizontal component $x$ and vertical component $y$.

**composite figure** (Lesson 11–2) A figure that can be separated into regions that are basic figures, such as triangles, rectangles, trapezoids, and circles.

**composite solid** (Lesson 11–4) A three-dimensional figure that is composed of simpler solids.

**composition of transformations** (Lesson 4–4) When a transformation is applied to a figure and then another transformation is applied to its image.

**compound event** (Lesson 12–5) Two or more simple events.

**compound statement** (Lesson 3–2) Two or more statements joined by the word *and* or *or*.

**concave polygon** (Lesson 2–3) A polygon with one or more interior angles with measures greater than 180°.

**concentric circles** (Lesson 10–1) Coplanar circles that have the same center.

**conclusion** (Lesson 3–2) The statement that immediately follows the word *then* in a conditional.

**concurrent lines** (Lesson 6–1) Three or more lines that intersect at a common point.

**conditional probability** (Lesson 12–7) The probability that an event will occur given that another event has already occurred.

**conditional statement** (Lesson 3–2) A statement that can be written in if-then form.

**cone** (Lesson 2–5) A solid figure with a circular base connected by a curved surface to a single vertex.

**congruence transformation** (Lesson 2–4) A transformation in which the position of the image may differ from that of the preimage, but the two figures remain congruent.

**congruent** (Lesson 1–3) Having the same size and shape.

**congruent angles** (Lesson 2–1) Two angles that have the same measure.

**forma de componente** Un vector escrito como $\langle x, y \rangle$, que describe el vector en términos de su componente horizontal $x$ y componente vertical $y$.

**figura compuesta** Una figura que se puede separar en regiones que son figuras básicas, tales como triángulos, rectángulos, trapezoides, y círculos.

**solido compuesta** Una figura tridimensional que se compone de figuras más simples.

**composición de transformaciones** Cuando una transformación se aplica a una figura y luego se aplica otra transformación a su imagen.

**evento compuesto** Dos o más eventos simples.

**enunciado compuesto** Dos o más declaraciones unidas por la palabra *y* o *o*.

**polígono cóncavo** Un polígono con uno o más ángulos interiores con medidas superiores a 180°.

**círculos concéntricos** Círculos coplanarios que tienen el mismo centro.

**conclusión** La declaración que inmediatamente sigue la palabra *entonces* en un condicional.

**líneas concurrentes** Tres o más líneas que se intersecan en un punto común.

**probabilidad condicional** La probabilidad de que un evento ocurra dado que otro evento ya ha ocurrido.

**enunciado condicional** Un enunciado escrito en la forma si-entonces.

**cono** Una figura sólida con una base circular conectada por una superficie curvada a un solo vértice.

**transformación de congruencia** Una transformación en la cual la posición de la imagen puede diferir de la de la preimagen, pero las dos figuras permanecen congruentes.

**congruente** Tener el mismo tamaño y forma.

**ángulo congruentes** Dos ángulos que tienen la misma medida.

**congruent arcs** (Lesson 10–2)   Arcs in the same or congruent circles that have the same measure.

**congruent polygons** (Lesson 5–2)   All of the parts of one polygon are congruent to the corresponding parts or matching parts of another polygon.

**congruent segments** (Lesson 1–3)   Line segments that are the same length.

**congruent solids** (Lesson 11–8)   Solid figures that have exactly the same shape, size, and a scale factor of 1:1.

**conic sections** (Lesson 11–5)   Cross sections of a right circular cone.

**conjecture** (Lesson 3–1)   An educated guess based on known information and specific examples.

**conjunction** (Lesson 3–2)   A compound statement using the word *and*.

**consecutive interior angles** (Lesson 3–7)   When two lines are cut by a transversal, interior angles that lie on the same side of the transversal.

**constructions** (Lesson 1–3)   Methods of creating figures without the use of measuring tools.

**contrapositive** (Lesson 3–2)   A statement formed by negating both the hypothesis and the conclusion of the converse of a conditional.

**converse** (Lesson 3–2)   A statement formed by exchanging the hypothesis and conclusion of a conditional statement.

**convex polygon** (Lesson 2–3)   A polygon with all interior angles measuring less than 180°.

**coordinate proofs** (Lesson 5–7)   Proofs that use figures in the coordinate plane and algebra to prove geometric concepts.

**coplanar** (Lesson 1–2)   Lying in the same plane.

**corollary** (Lesson 5–1)   A theorem with a proof that follows as a direct result of another theorem.

**arcos congruentes**   Arcos en los mismos círculos o congruentes que tienen la misma medida.

**poligonos congruentes**   Todas las partes de un polígono son congruentes con las partes correspondientes o partes coincidentes de otro polígono.

**segmentos congruentes**   Línea segmentos que son la misma longitud.

**sólidos congruentes**   Figuras sólidas que tienen exactamente la misma forma, tamaño y un factor de escala de 1:1.

**secciones cónicas**   Secciones transversales de un cono circular derecho.

**conjetura**   Una suposición educada basada en información conocida y ejemplos específicos.

**conjunción**   Una declaración compuesta usando la palabra *y*.

**ángulos internos consecutivos**   Cuando dos líneas se cortan por un ángulo transversal, interior que se encuentran en el mismo lado de la transversal.

**construcciones**   Métodos de creación de figuras sin el uso de herramientas de medición.

**antítesis**   Una afirmación formada negando tanto la hipótesis como la conclusión del inverso del condicional.

**recíproco**   Una declaración formada por el intercambio de la hipótesis y la conclusión de la declaración condicional.

**polígono convexo**   Un polígono con todos los ángulos interiores que miden menos de 180°.

**pruebas de coordenadas**   Pruebas que utilizan figuras en el plano de coordenadas y álgebra para probar conceptos geométricos.

**coplanar**   Acostado en el mismo plano.

**corolario**   Un teorema con una prueba que sigue como un resultado directo de otro teorema.

corresponding angles (Lesson 3–7)   When two lines are cut by a transversal, angles that lie on the same side of a transversal and on the same side of the two lines.

ángulos correspondientes   Cuando dos líneas se cortan transversalmente, los ángulos que se encuentran en el mismo lado de una transversal y en el mismo lado de las dos líneas.

corresponding parts (Lesson 5–2)   Corresponding angles and corresponding sides of two polygons.

partes correspondientes   Ángulos correspondientes y lados correspondientes.

cosine (Lesson 9–5)   The ratio of the length of the leg adjacent to an angle to the length of the hypotenuse.

coseno   Relación entre la longitud de la pierna adyacente a un ángulo y la longitud de la hipotenusa.

counterexample (Lesson 3–1)   A false example showing that a conjecture is not always true.

contraejemplo   Un ejemplo falso que muestra que una conjetura no siempre es verdadera.

cross section (Lesson 11–5)   The intersection of a solid and a plane.

sección transversal   Intersección de un sólido con un plano.

cylinder (Lesson 2–5)   A solid figure with two congruent and parallel circular bases connected by a curved surface.

cilindro   Una figura sólida con dos bases circulares congruentes y paralelas conectadas por una superficie curvada.

## D

decomposition (Lesson 11–1)   Separating a figure into two or more nonoverlapping parts.

descomposición   Separar una figura en dos o más partes que no se solapan.

deductive argument (Lesson 3–4)   An argument that guarantees the truth of the conclusion provided that its premises are true.

argumento deductivo   Un argumento que garantiza la verdad de la conclusión siempre que sus premisas sean verdaderas.

deductive reasoning (Lesson 3–3)   Reasoning that uses facts, rules, definitions, or properties to reach valid conclusions from given statements.

razonamiento deductivo   Razonamiento que utiliza hechos, reglas, definiciones o propiedades para llegar a conclusiones válidas de declaraciones dadas.

defined term (Lesson 1–1)   A term that has a definition and can be explained.

término definido   Un término que tiene una definición y se puede explicar.

definitions (Lesson 1–1)   An explanation that assigns properties to a mathematical object.

definiciones   Una explicación que asigna propiedades a un objeto matemático.

degree (Lesson 10–2)   $\frac{1}{360}$ of the circular rotation about a point.

grado   $\frac{1}{360}$ de la rotación circular alrededor de un punto.

density (Lesson 11–9)   A measure of the quantity of some physical property per unit of length, area, or volume.

densidad   Una medida de la cantidad de alguna propiedad física por unidad de longitud, área o volumen.

dependent events (Lesson 12–5)   Two or more events in which the outcome of one event affects the outcome of the other events.

eventos dependientes   Dos o más eventos en que el resultado de un evento afecta el resultado de los otros eventos.

diagonal (Lesson 7–1)   A segment that connects any two nonconsecutive vertices.

diameter of a circle or sphere (Lessons 10–1, 11–4)   A chord that passes through the center of a circle or sphere.

dilation (Lesson 8–1)   A nonrigid transformation that enlarges or reduces a geometric figure.

directed line segment (Lesson 1–5)   A line segment with an initial endpoint and a terminal endpoint.

directrix (Lesson 10–8)   An exterior line perpendicular to the line containing the foci of a curve.

disjunction (Lesson 3–2)   A compound statement using the word *or*.

distance (Lesson 1–4)   The length of the line segment between two points.

diagonal   Un segmento que conecta cualquier dos vértices no consecutivos.

diámetro de un círculo o esfera   Un acorde que pasa por el centro de un círculo o esfera.

dilatación   Una transformación no rígida que amplía o reduce una figura geométrica.

segment de línea dirigido   Un segmento de línea con un punto final inicial y un punto final terminal.

directriz   Una línea exterior perpendicular a la línea que contiene los focos de una curva.

disyunción   Una declaración compuesta usando la palabra *o*.

distancia   La longitud del segmento de línea entre dos puntos.

## E

edge of a polyhedron (Lesson 2–5)   A line segment where the faces of the polyhedron intersect.

enlargement (Lesson 8–1)   A dilation with a scale factor greater than 1.

equiangular polygon (Lesson 2–3)   A polygon with all angles congruent.

equidistant (Lesson 1–7)   A point is equidistant from other points if it is the same distance from them.

equidistant lines (Lesson 3–10)   Two lines for which the distance between the two lines, measured along a perpendicular line or segment to the two lines, is always the same.

equilateral polygon (Lesson 2–3)   A polygon with all sides congruent.

event (Lesson 12–1)   A subset of the sample space.

experiment (Lesson 12–1)   A situation involving chance.

exterior of an angle (Lesson 2–1)   The area outside of the two rays of an angle.

arista de un poliedro   Un segmento de línea donde las caras del poliedro se cruzan.

ampliación   Una dilatación con un factor de escala mayor que 1.

polígono equiangular   Un polígono con todos los ángulos congruentes.

equidistante   Un punto es equidistante de otros puntos si está a la misma distancia de ellos.

líneas equidistantes   Dos líneas para las cuales la distancia entre las dos líneas, medida a lo largo de una línea o segmento perpendicular a las dos líneas, es siempre la misma.

polígono equilátero   Un polígono con todos los lados congruentes.

evento   Un subconjunto cel espacio de muestra.

experimento   Una situación de riesgo.

exterior de un ángulo   El área fuera de los dos rayos de un ángulo.

exterior angle of a triangle  (Lesson 5–1)   An angle formed by one side of the triangle and the extension of an adjacent side.

ángulo exterior de un triángulo   Un ángulo formado por un lado del triángulo y la extensión de un lado adyacente.

exterior angles  (Lesson 3–7)   When two lines are cut by a transversal, any of the four angles that lie outside the region between the two intersected lines.

ángulos externos   Cuando dos líneas son cortadas por una transversal, cualquiera de los cuatro ángulos que se encuentran fuera de la región entre las dos líneas intersectadas.

## F

face of a polyhedron  (Lesson 2–5)   A flat surface of a polyhedron.

cara de un poliedro   Superficie plana de un poliedro.

factorial of $n$  (Lesson 12–4)   The product of the positive integers less than or equal to $n$.

factorial de $n$   El producto de los enteros positivos inferiores o iguales a $n$.

finite sample space  (Lesson 12–1)   A sample space that contains a countable number of outcomes.

espacio de muestra finito   Un espacio de muestra que contiene un número contable de resultados.

flow proof  (Lesson 3–4)   A proof that uses boxes and arrows to show the logical progression of an argument.

demostración de flujo   Una prueba que usa cajas y flechas para mostrar la progresión lógica de un argumento.

focus  (Lesson 10–8)   A point inside a parabola having the property that the distances from any point on the parabola to them and to a fixed line have a constant ratio for any points on the parabola.

foco   Un punto dentro de una parábola que tiene la propiedad de que las distancias desde cualquier punto de la parábola a ellos ya una línea fija tienen una relación constante para cualquier punto de la parábola.

formal proof  (Lesson 3–4)   A proof that contains statements and reasons organized in a two-column format.

prueba formal   Una prueba que contiene declaraciones y razones organizadas en un formato de dos columnas.

fractional distance  (Lesson 1–5)   An intermediary point some fraction of the length of a line segment.

distancia fraccionaria   Un punto intermediario de alguna fracción de la longitud de un segmento de línea.

## G

geometric mean  (Lesson 9–1)   The $n$th root, where $n$ is the number of elements in a set of numbers, of the product of the numbers.

media geométrica   La enésima raíz, donde $n$ es el número de elementos de un conjunto de números, del producto de los números.

geometric model  (Lesson 2–3)   A geometric figure that represents a real-life object.

modelo geométrico   Una figura geométrica que representa un objeto de la vida real.

geometric probability  (Lesson 12–3)   Probability that involves a geometric measure such as length or area.

probabilidad geométrica   Probabilidad que implica una medida geométrica como longitud o área.

**glide reflection** (Lesson 4–4) The composition of a translation followed by a reflection in a line parallel to the translation vector.

**reflexión del deslizamiento** La composición de una traducción seguida de una reflexión en una línea paralela al vector de traslación.

## H

**height of a parallelogram** (Lesson 11–1) The length of an altitude of the parallelogram.

**altura de un paralelogramo** La longitud de la altitud del paralelogramo.

**height of a solid** (Lesson 11–4) The length of the altitude of a solid figure.

**altura de un sólido** La longitud de la altitud de una figura sólida.

**height of a trapezoid** (Lesson 11–1) The perpendicular distance between the bases of a trapezoid.

**altura de un trapecio** La distancia perpendicular entre las bases de un trapecio.

**hypothesis** (Lesson 3–2) The statement that immediately follows the word *if* in a conditional.

**hipótesis** La declaración que sigue inmediatamente a la palabra *si* en un condicional.

## I

**if-then statement** (Lesson 3–2) A compound statement of the form *if p, then q*, where *p* and *q* are statements.

**enunciado si-entonces** Enunciado compuesto de la forma *si p, entonces q*, donde *p* y *q* son enunciados.

**image** (Lesson 2–4) The new figure in a transformation.

**imagen** La nueva figura en una transformación.

**incenter** (Lesson 6–2) The point of concurrency of the angle bisectors of a triangle.

**incentro** El punto de intersección de las bisectrices interiors de un triángulo.

**included angle** (Lesson 5–3) The interior angle formed by two adjacent sides of a triangle.

**ángulo incluido** El ángulo interior formado por dos lados adyacentes de un triángulo.

**included side** (Lesson 5–4) The side of a triangle between two angles.

**lado incluido** El lado de un triángulo entre dos ángulos.

**independent events** (Lesson 12–5) Two or more events in which the outcome of one event does not affect the outcome of the other events.

**eventos independientes** Dos o más eventos en los que el resultado de un evento no afecta el resultado de los otros eventos.

**indirect measurement** (Lesson 9–6) Using similar figures and proportions to measure an object.

**medición indirecta** Usando figuras y proporciones similares para medir un objeto.

**indirect proof** (Lesson 6–5) One assumes that the statement to be proven is false and then uses logical reasoning to deduce that a statement contradicts a postulate, theorem, or one of the assumptions.

**demostración indirecta** Se supone que la afirmación a ser probada es falsa y luego utiliza el razonamiento lógico para deducir que una afirmación contradice un postulado, teorema o uno de los supuestos.

**indirect reasoning** (Lesson 6–5) Reasoning that eliminates all possible conclusions but one so that the one remaining conclusion must be true.

**razonamiento indirecto** Razonamiento que elimina todas las posibles conclusiones, pero una de manera que la conclusión que queda una debe ser verdad.

inductive reasoning  (Lesson 3–1)   The process of reaching a conclusion based on a pattern of examples.

razonamiento inductivo   El proceso de llegar a una conclusión basada en un patrón de ejemplos.

infinite sample space  (Lesson 12–1)   A sample space with outcomes that cannot be counted.

espacio de muestra infinito   Un espacio de muestra con resultados que no pueden ser contados.

informal proof  (Lesson 3–4)   A paragraph that explains why the conjecture for a given situation is true.

prueba informal   Un párrafo que explica por qué la conjetura para una situación dada es verdadera.

inscribed angle  (Lesson 10–4)   An angle with its vertex on a circle and sides that contain chords of the circle.

ángulo inscrito   Un ángulo con su vértice en un círculo y lados que contienen acordes del círculo.

inscribed polygon  (Lesson 10–4)   A polygon inside a circle in which all of the vertices of the polygon lie on the circle.

polígono inscrito   Un polígono dentro de un círculo en el que todos los vértices del polígono se encuentran en el círculo.

intercepted arc  (Lesson 10–4)   The part of a circle that lies between the two lines intersecting it.

arco intersecado   La parte de un círculo que se encuentra entre las dos líneas que se cruzan.

interior of an angle  (Lesson 2–1)   The area between the two rays of an angle.

interior de un ángulo   El área entre los dos rayos de un ángulo.

interior angle of a triangle  (Lesson 5–1)   An angle at the vertex of a triangle.

ángulo interior de un triángulo   Un ángulo en el vértice de un triángulo.

interior angles  (Lesson 3–7)   When two lines are cut by a transversal, any of the four angles that lie inside the region between the two intersected lines.

ángulos interiores   Cuando dos líneas son cortadas por una transversal, cualquiera de los cuatro ángulos que se encuentran dentro de la región entre las dos líneas intersectadas.

intersection  (Lesson 1–2)   A set of points common to two or more geometric figures.

intersección   Un conjunto de puntos communes a dos o más figuras geométricas.

intersection of $A$ and $B$  (Lesson 12–2)   The set of all outcomes in the sample space of event $A$ that are also in the sample space of event $B$.

intersección de $A$ y $B$   El conjunto de todos los resultados en el espacio muestral del evento $A$ que también se encuentran en el espacio muestral del evento $B$.

inverse  (Lesson 3–2)   A statement formed by negating both the hypothesis and conclusion of a conditional statement.

inverso   Una declaración formada negando tanto la hipótesis como la conclusión de la declaración condicional.

inverse cosine  (Lesson 9–5)   The ratio of the length of the hypotenuse to the length of the leg adjacent to an angle.

inverso del coseno   Relación de la longitud de la hipotenusa con la longitud de la pierna adyacente a un ángulo.

inverse sine  (Lesson 9–5)   The ratio of the length of the hypotenuse to the length of the leg opposite an angle.

inverso del seno   Relación de la longitud de la hipotenusa con la longitud de la pierna opuesta a un ángulo.

inverse tangent  (Lesson 9–5)   The ratio of the length of the leg adjacent to an angle to the length of the leg opposite the angle.

inverso del tangente   Relación de la longitud de la pierna adyacente a un ángulo con la longitud de la pierna opuesta a un ángulo.

isometry (Lesson 2–4)   A transformation in which the position of the image may differ from that of the preimage, but the two figures remain congruent.

isometría   Una transformación en la cual la posición de la imagen puede diferir de la de la preimagen, pero las dos figuras permanecen congruentes.

isosceles trapezoid (Lesson 7–6)   A quadrilateral in which two sides are parallel and the legs are congruent.

trapecio isósceles   Un cuadrilátero en el que dos lados son paralelos y las patas son congruentes.

isosceles triangle (Lesson 5–6)   A triangle with at least two sides congruent.

triángulo isósceles   Un triángulo con al menos dos lados congruentes.

---

## J

joint frequencies (Lesson 12–8)   In a two-way frequency table, the frequencies in the interior of the table.

frecuencias articulares   En una tabla de frecuencia bidireccional, las frecuencias en el interior de la tabla.

---

## K

kite (Lesson 7–6)   A quadrilateral with exactly two distinct pairs of adjacent congruent sides.

cometa   Cuadrilátero que tiene exactamente dos pares diferentes do lados congruentes y adyacentes.

---

## L

lateral area (Lesson 11–4)   The sum of the areas of the lateral faces of the figure.

área lateral   La suma de las áreas de las caras laterales de la figura.

lateral edges (Lesson 11–4)   The intersection of two lateral faces.

aristas laterales   La intersección de dos caras laterales.

lateral faces (Lesson 11–4)   The faces that join the bases of a solid.

caras laterales   Las caras que unen las bases de un sólido.

lateral surface of a cone (Lesson 11–4)   The curved surface that joins the base of a cone to the vertex.

superficie lateral de un cono   La superficie curvada que une la base de un cono con el vértice.

lateral surface of a cylinder (Lesson 11–4)   The curved surface that joins the bases of a cylinder.

superficie lateral de un cilindro   La superficie curvada que une las bases de un cilindro.

legs of an isosceles triangle (Lesson 5–6)   The two congruent sides of an isosceles triangle.

patas de un triángulo isósceles   Los dos lados congruentes de un triángulo isósceles.

legs of a trapezoid (Lesson 7–6)   The nonparallel sides in a trapezoid.

patas de un trapecio   Los lados no paralelos en un trapezoide.

line (Lesson 1–2)   A line is made up of points, has no thickness or width, and extends indefinitely in both directions.

línea   Una línea está formada por puntos, no tiene espesor ni anchura, y se extiende indefinidamente en ambas direcciones.

line of reflection (Lesson 2–4)   A line midway between a preimage and an image.

línea de reflexión   Una línea a medio camino entre una preimagen y una imagen.

line of symmetry (Lesson 4–6)   An imaginary line that separates a figure into two congruent parts.

línea de simetría   Una línea imaginaria que separa una figura en dos partes congruentes.

line segment (Lesson 1–3)   A measurable part of a line that consists of two points, called endpoints, and all of the points between them.

segmento de línea   Una parte medible de una línea que consta de dos puntos, llamados extremos, y todos los puntos entre ellos.

line symmetry (Lesson 4–6)   Each half of a figure matches the other half exactly.

simetría de línea   Cada mitad de una figura coincide exactamente con la otra mitad.

linear pair (Lesson 2–1)   A pair of adjacent angles with noncommon sides that are opposite rays.

par lineal   Un par de ángulos adyacentes con lados no comunes que son rayos opuestos.

logically equivalent (Lesson 3–2)   Statements with the same truth value.

lógicamente equivalentes   Declaraciones con el mismo valor de verdad.

## M

magnitude (Lesson 4–2)   The length of a vector from the initial point to the terminal point.

magnitud   La longitud de un vector desde el punto inicial hasta el punto terminal.

magnitude of symmetry (Lesson 4–6)   The smallest angle through which a figure can be rotated so that it maps onto itself.

magnitud de la simetria   El ángulo más pequeño a través del cual una figura se puede girar para que se cargue sobre sí mismo.

major arc (Lesson 10–2)   An arc with measure greater than 180°.

arco mayor   Un arco con una medida superior a 180°.

marginal frequencies (Lesson 12–8)   In a two-way frequency table, the frequencies in the totals row and column.

frecuencias marginales   En una tabla de frecuencias de dos vías, las frecuencias en los totales de fila y columna.

median of a triangle (Lesson 6–3)   A line segment with endpoints that are a vertex of the triangle and the midpoint of the side opposite the vertex.

mediana de un triángulo   Un segmento de línea con extremos que son un vértice del triángulo y el punto medio del lado opuesto al vértice.

midpoint (Lesson 1–7)   The point on a line segment halfway between the endpoints of the segment.

punto medio   El punto en un segmento de línea a medio camino entre los extremos del segmento.

midsegment of a trapezoid (Lesson 7–6)   The segment that connects the midpoints of the legs of a trapezoid.

segment medio de un trapecio   El segmento que conecta los puntos medios de las patas de un trapecio.

midsegment of a triangle (Lesson 8–5)   The segment that connects the midpoints of the legs of a triangle.

segment medio de un triángulo   El segmento que conecta los puntos medios de las patas de un triángulo.

minor arc (Lesson 10–2)   An arc with measure less than 180°.

arco menor   Un arco con una medida inferior a 180°.

mutually exclusive (Lesson 12–6)   Events that cannot occur at the same time.

mutuamente exclusivos   Eventos que no pueden ocurrir al mismo tiempo.

## N

**negation** (Lesson 3–2)   A statement that has the opposite meaning, as well as the opposite truth value, of an original statement.

**net** (Lesson 2–6)   A two-dimensional figure that forms the surfaces of a three-dimensional object when folded.

**nonrigid motion** (Lesson 8–1)   A transformation that changes the dimensions of a given figure.

**negación**   Una declaración que tiene el significado opuesto, así como el valor de verdad opuesto, de una declaración original.

**red**   Una figura bidimensional que forma las superficies de un objeto tridimensional cuando se dobla.

**movimiento no rígida**   Una transformación que cambia las dimensiones de una figura dada.

## O

**octant** (Lesson 9–3)   One of the eight divisions of three-dimensional space.

**opposite rays** (Lesson 2–1)   Two collinear rays with a common endpoint.

**order of symmetry** (Lesson 4–6)   The number of times a figure maps onto itself.

**ordered triple** (Lesson 9–3)   Three numbers given in a specific order used to locate points in space.

**orthocenter** (Lesson 6–3)   The point of concurrency of the altitudes of a triangle.

**orthographic drawing** (Lesson 2–6)   The two-dimensional views of the top, left, front, and right sides of an object.

**outcome** (Lesson 12–1)   The result of a single performance or trial of an experiment.

**octante**   Una de las ocho divisiones del espacio tridimensional.

**rayos opuestos**   Dos rayos colineales con un punto final común.

**orden de la simetría**   El número de veces que una figura se asigna a sí misma.

**triple ordenado**   Tres números dados en un orden específico usado para localizar puntos en el espacio.

**ortocentro**   El punto de concurrencia de las altitudes de un triángulo.

**dibujo ortográfico**   Las vistas bidimensionales de los lados superior, izquierdo, frontal y derecho de un objeto.

**resultado**   El resultado de un solo rendimiento o ensayo de un experimento.

## P

**parabola** (Lesson 10–8)   A curved shape that results when a cone is cut at an angle by a plane that intersects the base.

**paragraph proof** (Lesson 3–4)   A paragraph that explains why the conjecture for a given situation is true.

**parallel lines** (Lesson 3–7)   Coplanar lines that do not intersect.

**parallel planes** (Lesson 3–7)   Planes that do not intersect.

**parábola**   Forma curvada que resulta cuando un cono es cortado en un ángulo por un plano que interseca la base.

**prueba de párrafo**   Un párrafo que explica por qué la conjetura para una situación dada es verdadera.

**líneas paralelas**   Líneas coplanares que no se intersecan.

**planos paralelas**   Planos que no se intersecan.

parallelogram (Lesson 7–2)  A quadrilateral with both pairs of opposite sides parallel.

paralelogramo  Un cuadrilátero con ambos pares de lados opuestos paralelos.

perimeter (Lesson 2–3)  The sum of the lengths of the sides of a polygon.

perimetro  La suma de las longitudes de los lados de un polígono.

permutation (Lesson 12–4)  An arrangement of objects in which order is important.

permutación  Un arreglo de objetos en el que el orden es importante.

perpendicular (Lesson 2–2)  Intersecting at right angles.

perpendicular  Intersección en ángulo recto.

perpendicular bisector (Lesson 6–1)  Any line, segment, or ray that passes through the midpoint of a segment and is perpendicular to that segment.

mediatriz  Cualquier línea, segmento o rayo que pasa por el punto medio de un segmento y es perpendicular a ese segmento.

pi (Lesson 10–1)  The ratio $\frac{\text{cricumference}}{\text{diameter}}$.

pi  Relación $\frac{\text{circunferencia}}{\text{diámetro}}$.

plane (Lesson 1–2)  A flat surface made up of points that has no depth and extends indefinitely in all directions.

plano  Una superficie plana compuesta de puntos que no tiene profundidad y se extiende indefinidamente en todas las direcciones.

plane symmetry (Lesson 11–5)  When a plane intersects a three-dimensional figure so one half is the reflected image of the other half.

simetría plana  Cuando un plano cruza una figura tridimensional, una mitad es la imagen reflejada de la otra mitad.

Platonic solid (Lesson 2–5)  One of five regular polyhedra.

sólido platónico  Uno de cinco poliedros regulares.

point (Lesson 1–2)  A location with no size, only position.

punto  Una ubicación sin tamaño, solo posición.

point of concurrency (Lesson 6–1)  The point of intersection of concurrent lines.

punto de concurrencia  El punto de intersección de líneas concurrentes.

point of symmetry (Lesson 4–6)  The point about which a figure is rotated.

punto de simetría  El punto sobre el que se gira una figura.

point of tangency (Lesson 10–5)  For a line that intersects a circle in one point, the point at which they intersect.

punto de tangencia  Para una línea que cruza un círculo en un punto, el punto en el que se cruzan.

point symmetry (Lesson 4–6)  A figure or graph has this when a figure is rotated 180° about a point and maps exactly onto the other part.

simetría de punto  Una figura o gráfica tiene esto cuando una figura se gira 180° alrededor de un punto y se mapea exactamente sobre la otra parte.

polygon (Lesson 2–3)  A closed plane figure with at least three straight sides.

polígono  Una figura plana cerrada con al menos tres lados rectos.

polyhedron (Lesson 2–5)  A closed three-dimensional figure made up of flat polygonal regions.

poliedros  Una figura tridimensional cerrada formada por regiones poligonales planas.

postulate (Lesson 1–1)  A statement that is accepted as true without proof.

postulado  Una declaración que se acepta como verdadera sin prueba.

precision (Lesson 2–7)   The repeatability, or reproducibility, of a measurement.

preimage (Lesson 2–4)   The original figure in a transformation.

principle of superposition (Lesson 5–2)   Two figures are congruent if and only if there is a rigid motion or series of rigid motions that maps one figure exactly onto the other.

prism (Lesson 2–5)   A polyhedron with two parallel congruent bases connected by parallelogram faces.

proof (Lesson 3–4)   A logical argument in which each statement is supported by a statement that is accepted as true.

proof by contradiction (Lesson 6–5)   One assumes that the statement to be proven is false and then uses logical reasoning to deduce that a statement contradicts a postulate, theorem, or one of the assumptions.

pyramid (Lesson 2–5)   A polyhedron with a polygonal base and three or more triangular faces that meet at a common vertex.

Pythagorean triple (Lesson 9–2)   A set of three nonzero whole numbers that make the Pythagorean Theorem true.

precisión   La repetibilidad, o reproducibilidad, de una medida.

preimagen   La figura original en una transformación.

principio de superposición   Dos figuras son congruentes si y sólo si hay un movimiento rígido o una serie de movimientos rígidos que traza una figura exactamente sobre la otra.

prisma   Un poliedro con dos bases congruentes paralelas conectadas por caras de paralelogramo.

prueba   Un argumento lógico en el que cada sentencia está respaldada por una sentencia aceptada como verdadera.

prueba por contradicción   Se supone que la afirmación a ser probada es falsa y luego utiliza el razonamiento lógico para deducir que una afirmación contradice un postulado, teorema o uno de los supuestos.

pirámide   Poliedro con una base poligonal y tres o más caras triangulares que se encuentran en un vértice común.

triplete Pitágorico   Un conjunto de tres números enteros distintos de cero que hacen que el Teorema de Pitágoras sea verdadero.

---

**R**

radian (Lesson 10–2)   A unit of angular measurement equal to $\frac{180°}{\pi}$ or about 57.296°.

radius of a circle or sphere (Lessons 10–1, 11–4)   A line segment from the center to a point on a circle or sphere.

radius of a regular polygon (Lesson 11–2)   The radius of the circle circumscribed about a regular polygon.

ray (Lesson 2–1)   Part of a line that starts at a point and extends to infinity.

rectangle (Lesson 7–4)   A parallelogram with four right angles.

reduction (Lesson 8–1)   A dilation with a scale factor between 0 and 1.

radián   Una unidad de medida angular igual o $\frac{180°}{\pi}$ alrededor de 57.296°.

radio de un círculo o esfera   Un segmento de línea desde el centro hasta un punto en un círculo o esfera.

radio de un polígono regular   El radio del círculo circunscrito alrededor de un polígono regular.

rayo   Parte de una línea que comienza en un punto y se extiende hasta el infinito.

rectángulo   Un paralelogramo con cuatro ángulos rectos.

reducción   Una dilatación con un factor de escala entre 0 y 1.

**reflection** (Lesson 2–4)   A function in which the preimage is reflected in the line of reflection.

**reflexión**   Función en la que la preimagen se refleja en la línea de reflexión.

**regular polygon** (Lesson 2–3)   A convex polygon that is both equilateral and equiangular.

**polígono regular**   Un polígono convexo que es a la vez equilátero y equiangular.

**regular polyhedron** (Lesson 2–5)   A polyhedron in which all of its faces are regular congruent polygons and all of the edges are congruent.

**poliedro regular**   Un poliedro en el que todas sus caras son polígonos congruentes regulares y todos los bordes son congruentes.

**regular pyramid** (Lesson 11–4)   A pyramid with a base that is a regular polygon.

**pirámide regular**   Una pirámide con una base que es un polígono regular.

**regular tessellation** (Lesson 4–5)   A tessellation formed by only one type of regular polygon.

**teselado regular**   Un teselado formado por un solo tipo de polígono regular.

**relative frequency** (Lesson 12–8)   In a two-way frequency table, the ratios of the number of observations in a category to the total number of observations.

**frecuencia relativa**   En una tabla de frecuencia bidireccional, las relaciones entre el número de observaciones en una categoría y el número total de observaciones.

**remote interior angles** (Lesson 5–1)   Interior angles of a triangle that are not adjacent to an exterior angle.

**ángulos internos no adyacentes**   Ángulos interiores de un triángulo que no están adyacentes a un ángulo exterior.

**rhombus** (Lesson 7–5)   A parallelogram with all four sides congruent.

**rombo**   Un paralelogramo con los cuatro lados congruentes.

**rigid motion** (Lesson 2–4)   A transformation in which the position of the image may differ from that of the preimage, but the two figures remain congruent.

**movimiento rígido**   Una transformación en la cual la posición de la imagen puede diferir de la de la preimagen, pero las dos figuras permanecen congruentes.

**rotation** (Lesson 2–4)   A function that moves every point of a preimage through a specified angle and direction about a fixed point.

**rotación**   Función que mueve cada punto de una preimagen a través de un ángulo y una dirección especificados alrededor de un punto fijo.

**rotational symmetry** (Lesson 4–6)   A figure can be rotated less than 360° about a point so that the image and the preimage are indistinguishable.

**simetría rotacional**   Una figura puede girar menos de 360° alrededor de un punto para que la imagen y la preimagen sean indistinguibles.

## S

**sample space** (Lesson 12–1)   The set of all possible outcomes.

**espacio muestral**   El conjunto de todos los resultados posibles.

**scale factor of a dilation** (Lesson 8–1)   The ratio of a length on an image to a corresponding length on the preimage.

**factor de escala de una dilatación**   Relación de una longitud en una imagen con una longitud correspondiente en la preimagen.

**secant** (Lesson 10–6)   Any line or ray that intersects a circle in exactly two points.

**secante**   Cualquier línea o rayo que cruce un círculo en exactamente dos puntos.

**sector** (Lesson 11–3)   A region of a circle bounded by a central angle and its intercepted arc.

**segment bisector** (Lesson 1–7)   Any segment, line, plane, or point that intersects a line segment at its midpoint.

**semicircle** (Lesson 10–2)   An arc that measures exactly 180°.

**semiregular tessellation** (Lesson 4–5)   A tessellation formed by two or more regular polygons.

**sides of an angle** (Lesson 2–1)   The rays that form an angle.

**significant figures** (Lesson 2–8)   The digits that contribute to a number's precision in a measurement.

**similar polygons** (Lesson 8–2)   Two figures are similar polygons if one can be obtained from the other by a dilation or a dilation with one or more rigid transformations.

**similar solids** (Lesson 11–8)   Solid figures with the same shape but not necessarily the same size.

**similar triangles** (Lesson 8–3)   Triangles in which all of the corresponding angles are congruent and all of the corresponding sides are proportional.

**similarity ratio** (Lesson 8–2)   The scale factor between two similar polygons.

**similarity transformation** (Lesson 8–2)   A transformation in which a figure and its transformation image have the same shape.

**sine** (Lesson 9–5)   The ratio of the length of the leg opposite an angle to the length of the hypotenuse.

**skew lines** (Lesson 3–7)   Noncoplanar lines that do not intersect.

**slant height of a pyramid or right cone** (Lesson 11–4) The length of a segment with one endpoint on the base edge of the figure and the other at the vertex.

**slope** (Lesson 3–8)   The ratio of the change in the y-coordinates (rise) to the corresponding change in the x-coordinates (run) as you move from one point to another along a line.

**sector**   Una región de un círculo delimitada por un ángulo central y su arco interceptado.

**bisectriz del segmento**   Cualquier segmento, línea, plano o punto que interseca un segmento de línea en su punto medio.

**semicírculo**   Un arco que mide exactamente 180°.

**teselado semiregular**   Un teselado formado por dos o más polígonos regulares.

**lados de un ángulo**   Los rayos que forman un ángulo.

**dígitos significantes**   Los dígitos que contribuyen a la precisión de un número en una medición.

**polígonos similares**   Dos figuras son polígonos similares si uno puede ser obtenido del otro por una dilatación o una dilatación con una o más transformaciones rígidas.

**sólidos similares**   Figuras sólidas con la misma forma pero no necesariamente del mismo tamaño.

**triángulos similares**   Triángulos en los cuales todos los ángulos correspondientes son congruentes y todos los lados correspondientes son proporcionales.

**relación de similitud**   El factor de escala entre dos polígonos similares.

**transformación de similitud**   Una transformación en la que una figura y su imagen de transformación tienen la misma forma.

**seno**   La relación entre la longitud de la pierna opuesta a un ángulo y la longitud de la hipotenusa.

**líneas alabeadas**   Líneas no coplanares que no se cruzan.

**altura inclinada de una pirámide o cono derecho**   La longitud de un segmento con un punto final en el borde base de la figura y el otro en el vértice.

**pendiente**   La relación entre el cambio en las coordenadas y (subida) y el cambio correspondiente en las coordenadas x (ejecución) a medida que se mueve de un punto a otro a lo largo de una línea.

**slope criteria** (Lesson 3–8)  Outlines a method for proving the relationship between lines based on a comparison of the slopes of the lines.

**criterios de pendiente**  Describe un método para probar la relación entre líneas basado en una comparación de las pendientes de las líneas.

**solid of revolution** (Lesson 11–5)  A solid figure obtained by rotating a shape around an axis.

**sólido de revolución**  Una figura sólida obtenida girando una forma alrededor de un eje.

**solving a triangle** (Lesson 9–7)  When you are given measurements to find the unknown angle and side measures of a triangle.

**resolver un triángulo**  Cuando se le dan mediciones para encontrar el ángulo desconocido y las medidas laterales de un triángulo.

**space** (Lesson 1–2)  A boundless three-dimensional set of all points.

**espacio**  Un conjunto tridimensional ilimitado de todos los puntos.

**sphere** (Lesson 2–5)  A set of all points in space equidistant from a given point called the center of the sphere.

**esfera**  Un conjunto de todos los puntos del espacio equidistantes de un punto dado llamado centro de la esfera.

**square** (Lesson 7–5)  A parallelogram with all four sides and all four angles congruent.

**cuadrado**  Un paralelogramo con los cuatro lados y los cuatro ángulos congruentes.

**statement** (Lesson 3–2)  Any sentence that is either true or false, but not both.

**enunciado**  Cualquier oración que sea verdadera o falsa, pero no ambas.

**straight angle** (Lesson 2–1)  An angle that measures 180°.

**ángulo recto**  Un ángulo que mide 180°.

**supplementary angles** (Lesson 2–2)  Two angles with measures that have a sum of 180°.

**ángulos suplementarios**  Dos ángulos con medidas que tienen una suma de 180°.

**surface area** (Lesson 2–5)  The sum of the areas of all faces and side surfaces of a three-dimensional figure.

**área de superficie**  La suma de las áreas de todas las caras y superficies laterales de una figura tridimensional.

**symmetry** (Lesson 4–6)  A figure has this if there exists a rigid motion—reflection, translation, rotation, or glide reflection—that maps the figure onto itself.

**simetría**  Una figura tiene esto si existe un movimiento rígido—reflexión, una traducción, una rotación o una reflexión de deslizamiento rígida—que mapea la figura sobre sí misma.

**synthetic geometry** (Lesson 1–1)  The study of geometric figures without the use of coordinates.

**geometría sintética**  El estudio de figuras geométricas sin el uso de coordenadas.

**T**

**tangent** (Lesson 9–5)  The ratio of the length of the leg opposite an angle to the length of the leg adjacent to the angle.

**tangente**  La relación entre la longitud de la pata opuesta a un ángulo y la longitud de la pata adyacente al ángulo.

**tangent to a circle** (Lesson 10–5)  A line or segment in the plane of a circle that intersects the circle in exactly one point and does not contain any points in the interior of the circle.

**tangente a un círculo**  Una línea o segmento en el plano de un círculo que interseca el círculo en exactamente un punto y no contiene ningún punto en el interior del círculo.

**tangent to a sphere** (Lesson 11–4)  A line that intersects the sphere in exactly one point.

**tessellation** (Lesson 4–5)  A repeating pattern of one or more figures that covers a plane with no overlapping or empty spaces.

**theorem** (Lesson 1–1)  A statement that can be proven true using undefined terms, definitions, and postulates.

**transformation** (Lesson 2–4)  A function that takes points in the plane as inputs and gives other points as outputs.

**translation** (Lesson 2–4)  A function in which all of the points of a figure move the same distance in the same direction.

**translation vector** (Lesson 2–4)  A directed line segment that describes both the magnitude and direction of the slide if the magnitude is the length of the vector from its initial point to its terminal point.

**transversal** (Lesson 3–7)  A line that intersects two or more lines in a plane at different points.

**trapezoid** (Lesson 7–6)  A quadrilateral with exactly one pair of parallel sides.

**trigonometric ratio** (Lesson 9–5)  A ratio of the lengths of two sides of a right triangle.

**trigonometry** (Lesson 9–5)  The study of the relationships between the sides and angles of triangles.

**truth value** (Lesson 3–2)  The truth or falsity of a statement.

**two-column proof** (Lesson 3–4)  A proof that contains statements and reasons organized in a two-column format.

**two-way frequency table** (Lesson 12–8)  A table used to show the frequencies of data from a survey or experiment classified according to two variables, with the rows indicating one variable and the columns indicating the other.

**tangente a una esfera**  Una línea que interseca la esfera exactamente en un punto.

**teselado**  Patrón repetitivo de una o más figuras que cubre un plano sin espacios superpuestos o vacíos.

**teorema**  Una afirmación o conjetura que se puede probar verdad utilizando términos, definiciones y postulados indefinidos.

**transformación**  Función que toma puntos en el plano como entradas y da otros puntos como salidas.

**traslación**  Función en la que todos los puntos de una figura se mueven en la misma dirección.

**vector de traslación**  Un segmento de línea dirigido que describe tanto la magnitud como la dirección de la diapositiva si la magnitud es la longitud del vector desde su punto inicial hasta su punto terminal.

**transversal**  Una línea que interseca dos o más líneas en un plano en diferentes puntos.

**trapecio**  Un cuadrilátero con exactamente un par de lados paralelos.

**relación trigonométrica**  Una relación de las longitudes de dos lados de un triángulo rectángulo.

**trigonometría**  El estudio de las relaciones entre los lados y los ángulos de los triángulos.

**valor de verdad**  La verdad o la falsedad de una declaración.

**prueba de dos columnas**  Una prueba que contiene declaraciones y razones organizadas en un formato de dos columnas.

**tabla de frecuencia bidireccional**  Una tabla utilizada para mostrar las frecuencias de datos de una encuesta o experimento clasificados de acuerdo a dos variables, con las filas indicando una variable y las columnas que indican la otra.

**undefined terms** (Lesson 1–1)  Words that are not formally explained by means of more basic words and concepts.

**términos indefinidos**  Palabras que no se explican formalmente mediante palabras y conceptos más básicos.

**uniform tessellation** (Lesson 4–5)  A tessellation that contains the same arrangement of shapes and angles at each vertex.

**teselado uniforme**  Un teselado que contiene la misma disposición de formas y ángulos en cada vértice.

**union of *A* and *B*** (Lesson 12–2)  The set of all outcomes in the sample space of event *A* combined with all outcomes in the sample space of event *B*.

**unión de *A* y *B***  El conjunto de todos los resultados en el espacio muestral del evento *A* combinado con todos los resultados en el espacio muestral del evento *B*.

---

**V**

---

**valid argument** (Lesson 3–3)  An argument is valid if it is impossible for all of the premises, or supporting statements, of the argument to be true and its conclusion false.

**argumento válido**  Un argumento es válido si es imposible que todas las premisas o argumentos de apoyo del argumento sean verdaderos y su conclusión sea falsa.

**vertex of an angle** (Lesson 2–1)  The common endpoint of the two rays that form an angle.

**vértice de un ángulo**  El punto final común de los dos rayos que forman un ángulo.

**vertex angle** (Lesson 5–6)  The angle between the sides that are the legs of an isosceles triangle.

**ángulo del vértice**  El ángulo entre los lados que son las patas de un triángulo isósceles.

**vertex of a polyhedron** (Lesson 2–5)  The intersection of three edges of a polyhedron.

**vértice de un polígono**  La intersección de tres bordes de un poliedro.

**vertical angles** (Lesson 2–1)  Two nonadjacent angles formed by two intersecting lines.

**ángulos verticales**  Dos ángulos no adyacentes formados por dos líneas de intersección.

**volume** (Lesson 2–5)  The measure of the amount of space enclosed by a three-dimensional figure.

**volumen**  La medida de la cantidad de espacio encerrada por una figura tridimensional.

# Index